P9-CBK-906

Pitman Research Notes in Mathematics Series

Main Editors
H. Brezis, Université de Paris
R. G. Douglas, State University of New York at Stony Brook
A. Jeffrey, University of Newcastle-upon-Tyne *(Founding Editor)*

Editorial Board
R. Aris, University of Minnesota
A. Bensoussan, INRIA, France
S. Bloch, University of Chicago
B. Bollobás, University of Cambridge
W. Bürger, Universität Karlsruhe
S. Donaldson, University of Oxford
J. Douglas Jr, Purdue University
R. J. Elliott, University of Alberta
G. Fichera, Università di Roma
R. P. Gilbert, University of Delaware
R. Glowinski, Université de Paris
K. P. Hadeler, Universität Tübingen
K. Kirchgässner, Universität Stuttgart
B. Lawson, State University of New York at Stony Brook
W. F. Lucas, Claremont Graduate School
R. E. Meyer, University of Wisconsin-Madison
S. Mori, Nagoya University
L. E. Payne, Cornell University
G. F. Roach, University of Strathclyde
J. H. Seinfeld, California Institute of Technology
B. Simon, California Institute of Technology
I. N. Stewart, University of Warwick
S. J. Taylor, University of Virginia

Submission of proposals for consideration
Suggestions for publication, in the form of outlines and representative samples, are invited by the Editorial Board for assessment. Intending authors should approach one of the main editors or another member of the Editorial Board, citing the relevant AMS subject classifications. Alternatively, outlines may be sent directly to the publisher's offices. Refereeing is by members of the board and other mathematical authorities in the topic concerned, throughout the world.

Preparation of accepted manuscripts
On acceptance of a proposal, the publisher will supply full instructions for the preparation of manuscripts in a form suitable for direct photo-lithographic reproduction. Specially printed grid sheets are provided and a contribution is offered by the publisher towards the cost of typing. Word processor output, subject to the publisher's approval, is also acceptable.

Illustrations should be prepared by the authors, ready for direct reproduction without further improvement. The use of hand-drawn symbols should be avoided wherever possible, in order to maintain maximum clarity of the text.

The publisher will be pleased to give any guidance necessary during the preparation of a typescript, and will be happy to answer any queries.

Important note
In order to avoid later retyping, intending authors are strongly urged not to begin final preparation of a typescript before receiving the publisher's guidelines and special paper. In this way it is hoped to preserve the uniform appearance of the series.

Longman Scientific & Technical
Longman House
Burnt Mill
Harlow, Essex, UK
(tel (0279) 426721)

Titles in this series

1 Improperly posed boundary value problems
A Carasso and A P Stone

2 Lie algebras generated by finite dimensional ideals
I N Stewart

3 Bifurcation problems in nonlinear elasticity
R W Dickey

4 Partial differential equations in the complex domain
D L Colton

5 Quasilinear hyperbolic systems and waves
A Jeffrey

6 Solution of boundary value problems by the method of integral operators
D L Colton

7 Taylor expansions and catastrophes
T Poston and I N Stewart

8 Function theoretic methods in differential equations
R P Gilbert and R J Weinacht

9 Differential topology with a view to applications
D R J Chillingworth

10 Characteristic classes of foliations
H V Pittie

11 Stochastic integration and generalized martingales
A U Kussmaul

12 Zeta-functions: An introduction to algebraic geometry
A D Thomas

13 Explicit *a priori* inequalities with applications to boundary value problems
V G Sigillito

14 Nonlinear diffusion
W E Fitzgibbon III and H F Walker

15 Unsolved problems concerning lattice points
J Hammer

16 Edge-colourings of graphs
S Fiorini and R J Wilson

17 Nonlinear analysis and mechanics: Heriot-Watt Symposium Volume I
R J Knops

18 Actions of fine abelian groups
C Kosniowski

19 Closed graph theorems and webbed spaces
M De Wilde

20 Singular perturbation techniques applied to integro-differential equations
H Grabmüller

21 Retarded functional differential equations: A global point of view
S E A Mohammed

22 Multiparameter spectral theory in Hilbert space
B D Sleeman

24 Mathematical modelling techniques
R Aris

25 Singular points of smooth mappings
C G Gibson

26 Nonlinear evolution equations solvable by the spectral transform
F Calogero

27 Nonlinear analysis and mechanics: Heriot-Watt Symposium Volume II
R J Knops

28 Constructive functional analysis
D S Bridges

29 Elongational flows: Aspects of the behaviour of model elasticoviscous fluids
C J S Petrie

30 Nonlinear analysis and mechanics: Heriot-Watt Symposium Volume III
R J Knops

31 Fractional calculus and integral transforms of generalized functions
A C McBride

32 Complex manifold techniques in theoretical physics
D E Lerner and P D Sommers

33 Hilbert's third problem: scissors congruence
C-H Sah

34 Graph theory and combinatorics
R J Wilson

35 The Tricomi equation with applications to the theory of plane transonic flow
A R Manwell

36 Abstract differential equations
S D Zaidman

37 Advances in twistor theory
L P Hughston and R S Ward

38 Operator theory and functional analysis
I Erdelyi

39 Nonlinear analysis and mechanics: Heriot-Watt Symposium Volume IV
R J Knops

40 Singular systems of differential equations
S L Campbell

41 N-dimensional crystallography
R L E Schwarzenberger

42 Nonlinear partial differential equations in physical problems
D Graffi

43 Shifts and periodicity for right invertible operators
D Przeworska-Rolewicz

44 Rings with chain conditions
A W Chatters and C R Hajarnavis

45 Moduli, deformations and classifications of compact complex manifolds
D Sundararaman

46 Nonlinear problems of analysis in geometry and mechanics
M Atteia, D Bancel and I Gumowski

47 Algorithmic methods in optimal control
W A Gruver and E Sachs

48 Abstract Cauchy problems and functional differential equations
F Kappel and W Schappacher

49 Sequence spaces
W H Ruckle

50 Recent contributions to nonlinear partial differential equations
H Berestycki and H Brezis

51 Subnormal operators
J B Conway

52 Wave propagation in viscoelastic media
F Mainardi
53 Nonlinear partial differential equations and their applications: Collège de France Seminar. Volume I
H Brezis and J L Lions
54 Geometry of Coxeter groups
H Hiller
55 Cusps of Gauss mappings
T Banchoff, T Gaffney and C McCrory
56 An approach to algebraic K-theory
A J Berrick
57 Convex analysis and optimization
J-P Aubin and R B Vintner
58 Convex analysis with applications in the differentiation of convex functions
J R Giles
59 Weak and variational methods for moving boundary problems
C M Elliott and J R Ockendon
60 Nonlinear partial differential equations and their applications: Collège de France Seminar. Volume II
H Brezis and J L Lions
61 Singular systems of differential equations II
S L Campbell
62 Rates of convergence in the central limit theorem
Peter Hall
63 Solution of differential equations by means of one-parameter groups
J M Hill
64 Hankel operators on Hilbert space
S C Power
65 Schrödinger-type operators with continuous spectra
M S P Eastham and H Kalf
66 Recent applications of generalized inverses
S L Campbell
67 Riesz and Fredholm theory in Banach algebra
B A Barnes, G J Murphy, M R F Smyth and T T West
68 Evolution equations and their applications
F Kappel and W Schappacher
69 Generalized solutions of Hamilton-Jacobi equations
P L Lions
70 Nonlinear partial differential equations and their applications: Collège de France Seminar. Volume III
H Brezis and J L Lions
71 Spectral theory and wave operators for the Schrödinger equation
A M Berthier
72 Approximation of Hilbert space operators I
D A Herrero
73 Vector valued Nevanlinna Theory
H J W Ziegler
74 Instability, nonexistence and weighted energy methods in fluid dynamics and related theories
B Straughan
75 Local bifurcation and symmetry
A Vanderbauwhede
76 Clifford analysis
F Brackx, R Delanghe and F Sommen
77 Nonlinear equivalence, reduction of PDEs to ODEs and fast convergent numerical methods
E E Rosinger
78 Free boundary problems, theory and applications. Volume I
A Fasano and M Primicerio
79 Free boundary problems, theory and applications. Volume II
A Fasano and M Primicerio
80 Symplectic geometry
A Crumeyrolle and J Grifone
81 An algorithmic analysis of a communication model with retransmission of flawed messages
D M Lucantoni
82 Geometric games and their applications
W H Ruckle
83 Additive groups of rings
S Feigelstock
84 Nonlinear partial differential equations and their applications: Collège de France Seminar. Volume IV
H Brezis and J L Lions
85 Multiplicative functionals on topological algebras
T Husain
86 Hamilton-Jacobi equations in Hilbert spaces
V Barbu and G Da Prato
87 Harmonic maps with symmetry, harmonic morphisms and deformations of metrics
P Baird
88 Similarity solutions of nonlinear partial differential equations
L Dresner
89 Contributions to nonlinear partial differential equations
C Bardos, A Damlamian, J I Díaz and J Hernández
90 Banach and Hilbert spaces of vector-valued functions
J Burbea and P Masani
91 Control and observation of neutral systems
D Salamon
92 Banach bundles, Banach modules and automorphisms of C*-algebras
M J Dupré and R M Gillette
93 Nonlinear partial differential equations and their applications: Collège de France Seminar. Volume V
H Brezis and J L Lions
94 Computer algebra in applied mathematics: an introduction to MACSYMA
R H Rand
95 Advances in nonlinear waves. Volume I
L Debnath
96 FC-groups
M J Tomkinson
97 Topics in relaxation and ellipsoidal methods
M Akgül
98 Analogue of the group algebra for topological semigroups
H Dzinotyiweyi
99 Stochastic functional differential equations
S E A Mohammed

100 Optimal control of variational inequalities
V Barbu
101 Partial differential equations and dynamical systems
W E Fitzgibbon III
102 Approximation of Hilbert space operators. Volume II
C Apostol, L A Fialkow, D A Herrero and D Voiculescu
103 Nondiscrete induction and iterative processes
V Ptak and F-A Potra
104 Analytic functions – growth aspects
O P Juneja and G P Kapoor
105 Theory of Tikhonov regularization for Fredholm equations of the first kind
C W Groetsch
106 Nonlinear partial differential equations and free boundaries. Volume I
J I Díaz
107 Tight and taut immersions of manifolds
T E Cecil and P J Ryan
108 A layering method for viscous, incompressible L_p flows occupying R^n
A Douglis and E B Fabes
109 Nonlinear partial differential equations and their applications: Collège de France Seminar. Volume VI
H Brezis and J L Lions
110 Finite generalized quadrangles
S E Payne and J A Thas
111 Advances in nonlinear waves. Volume II
L Debnath
112 Topics in several complex variables
E Ramírez de Arellano and D Sundararaman
113 Differential equations, flow invariance and applications
N H Pavel
114 Geometrical combinatorics
F C Holroyd and R J Wilson
115 Generators of strongly continuous semigroups
J A van Casteren
116 Growth of algebras and Gelfand–Kirillov dimension
G R Krause and T H Lenagan
117 Theory of bases and cones
P K Kamthan and M Gupta
118 Linear groups and permutations
A R Camina and E A Whelan
119 General Wiener–Hopf factorization methods
F-O Speck
120 Free boundary problems: applications and theory, Volume III
A Bossavit, A Damlamian and M Fremond
121 Free boundary problems: applications and theory, Volume IV
A Bossavit, A Damlamian and M Fremond
122 Nonlinear partial differential equations and their applications: Collège de France Seminar. Volume VII
H Brezis and J L Lions
123 Geometric methods in operator algebras
H Araki and E G Effros
124 Infinite dimensional analysis–stochastic processes
S Albeverio
125 Ennio de Giorgi Colloquium
P Krée
126 Almost-periodic functions in abstract spaces
S Zaidman
127 Nonlinear variational problems
A Marino, L Modica, S Spagnolo and M Degiovanni
128 Second-order systems of partial differential equations in the plane
L K Hua, W Lin and C-Q Wu
129 Asymptotics of high-order ordinary differential equations
R B Paris and A D Wood
130 Stochastic differential equations
R Wu
131 Differential geometry
L A Cordero
132 Nonlinear differential equations
J K Hale and P Martinez-Amores
133 Approximation theory and applications
S P Singh
134 Near-rings and their links with groups
J D P Meldrum
135 Estimating eigenvalues with *a posteriori/a priori* inequalities
J R Kuttler and V G Sigillito
136 Regular semigroups as extensions
F J Pastijn and M Petrich
137 Representations of rank one Lie groups
D H Collingwood
138 Fractional calculus
G F Roach and A C McBride
139 Hamilton's principle in continuum mechanics
A Bedford
140 Numerical analysis
D F Griffiths and G A Watson
141 Semigroups, theory and applications. Volume I
H Brezis, M G Crandall and F Kappel
142 Distribution theorems of L-functions
D Joyner
143 Recent developments in structured continua
D De Kee and P Kaloni
144 Functional analysis and two-point differential operators
J Locker
145 Numerical methods for partial differential equations
S I Hariharan and T H Moulden
146 Completely bounded maps and dilations
V I Paulsen
147 Harmonic analysis on the Heisenberg nilpotent Lie group
W Schempp
148 Contributions to modern calculus of variations
L Cesari
149 Nonlinear parabolic equations: qualitative properties of solutions
L Boccardo and A Tesei
150 From local times to global geometry, control an physics
K D Elworthy

151 A stochastic maximum principle for optimal control of diffusions
U G Haussmann
152 Semigroups, theory and applications. Volume II
H Brezis, M G Crandall and F Kappel
153 A general theory of integration in function spaces
P Muldowney
154 Oakland Conference on partial differential equations and applied mathematics
L R Bragg and J W Dettman
155 Contributions to nonlinear partial differential equations. Volume II
J I Díaz and P L Lions
156 Semigroups of linear operators: an introduction
A C McBride
157 Ordinary and partial differential equations
B D Sleeman and R J Jarvis
158 Hyperbolic equations
F Colombini and M K V Murthy
159 Linear topologies on a ring: an overview
J S Golan
160 Dynamical systems and bifurcation theory
M I Camacho, M J Pacifico and F Takens
161 Branched coverings and algebraic functions
M Namba
162 Perturbation bounds for matrix eigenvalues
R Bhatia
163 Defect minimization in operator equations: theory and applications
R Reemtsen
164 Multidimensional Brownian excursions and potential theory
K Burdzy
165 Viscosity solutions and optimal control
R J Elliott
166 Nonlinear partial differential equations and their applications. Collège de France Seminar. Volume VIII
H Brezis and J L Lions
167 Theory and applications of inverse problems
H Haario
168 Energy stability and convection
G P Galdi and B Straughan
169 Additive groups of rings. Volume II
S Feigelstock
170 Numerical analysis 1987
D F Griffiths and G A Watson
171 Surveys of some recent results in operator theory. Volume I
J B Conway and B B Morrel
172 Amenable Banach algebras
J-P Pier
173 Pseudo-orbits of contact forms
A Bahri
174 Poisson algebras and Poisson manifolds
K H Bhaskara and K Viswanath
175 Maximum principles and eigenvalue problems in partial differential equations
P W Schaefer
176 Mathematical analysis of nonlinear, dynamic processes
K U Grusa
177 Cordes' two-parameter spectral representation theory
D F McGhee and R H Picard
178 Equivariant K-theory for proper actions
N C Phillips
179 Elliptic operators, topology and asymptotic methods
J Roe
180 Nonlinear evolution equations
J K Engelbrecht, V E Fridman and E N Pelinovski
181 Nonlinear partial differential equations and their applications. Collège de France Seminar. Volume IX
H Brezis and J L Lions
182 Critical points at infinity in some variational problems
A Bahri
183 Recent developments in hyperbolic equations
L Cattabriga, F Colombini, M K V Murthy and S Spagnolo
184 Optimization and identification of systems governed by evolution equations on Banach space
N U Ahmed
185 Free boundary problems: theory and applications. Volume I
K H Hoffmann and J Sprekels
186 Free boundary problems: theory and applications. Volume II
K H Hoffmann and J Sprekels
187 An introduction to intersection homology theory
F Kirwan
188 Derivatives, nuclei and dimensions on the frame of torsion theories
J S Golan and H Simmons
189 Theory of reproducing kernels and its applications
S Saitoh
190 Volterra integrodifferential equations in Banach spaces and applications
G Da Prato and M Iannelli
191 Nest algebras
K R Davidson
192 Surveys of some recent results in operator theory. Volume II
J B Conway and B B Morrel
193 Nonlinear variational problems. Volume II
A Marino and M K Murthy
194 Stochastic processes with multidimensional parameter
M E Dozzi
195 Prestressed bodies
D Iesan
196 Hilbert space approach to some classical transforms
R H Picard
197 Stochastic calculus in application
J R Norris
198 Radical theory
B J Gardner
199 The C* – algebras of a class of solvable Lie groups
X Wang

200 Stochastic analysis, path integration and dynamics
D Elworthy

201 Riemannian geometry and holonomy groups
S Salamon

202 Strong asymptotics for extremal errors and polynomials associated with Erdös type weights
D S Lubinsky

203 Optimal control of diffusion processes
V S Borkar

204 Rings, modules and radicals
B J Gardner

205 Two-parameter eigenvalue problems in ordinary differential equations
M Faierman

206 Distributions and analytic functions
R D Carmichael and D Mitrović

207 Semicontinuity, relaxation and integral representation in the calculus of variations
G Buttazzo

208 Recent advances in nonlinear elliptic and parabolic problems
P Bénilan, M Chipot, L Evans and M Pierre

209 Model completions, ring representations and the topology of the Pierce sheaf
A Carson

210 Retarded dynamical systems
G Stepan

211 Function spaces, differential operators and nonlinear analysis
L Paivarinta

212 Analytic function theory of one complex variable
C C Yang, Y Komatu and K Niino

213 Elements of stability of visco-elastic fluids
J Dunwoody

214 Jordan decompositions of generalised vector measures
K D Schmidt

215 A mathematical analysis of bending of plates with transverse shear deformation
C Constanda

216 Ordinary and partial differential equations Vol II
B D Sleeman and R J Jarvis

217 Hilbert modules over function algebras
R G Douglas and V I Paulsen

218 Graph colourings
R Wilson and R Nelson

219 Hardy-type inequalities
A Kufner and B Opic

220 Nonlinear partial differential equations and their applications. College de France Seminar Volume X
H Brezis and J L Lions

221 Workshop on dynamical systems
E Shiels and Z Coelho

222 Geometry and analysis in nonlinear dynamics
H W Broer and F Takens

223 Fluid dynamical aspects of combustion theory
M Onofri and A Tesei

224 Approximation of Hilbert space operators. Volume I. 2nd edition
D Herrero

225 Operator Theory: Proceedings of the 1988 GPOTS–Wabash conference
J B Conway and B B Morrel

226 Local cohomology and localization
J L Bueso Montero, B Torrecillas Jover and A Verschoren

227 Nonlinear waves and dissipative effects
D Fusco and A Jeffrey

228 Numerical analysis. Volume III
D F Griffiths and G A Watson

229 Recent developments in structured continua. Volume III
D De Kee and P Kaloni

230 Boolean methods in interpolation and approximation
F J Delvos and W Schempp

231 Further advances in twistor theory, Volume 1
L J Mason and L P Hughston

232 Further advances in twistor theory, Volume 2
L J Mason and L P Hughston

233 Geometry in the neighborhood of invariant manifolds of maps and flows and linearization
U Kirchgraber and K Palmer

234 Quantales and their applications
K I Rosenthal

235 Integral equations and inverse problems
V Petkov and R Lazarov

236 Pseudo-differential operators
S R Simanca

237 A functional analytic approach to statistical experiments
I M Bomze

238 Quantum mechanics, algebras and distributions
D Dubin and M Hennings

239 Hamilton flows and evolution semigroups
J Gzyl

240 Topics in controlled Markov chains
V S Borkar

241 Invariant manifold theory for hydrodynamic transition
S Sritharan

242 Lectures on the spectrum of $L^2(\Gamma\backslash G)$
F L Williams

D Fusco (Editor)

Universita di Messina, Italy
and

A Jeffrey (Editor)

University of Newcastle upon Tyne, England

Nonlinear waves and dissipative effects

Copublished in the United States with
John Wiley & Sons, Inc., New York

Longman Scientific & Technical,
Longman Group UK Limited,
Longman House, Burnt Mill, Harlow,
Essex CM20 2JE, England
and Associated Companies throughout the world.

Copublished in the United States with
John Wiley & Sons, Inc., 605 Third Avenue, New York, NY 10158

© Longman Group UK Limited 1991

All rights reserved; no part of this publication
may be reproduced, stored in a retrieval system,
or transmitted in any form or by any means, electronic,
mechanical, photocopying, recording, or otherwise,
without either the prior written permission of the Publishers
or a licence permitting restricted copying in the United Kingdom
issued by the Copyright Licensing Agency Ltd,
90 Tottenham Court Road, London W1P 9HE

First published 1991

AMS Subject Classification: 73BXX, 73DXX, 35LXX, 76XX, 80A20, 35BXX

ISSN 0269-3674

British Library Cataloguing in Publication Data
A catalogue record for this book is
available from the British Library

Library of Congress Cataloging-in-Publication Data
Nonlinear waves and dissipative effects / D. Fusco (editor) and A. Jeffrey (editor).
p. cm. – (Pitman research notes in mathematics series, ISSN
ISSN 0269-3674; 227)
1. Nonlinear waves. 2. Wave-motion, Theory of. I. Fusco, D.
(Domenico) II. Jeffrey, Alan. III. Series.
QA927.N667 1991
530. 1'4--dc20
91-28956
CIP

Printed and bound in Great Britain
by Biddles Ltd, Guildford and King's Lynn

Contents

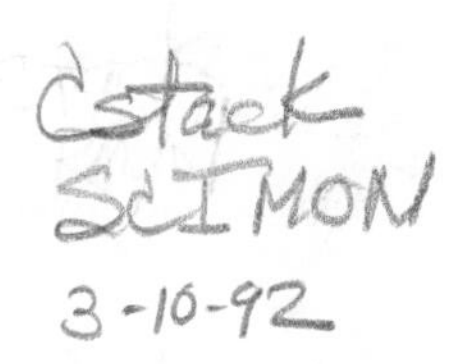

Part I General Topics on Dissipative Waves

Part II Analytical and Numerical Methods for the Solution of Nonlinear Evolutionary Models

Part III Hyperbolic Dissipative Models from Kinetic Theory and Extended Thermodynamics

Preface

Dissipative wave processes are of great interest in both theoretical and experimental investigations. The mathematical models which are usually considered in this context are quasilinear systems of partial differential equations representing balance nonconservative laws (i.e. involving source-like terms). Models of this type are encountered in the study of flows in fluid filled tubes or channels, in irreversible thermodynamics, and also in the study of nerve pulse transmission and diffusion in solids. In many cases the resulting systems of field equations are hyperbolic, so they allow disturbances to propagate with finite wave speeds. In the last decade a great deal of attention has been focussed on the study of dissipation in wave dynamic problems governed by quasilinear hyperbolic models. Euromech Colloquium 270 held in Reggio Calabria, Italy, 25-28 September 1990 was devoted to the study of this general area of research. As Co-chairmen of the Colloquium our task was to bring together scientists working in different areas of wave propagation with the object of comparing their methods of approach and results in the belief that this would lead to a better understanding of hyperbolic dissipative theories. Most of the participants at this Colloquium have published their contributions in the present volume.

The papers in this volume have been grouped into four parts.

Part I presents papers whose main aim is to develop general approaches which allow the characterization of wave amplitude evolution laws in various physical situations in which dissipative effects are significant.

Part II contains a survey of various procedures for the analysis and numerical solution of nonlinear models of interest in wave propagation. Most of these papers deal with analytical methods, though some develop approximate and/or numerical techniques for determining wave-like solutions to nonlinear evolution equations.

Part III concerns hyperbolic models obtained from within the framework of Kinetic Theory and Extended Thermodynamics. The papers encompass a number of fields such as radiative transfer, the Boltzmann equation, relativistic Magnetohydrodynamics, light scattering and nonlinear heat conduction.

Part IV is devoted to the mathematical modeling of some experimental tests performed on waves in the sea, in fluid-filled tubes, in viscous systems and in beams under torsion.

This volume is intended mainly for research workers interested both in constructing evolutionary models for dissipative media on the basis of experimental evidence, and in developing mathematical tools for determining nonlinear wave behaviour. We hope readers will find this collection of use when undertaking research in the many fascinating and varied aspects of the study of wave phenomena.

We wish to thank all the participants for attending the Colloquium and the contributors to this volume for their collaboration when preparing their paper for publication.

We express our gratitude to the Committee for Mathematics, and to the National Group for Mathematical Physics of the National Council for Research of Italy for their financial support which enabled us to run the Colloquium. We address our special thanks to the University of Reggio Calabria and to the local faculties of Engineering and Architecture for providing additional financial support together with a wide range of help and facilities, all of which contributed so much to the success and pleasure of the meeting.

Messina, Newcastle upon Tyne, March 1991

D. Fusco

A. Jeffrey

List of contributors

Anile M.
Dipartimento di Matematica, Università di Catania
viale A. Doria, 6, 95125 Catania
ITALY

Bampi F.
Istituto Matematico di Ingegneria, Universita' di Genova
Piazzale J.F. Kennedy, 16129 Genova
ITALY

Barbaro G.
Istituto di Ingegneria Civile, Universita' di Reggio Calabria
Via V. Veneto, 69, 89100 Reggio Calabria
ITALY

Bellomo N.
Politecnico di Torino, Dipartimento di Matematica
C. Duca Degli Abruzzi, 24, 10129 Torino
ITALY

Boccotti P.
Istituto di Ingegneria Civile, Universita' di Reggio Calabria
Via V. Veneto, 69, 89100 Reggio Calabria
ITALY

Boffi V.
Dipartimento di Metodi e Modelli Mtematici
per le Scienze Applicate, Universita' di Roma " la Sapienza"
Via A. Scarpa, 10, 00161 Roma
ITALY

Bonzani I.
Politecnico di Torino, Dipartimento di Matematica
C. Duca Degli Abruzzi, 24, 10129 Torino
ITALY

Boulanger P.
Department de Mathematique
Universite' Libre de Bruxelles
Campus Plaine, C.P. 218/1, 1050 Bruxelles
BELGIUM

Camera-Roda G.
Dipartimento di Ingegneria Chimica e di Processo
Universitá di Bologna
viale Risorgimento 2
40136 Bologna
ITALY

Caorsi S.
Dipartimento di Ingegneria, Biofisica ed Elettronica, Universita' di Genova
via All'Opera Pia, 11/A, 16145 Genova
ITALY

Caviglia G.
Dipartimento di Matematica, Università di Genova
via Alberti, 4, 16132 Genova
ITALY

Charru F.
Institut de Mécanique des Fluides, C.N.R.S. / U.R.A. 005,
Avenue C. Soula, 31400 Toulouse
FRANCE

Ciancio V.
Department of Mathematics, University of Messina
Contrada Papardo, Salita Sperone 31, 98166 S. Agata, Messina
ITALY

Doghieri F.
Dipartimento di Ingegneria Chimica e di Processo
Universitá di Bologna
viale Risorgimento 2
40136 Bologna
ITALY

Donato A.
Department of Mathematics, University of Messina
Contrada Papardo, Salita Sperone 31, 98166 S. Agata, Messina
ITALY

Engelbrecht J.
Institute of Cybernetics, Estonian Academy of Sciences
Akadeemia 21, Tallinn, SU 200108
ESTONIA - USSR

Fabre J.
Institut de Mécanique des Fluides, C.N.R.S. / U.R.A. 005,
Avenue C. Soula, 31400 Toulouse
FRANCE

Fazio R.
Department of Mathematics, University of Messina
Contrada Papardo, Salita Sperone 31, 98166 S. Agata, Messina
ITALY

Giambò S.
Department of Mathematics, University of Messina
Contrada Papardo, Salita Sperone 31, 98166 S. Agata, Messina
ITALY

Hayes M.
Department of Mathematical Physics, University College, Belfield
Dublin, 4
IRELAND

Hindelang F.
Fakultat LRT, Universität der B. W. Munchen
D - 8014 Neubiberg
GERMANY

Hornung K.
Fakultat LRT, Universität der B. W. Munchen
D - 8014 Neubiberg
GERMANY

Lebon G.
Universitè de Liege
B 6, Sart Tilman B-4000 Liege,
BELGIUM

Leevers P.S.
Department of Mechanical Engineering, Imperial College
Exhibition Road, SW7 2BX London
GREAT BRITAIN

Liu I-Shih
Instituto de Matematica
Universidade Federal do Rio de Janeiro
Caixa Postal 68530, 21944 Rio de Janeiro
BRAZIL

Luger P.
Fakultat LRT, Universität der B. W. Munchen
D - 8014 Neubiberg
GERMANY

Mallamace F.

Department of Physics, University of Messina

Contrada Papardo, Salita Sperone 31, 98166 S. Agata, Messina

ITALY

Manganaro N.

Department of Mathematics, University of Messina

Contrada Papardo, Salita Sperone 31, 98166 S. Agata, Messina

ITALY

Mannino L.

Istituto di Ingegneria Civile, Universita' di Reggio Calabria

Via V. Veneto, 69, 89100 Reggio Calabria

ITALY

Marzocchi A.

Dipartimento di Matematica, Università Cattolica del S. Cuore

Via Trieste 17, 25121 Brescia

ITALY

Mayer A. P.

Department of Mathematics, University of Edinburgh

The King's Buildings - Mayfield Road, Edinburgh EH9 3J2

GREAT BRITAIN

Micali N.

Department of Physics, University of Messina

Contrada Papardo, Salita Sperone 31, 98166 S. Agata, Messina

ITALY

Michel M.

Institut de Mécanique des Fluides, C.N.R.S. / U.R.A. 005,

Avenue C. Soula, 31400 Toulouse

FRANCE

Moodie B.
Applied Mathematics Institute
Department of Mathematics - University of Alberta
533 Central Academic Building, monton, T6G 2G1
CANADA

Müller I.
Hermann-Föttinger-Institut, Technische Universität Berlin
Strasse des 17 Juni 135, D-1000 Berlin 12
GERMANY

Muracchini A.
Dipartimento di Matematica - CIRAM
via Saragozza, 8, 40123 Bologna
ITALY

Oliveri F.
Department of Mathematics, University of Messina
Contrada Papardo, Salita Sperone 31, 98166 S. Agata, Messina
ITALY

Palumbo A.
Istituto del Biennio, Facolta' di Ingegneria
Via E. Cuzzocrea, 48, 89128 Reggio Calabria
ITALY

Parker D. F.
Department of Mathematics, University of Edinburgh
The King's Buildings - Mayfield Road, Edinburgh EH9 3JZ
GREAT BRITAIN

Peipman T.
Dept. of Structural Mech., Technical University of Tallinn
Ehitajate 5, SU 200108 Tallinn
ESTONIA - USSR

Ruggeri T.
Dipartimento di Matematica - CIRAM
Universitá di Bologna
via Saragozza, 8 - 40123 Bologna
ITALY

Sammartino M.
Dipartimento di Matematica, Università di Catania
viale A. Doria, 6, 95125 Catania
ITALY

Samsonov A. M.
A. F. Ioffe Physical Technical Institute of the USSR Academy of Sciences
Leningrad 194021
USSR

Sarti G. C.
Dipartimento di Ingegneria Chimica e di Processo
Universitá di Bologna
viale Risorgimento 2
40136 Bologna
ITALY

Schierwagen A. K.
Fachbereich Informatik der Universitat Leipzig
Augustusplatz 10/11, 7010 Leipzig
GERMANY

Seccia L.
Dipartimento di Matematica - CIRAM
via Saragozza, 8
40123 BOLOGNA

Stronge W. J.
Department of Engineering, University of Cambridge
Trumpington Street CB2 1PZ Cambridge
GREAT BRITAIN

Strumia A.
Dipartimento di Matematica - CIRAM
via Saragozza, 8 40123 Bologna
ITALY

Suvorova J.
Mechanical Engineering
Research Institute
101830, Moscow
USSR

Valenti G.
Istituto del Biennio, Facolta' di Ingegneria
Via E. Cuzzocrea, 48, 89128 Reggio Calabria
ITALY

Vasi C.
Department of Physics, University of Messina
Contrada Papardo, Salita Sperone 31, 98166 S. Agata, Messina
ITALY

Wheel M.
Department of Mechanical Engineering, Imperial College
Exhibition Road, SW7 2BX London
GREAT BRITAIN

Yuanping He
Applied Mathematics Institute
Department of Mathematics - University of Alberta
533 Central Academic Building, monton, T6G 2G1
CANADA

Zordan C.
Dipartimento di Ingegneria, Biofisica ed Elettronica, Università di Genova
via Opera Pia, 11/A, 16145 GENOVA
ITALY

Part I

General Topics on Dissipative Waves

F BAMPI AND C ZORDAN

Evanescent waves and dissipative systems

ABSTRACT

Evanescent (or inhomogeneous) waves appear naturally as possible emergent waves generated by a collision of a plane wave against an interface between two different media. In conjunction with dissipative systems, this note points out a remarkable feature of evanescent waves. Specifically, a definition of dissipative hyperbolicity is proposed and it is shown that, unlike standard hyperbolic systems, dissipation is consistent with evanescent waves which attenuate along the propagation direction. A physical model which is dissipative hyperbolic is also exhibited.

1. INTRODUCTION

Plane waves and evanescent waves—sometimes called inhomogeneous plane waves [1,2]—are the relevant ingredients entering many physical problems described by linear systems of partial differential equations. A well known example in which evanescent waves are of vital importance is provided by the total reflection of electromagnetic waves in a wave guide, as mentioned also in textbooks [3]. In general, a typical situation mirroring the cited example is the problem of oblique incidence of plane waves on a boundary between two different media. As shown in [4,5], calculation of emergent waves necessarily implies the presence of evanescent waves.

The nice features of evanescent waves are not confined to the field of wave generation at an interface. Indeed, as we shall show in this note, appeal to the properties of evanescent-wave solutions allows a significant characterization of dissipation in physical systems described by linear differential equations. To carry out this program, Sect. 2 provides first the formal definition of plane and evanescent waves, while Sect. 3 draws the consequences of applying such definitions to hyperbolic systems. In full generality, it turns out that plane waves are characterized by one arbitrary function; on the contrary, the knowledge of evanescent waves depends on the choice of two arbitrary functions. As a peculiar result, evanescent waves compatible with hyperbolic systems cannot attenuate along the propagation direction.

To proceed further, notice that the oblique-incidence problems establish that the arbitrary functions involved in evanescent-wave solutions must be the Hilbert transforms of each other [6]. This fact supports the wide use of the complex exponential function as an effective tool for analyzing wave behaviors in linear problems. The complex exponential function is employed in Sect. 4 for defining dissipative hyperbolicity. The main result is that dissipative hyperbolic systems allow evanescent waves to attenuate along the propagation direction. In conclusion evanescent waves serve effectively for detecting dissipation in linear systems. Significantly, Sect. 5 proves that the differential system governing linear viscous fluids is dissipative hyperbolic.

2. EVANESCENT WAVES

When the behavior of a physical system is described by linear partial differential equations, there exist often distinguished solutions called plane waves. Precisely, on denoting by U the N-dimensional column vector representing the relevant field variables, we have the following

DEFINITION 2.1. *Consider a linear system of partial differential equations; a solution of the form*

$$U = U(\varphi), \tag{1}$$

which, for any real frequency ω, depends on space coordinates $\mathbf{x}$ and time t through the single phase variable

$$\varphi(\mathbf{x}, t) = \omega t - \mathbf{k} \cdot \mathbf{x}, \tag{2}$$

$\mathbf{k}$ *being the wavenumber vector, is called a plane wave.*

It is apparent that the solution (1) represents an assigned profile $U = U(\varphi)$ traveling unaltered along the direction

$$\mathbf{n} = \mathbf{k}/|\mathbf{k}| \tag{3}$$

at a propagation speed

$$c = \omega/k, \qquad k = |\mathbf{k}|. \tag{4}$$

Of course, plane-wave solutions not only do not exhaust the class of solutions to linear differential equations, but there are differential systems which admit no plane-wave solutions at all; a well known example of such systems is provided by the linear Navier-Stokes equations governing viscous fluid-dynamics (cf. Sect. 5). Also, as the literature shows [4,6,7,8], plane-wave solutions do not suffice for providing a consistent evaluation of the reflection-transmission pattern generated by the oblique incidence of a plane wave on a boundary. Accordingly, Definition 2.1 is generalized as follows.

DEFINITION 2.2. *Consider a linear system of partial differential equations; a solution of the form*

$$U = U(\varphi_1, \varphi_2) \tag{5}$$

where

$$\varphi_1 = \omega t - \mathbf{k}_1 \cdot \mathbf{x}, \qquad \varphi_2 = \mathbf{k}_2 \cdot \mathbf{x}, \tag{6}$$

is called an evanescent wave.

Sometimes in the literature [1,2], evanescent waves are referred to as inhomogeneous plane waves. From a physical viewpoint, an evanescent wave is such that its profile (5), as evaluated on the planes of constant amplitudes φ_2 = const, propagates without changes along the direction $\mathbf{n}_1 = \mathbf{k}_1/|\mathbf{k}_1|$ at a speed $V_{\text{phase}} = \omega/|\mathbf{k}_1|$, whereas on the planes of constant phase φ_1 = const the amplitude attenuates along the direction $\mathbf{n}_2 = \mathbf{k}_2/|\mathbf{k}_2|$ at a rate measured by the quantity $|\mathbf{k}_2|$. Naturally, attenuation takes place provided that the column vector U vanishes as $\mathbf{n}_{\text{att}} \to \infty$.

Detailed discussions on evanescent (or inhomogeneous) waves can be found in References [1,8].

3. HYPERBOLIC SYSTEMS

Consider a physical system whose behavior is described by the following N linear differential equations

$$\frac{\partial U}{\partial t} + A^p \frac{\partial U}{\partial x^p} = 0, \tag{7}$$

where A^p, $p = 1,2,3$, are constant $N \times N$ matrices, and suppose that the system (7) is hyperbolic according to the usual definition—see, e.g., [9].

It is an easy matter to prove, by substitution, that a plane-wave solution (1) to the system (7) splits as

$$U = U(\varphi) = \mathcal{U}[\omega(t - \mathbf{n} \cdot \mathbf{x}/c)]\Pi, \tag{8}$$

where $\mathcal{U}$ (the amplitude) is an arbitrary function while the propagation speed c and the polarization vector Π satisfy the following well-known algebraic conditions [10]

$$\det |A_n - cI| = 0, \qquad (A_n - cI)\Pi = 0, \tag{9}$$

the symbol I standing for the N dimensional identity matrix and $A_n = A^p n_p$. We remind the reader that hyperbolicity implies that all the propagation speeds c, eigenvalues of A_n, are real and that the set of polarization vectors Π, eigenvectors of A_n, is a basis for $\mathbb{R}^N$.

Look now at evanescent waves. Substitution of (5) into (7) shows that the function $U(\varphi_1, \varphi_2)$ satisfies the system

$$\mathcal{A}\frac{\partial U}{\partial \varphi_1} + \mathcal{B}\frac{\partial U}{\partial \varphi_2} = 0, \tag{10}$$

where

$$\mathcal{A} = \omega I - k_{1p}A^p, \qquad \mathcal{B} = k_{2p}A^p. \tag{11}$$

Experience teaches us that evanescent waves arise when, looking for plane-wave solutions to (7), the phase (2) becomes a complex quantity [6,7]. Fruitfully, we profit of such a circumstance by using the complex formalism as a formal tool for calculating explicitly the evanescent-wave solution (5). To avoid attenuation in time, we assume that the frequency ω is always a real quantity, whereas, we let, if necessary the wavenumber k take on complex values. Also, we denote the real and imaginary part of a complex quantity ψ in accordance with the formula

$$\psi = \psi_1 + i\psi_2, \qquad i = \sqrt{-1}.$$

Hence, evanescent waves can be calculated from plane waves by letting the phase φ take on the expression

$$\varphi = \varphi_1 - i\varphi_2, \tag{12}$$

with φ_1 and φ_2 given by (6). This means that the complex wavenumber vector is $\mathbf{k} = \mathbf{k}_1 - i\mathbf{k}_2$ which in turn makes $\mathbf{n}$ into a complex direction via (3). As a consequence, the polarization vector becomes complex too and, owing to (9) and (11), its real and imaginary parts satisfy the system

$$\begin{cases} \mathcal{A}\Pi_1 + \mathcal{B}\Pi_2 = 0, \\ \mathcal{A}\Pi_2 - \mathcal{B}\Pi_1 = 0. \end{cases} \tag{13}$$

Therefore, a general solution to (10) can easily be found as follows. Let $\mathcal{U}(\varphi)$ be an arbitrary function of the complex phase (12); we have

$$\mathcal{U}(\varphi_1 - i\varphi_2) = \mathcal{U}_1(\varphi_1, \varphi_2) + i\mathcal{U}_2(\varphi_1, \varphi_2) \tag{14}$$

Owing to the system (13) and to the Cauchy-Riemann conditions for the complex function (14), it is a simple matter to ascertain that the real and imaginary part of the complex N-dimensional vector

$$U = (\mathcal{U}_1 + i\mathcal{U}_2)(\Pi_1 + i\Pi_2) = (\mathcal{U}_1\Pi_1 - \mathcal{U}_2\Pi_2) + i(\mathcal{U}_1\Pi_2 + \mathcal{U}_2\Pi_1) \tag{15}$$

are separately solutions to the system (10)—see [5]. Accordingly, the general evanescent-wave solution to (10) is a linear combination of the real and imaginary part of (15). In view of this, it is evident that there is no reason for employing the same function $\mathcal{U}$ both for the real and imaginary part of (15). In conclusion, the general evanescent-wave solution to (7) can be given the form

$$U(\varphi_1, \varphi_2) = (\mathcal{U}_1 + \mathcal{V}_2)\Pi_1 - (\mathcal{U}_2 - \mathcal{V}_1)\Pi_2, \tag{16}$$

where the functions $\mathcal{U}_1$, $\mathcal{U}_2$ and $\mathcal{V}_1$, $\mathcal{V}_2$ are evaluated through relation (14) by means of two totally arbitrary functions $\mathcal{U}$ and $\mathcal{V}$, respectively.

As shown in [6], when waves are generated at a boundary via oblique incidence, the Rankine-Hugoniot conditions and the requisite of well-behavior at infinity demand that the functions $\mathcal{U}$ and $\mathcal{V}$ are conjugate functions; in other words $\mathcal{U}$ and $\mathcal{V}$ must be the Hilbert transforms of each other. Since the functions $\cos\theta$ and $-\sin\theta$ are conjugate functions [11], frequently the formal machinery of Hilbert transforms is effectively avoided by utilizing from the start complex exponential functions. For simplicity, from now on we shall use the complex exponential function formalism.

As proved in [8], there holds a theorem which summarizes the situation.

THEOREM 3.1. *Evanescent waves compatible with hyperbolic systems attenuate in accordance with the following rules*

(a) the amplitude attenuation and the propagation occur along orthogonal directions if and only if either c is real or c is purely imaginary;

(b) the amplitude attenuation cannot occur along the propagation direction, in formula $\mathbf{k}_1 \times \mathbf{k}_2 \neq \mathbf{0}$ always.

The proof relies on a result that hyperbolic systems allow for evanescent waves provided that the vector $\mathbf{n}$ has both the real and the imaginary part nonvanishing.

4. DISSIPATIVE HYPERBOLIC SYSTEMS

We have just seen that evanescent waves compatible with hyperbolic systems enjoy peculiar properties. Unfortunately, hyperbolic systems rule out the possibility of accounting intrinsically for dissipation. Perhaps, the most convincing result is that, according to Theorem 3.1 (b), hyperbolicity forbids the existence of evanescent waves which attenuate along the propagation direction whereas, from the physical point of view, attenuation along the propagation direction seems unavoidable when dissipation occurs in linear systems.

The simplest way that dissipation can enter hyperbolic systems is via an inhomogeneous term on the r.h.s. of (7), namely via a source term. Also, dissipation may be intrinsically present in the physical model, as it happens in the case of viscoelasticity, viscous fluid dynamics, electrodynamics in lossy media, to mention only a few. This suggests that we propose an enlarged definition of hyperbolicity we shall call dissipative hyperbolicity.

To this end, we consider a general linear system of N differential equations in the form

$$\frac{\partial U}{\partial t} + \mathcal{L}U = 0, \tag{17}$$

$\mathcal{L}$ being a linear integro-differential operator which does not involve time derivatives of U. On looking for solutions to (17) in the form of exponential plane waves, viz

$$U = \exp(i\varphi)\Pi, \tag{18}$$

we arrive at the following algebraic condition

$$[L(k\mathbf{n}) - \omega I]\Pi = 0, \tag{19}$$

where the $L(k\mathbf{n})$ denotes the constant matrix defined as

$$L(k\mathbf{n})\Pi = i\exp(-i\varphi)\mathcal{L}[\Pi\exp(i\varphi)]. \tag{20}$$

We are now in a position to make the following definition [8].

DEFINITION 4.1. *The system* (17) *is said dissipative hyperbolic in the t-direction if and only if*

(a) *there exist physical parameters (the dissipation parameters) whose vanishing renders the system* (17) *hyperbolic in the usual sense;*

(b) *for any real unit vector* $\mathbf{n}$, *the set of right eigenvector* Π, *as calculated from* (19), *constitutes a basis for* $\mathbb{C}^N$.

Note that condition (a) inhibits elliptic, but not parabolic, systems to satisfy the previous definition. Of course suitable (thermodynamic) restrictions are to be imposed on the dissipation parameters in order that energy is really dissipated. Many of the physical models for dissipative systems are in fact dissipative hyperbolic models. As an example, in next Section we deal with the case of linear viscous fluids; a detailed discussion on inhomogeneous hyperbolic systems, isotropic viscoelasticity, and electromagnetic waves in a conducting medium as examples of dissipative hyperbolic systems can be found in [8].

A dissipative hyperbolic system typically differs from a hyperbolic system in that condition (19), along with definition $(4)_1$, may render the propagation speed c complex even when $\mathbf{n}$ is a real vector, whereas this is forbidden for standard hyperbolic systems by their very definition. Hence Theorem 3.1 modifies as follows.

THEOREM 4.1. *Evanescent waves compatible with dissipative hyperbolic systems are characterized in the following way:*

(a) *the quantity* c *is a real or a purely imaginary quantity, then*

$$\begin{cases} c = c_1 \quad or \quad c = ic_2 \\ \mathbf{n} = \mathbf{n}_1 + i\mathbf{n}_2 \end{cases} \implies \mathbf{k}_1 \cdot \mathbf{k}_2 = 0;$$

(b) *the quantity* c *is a complex quantity, then either*

$$\begin{cases} c = c_1 + ic_2 \\ \mathbf{n} = \mathbf{n}_1 \end{cases} \implies \mathbf{k}_1 \times \mathbf{k}_2 = \mathbf{0}$$

or

$$\begin{cases} c = c_1 + ic_2 \\ \mathbf{n} = \mathbf{n}_1 + i\mathbf{n}_2 \end{cases} \quad \textit{(the general case).}$$

We stress that, as condition (b) shows, dissipative hyperbolic systems do admit evanescent waves which attenuate along the propagation direction, possibility which is ruled out by standard hyperbolicity. Thus the presence of evanescent waves satisfying condition (b) must be regarded as the peculiar feature which fully characterizes dissipative hyperbolicity.

5. AN EXAMPLE OF DISSIPATIVE HYPERBOLICITY

As an outstanding example of dissipative hyperbolicity, we consider the equations governing viscous fluid-dynamics. It is well known—see, e.g., [12]—that a viscous fluid of mass density ρ, subject to a pressure p, and traveling at velocity $\mathbf{v}$ is governed by the Navier-Stokes equations

$$\begin{cases} \dfrac{\partial \rho}{\partial t} + \rho \nabla \cdot \mathbf{v} + (\mathbf{v} \cdot \nabla)\rho = 0, \\ \rho\Big(\dfrac{\partial \mathbf{v}}{\partial t} + (\mathbf{v} \cdot \nabla)\mathbf{v}\Big) = -\nabla p + \mu \nabla^2 \mathbf{v} + (\mu + \lambda)\nabla(\nabla \cdot \mathbf{v}), \end{cases} \tag{21}$$

where μ and λ are the viscosity coefficients. Also, for simplicity, we make the constitutive assumption $p = p(\rho)$. To obtain a linear system, we linearize the balance equations (21) around the constant state $\rho = \rho_0$, $\mathbf{v} = \mathbf{0}$. Since $\nabla p = dp/d\rho \, \nabla \rho$, on introducing the propagation speed c_0 for inviscid fluids defined by the formula $c_0 = (dp/d\rho)_{\rho=\rho_0}$, we get

$$\begin{cases} \dfrac{\partial \rho}{\partial t} + \rho_0 \nabla \cdot \mathbf{v} = 0, \\ \dfrac{\partial \mathbf{v}}{\partial t} + \dfrac{c_0}{\rho_0}\nabla \rho - \dfrac{\mu}{\rho_0}\nabla^2 \mathbf{v} - \dfrac{\mu+\lambda}{\rho_0}\nabla(\nabla \cdot \mathbf{v}) = 0. \end{cases} \tag{22}$$

On setting $U = (\rho \quad \mathbf{v})^T$, we search solutions to (22) of the form (18); since the fluid is isotropic, we lose no generality by assuming that the wave travels along the z axis. Accordingly, condition (19) implies

$$\det \begin{vmatrix} \omega & 0 & 0 & -\rho_0 k \\ 0 & \omega - ib_T^2 k^2 & 0 & 0 \\ 0 & 0 & \omega - ib_T^2 k^2 & 0 \\ -(c_0/\rho_0)k & 0 & 0 & \omega - ib_L^2 k^2 \end{vmatrix} = 0, \tag{23}$$

having employed the notation

$$b_T^2 = \frac{\mu}{\rho_0}, \qquad b_L^2 = \frac{2\mu + \lambda}{\rho_0}.$$

As a calculation shows, condition (23) explicitly reads

$$(\omega - ib_T^2 k^2)^2 [\omega^2 - (c_0 + i\omega b_L^2)k^2] = 0; \tag{24}$$

whose solutions are

$$k_T^2 = \frac{\omega}{ib_T^2}, \quad \text{multiplicity} = 2; \tag{25}$$

$$k_L^2 = \frac{\omega^2}{c_0 + i\omega b_L^2}. \tag{26}$$

We observe that waves in viscous fluids are dispersive. The eigenvectors associated with the solutions (25) and (26) are found to be

$$k = k_L^{\pm} \Longrightarrow \Pi = (\rho_0 k_L^{\pm} \quad 0 \quad 0 \quad \omega)^T ;$$
$$k = k_T^{\pm} \Longrightarrow \begin{cases} \Pi = (0 \quad 1 \quad 0 \quad 0)^T, \\ \Pi = (0 \quad 0 \quad 1 \quad 0)^T. \end{cases}$$

Apparently, these four eigenvectors constitute a basis for $\mathbb{C}^4$: therefore the system (22) is dissipative hyperbolic. It should come as no surprise that the dissipation parameters are the viscosities μ and λ.

Note that somewhat different results on inhomogeneous wave propagation in viscous fluids have been obtained by Boulanger and Hayes in [13] simply because they have considered incompressible viscous fluids.

A further comment is in order. The propagation speeds relative to (linear) inviscid fluids can be obtained by (25) and (26) by letting the dissipation parameters μ and λ vanish; using (4) we can write

$$c = \pm c_0, \qquad c = 0, \quad \text{multiplicity} = 2.$$

We point out that inviscid fluids admit four propagation speeds, two of them concerning material waves. Now viscosity causes a splitting of the zero speeds thereby making viscous fluids possess six propagation speeds. Although this could seem a strange result, the explanation is quite easy. Owing to the isotropy of the fluid, as soon as a wave can propagate in the direction $\mathbf{n}$ with speed $\tilde{c}$, another wave must exist which propagate in the same direction $\mathbf{n}$ with opposite speed $-\tilde{c}$. Accordingly, since viscous fluids do not admit any material wave, the material wave of inviscid fluids must correspond to two opposite speeds, thereby rendering higher the number of propagation speeds in viscous fluids.

REFERENCES

1. M. HAYES, Inhomogeneous plane waves. Arch. Ration. Mech. Anal. **85** (1984) 41–79.
2. M. HAYES, M.J.P. MUSGRAVE, On energy flux and group velocity. Wave Motion **1** (1979) 75–82.
3. J.D. JACKSON, The Theory of Electromagnetism (Pergamon Oxford 1964).
4. F. BAMPI, C. ZORDAN, Solving Snell's law. Mech. Res. Comm. (in print).
5. F. BAMPI, C. ZORDAN, Discontinuity waves cannot describe evanescent waves. (in preparation).
6. F. BAMPI, C. ZORDAN, Hilbert transforms in wave propagation theory. (submitted for publication).

7. E. DIEULESAINT, D. ROYER, Elastic Waves in Solids (Wiley New York 1980).
8. C. ZORDAN, Wave generation at an interface. Atti Accad. Peloritana, Classe Sci. Mat. Fis. Nat. (in print).
9. A. JEFFREY, Quasilinear Hyperbolic Systems and Waves (Pitman London 1976).
10. G. BOILLAT, La Propagation des Ondes (Gauthier Villars Paris 1965).
11. E.C. TITCHMARSH, Introduction to the Theory of Fourier Integrals (Clarendon Oxford 1937).
12. H. LAMB, Hydrodynamics (Cambridge University Press Cambridge 1932).
13. PH. BOULANGER, M. HAYES, Inhomogeneous plane waves in viscous fluids. Continuum Mech. Thermodyn. **2** (1990) 1–16.

Franco Bampi
Istituto Matematico di Ingegneria
Università di Genova
Piazzale J.F. Kennedy, Pad. D
16129 Genova

Clara Zordan
Dip. Ing. Biofisica ed Elettronica
Università di Genova
Via all'Opera Pia 11a
16145 Genova

Ph BOULANGER AND M HAYES

On the energy flux for finite amplitude waves in Navier-Stokes and second grade fluids

Abstract

In this note, two finite amplitude inhomogeneous waves are considered. One is a circularly polarized wave which may propagate in a second grade fluid. The second is a unidirectional motion which is possible in a Navier-Stokes fluid. For both waves we obtain moduli independent relations between the mean energy flux, energy density and dissipation. These are the same as those obtained previously for small amplitude waves in classical linearly viscous fluids. However, for a second grade fluid the energy density is the sum of the kinetic energy and a stored energy density.

1 Introduction

In [1], the propagation of small amplitude inhomogeneous waves in a classical linearly viscous incompressible fluid has been considered. For such waves, the velocity field is of the form $\mathbf{v} = \{\mathbf{A} \exp \iota(\mathbf{K}.\mathbf{x} - \omega t)\}^+$, where $\mathbf{K} = \mathbf{K}^+ + \iota\mathbf{K}^-$ is the complex wave vector, $\mathbf{A} = \mathbf{A}^+ + \iota\mathbf{A}^-$ the complex amplitude, and $\omega = \omega^+ + \iota\omega^-$ the complex angular frequency. As the waves are assumed to be of small amplitude, the non-linear term $\mathbf{v}.\text{grad}\,\mathbf{v}$ has been neglected in the expression for the acceleration. It has then been shown that, for all inhomogeneous wave solutions obtained, the "weighted mean" energy flux vector $\hat{\mathbf{R}}$, the "weighted mean" kinetic energy $\hat{E}$, and the "weighted mean" dissipation $\hat{D}$ are related through $\hat{\mathbf{R}}.\mathbf{K}^+ = \omega^+\hat{E}$, $\hat{\mathbf{R}}.\mathbf{K}^- = \omega^-\hat{E} + (1/2)\hat{D}$.

The purpose of this note is to consider the energy flux vector, the energy density and energy dissipation for finite amplitude motions in Navier-Stokes and second grade fluids. Two finite amplitude inhomogeneous waves have been obtained in [2]. The first is a circularly polarized inhomogeneous wave which may propagate in a second grade fluid and therefore also in a Navier-Stokes fluid. Its velocity field does not depend on the material moduli. The second is a linearly polarized wave motion which is possible in a Navier-Stokes fluid. The flow is unidirectional.

For these finite amplitude waves in a Navier-Stokes fluid we obtain the same modulus independent relations between $\hat{\mathbf{R}}, \hat{E}, \hat{D}$ as in the case of small amplitude waves.

For the finite amplitude wave obtained in the case of a second grade fluid, these relations also remain valid with an appropriate definition of the energy density E as the sum of the kinetic energy and a stored energy, and a corresponding appropriate definition of the dissipation D.

2 Basic equations

The constitutive equation for a second grade fluid is

$$\mathbf{t} = -p\mathbf{1} + \mu\mathbf{A}_1 + \alpha_1\mathbf{A}_2 + \alpha_2\mathbf{A}_1^2, \tag{1}$$

with the constraint of incompressibility

$$\text{div } \mathbf{v} = 0, \tag{2}$$

where $\mathbf{v}$ is the velocity field, p is the pressure, and μ, α_1, α_2 are material moduli. The Rivlin-Ericksen tensors $\mathbf{A}_1$ and $\mathbf{A}_2$ are defined through

$$\mathbf{A}_1 = (\text{grad } \mathbf{v}) + (\text{grad } \mathbf{v})^T, \tag{3}$$

$$\mathbf{A}_2 = \frac{d}{dt}\mathbf{A}_1 + \mathbf{A}_1(\text{grad } \mathbf{v}) + (\text{grad } \mathbf{v})^T\mathbf{A}_1, \tag{4}$$

where $\frac{d}{dt}$ denotes the material time derivative. When $\alpha_1 = \alpha_2 = 0$, we recover the Navier-Stokes fluid (μ is the coefficient of viscosity).

Neglecting body forces, the equations of motion are

$$\rho\frac{d\mathbf{v}}{dt} = \text{div } \mathbf{t}. \tag{5}$$

Substituting the constitutive equation (1) into the equations of motion (5) gives

$$\begin{aligned} \mu\Delta\mathbf{v} &+ \alpha_1\Delta\mathbf{v}_t + \alpha_1\Delta\omega \times \mathbf{v} \\ &+ (\alpha_1 + \alpha_2)\{\mathbf{A}_1\Delta\mathbf{v} + 2\text{div } [(\text{grad } \mathbf{v})(\text{grad } \mathbf{v})^T]\} \\ &- \rho\mathbf{v}_t - \rho\omega \times \mathbf{v} = \text{grad } \hat{p}, \end{aligned} \tag{6}$$

where $\omega = curl\mathbf{v}$ is the vorticity, and $\hat{p}$ is defined by

$$\hat{p} = p + \frac{1}{2}\rho\mathbf{v}.\mathbf{v} - \alpha_1\mathbf{v}.\Delta\mathbf{v} - \frac{1}{4}(2\alpha_1 + \alpha_2)tr(\mathbf{A}_1^2). \tag{7}$$

The subscript t denotes the partial time derivative, and Δ denotes the Laplacian.

The finite amplitude inhomogeneous waves we consider [2] are the following.

(1) Circularly Polarized Wave (Second Grade Fluid)

The components u, v, w of the velocity field $\mathbf{v}$ in Cartesian coordinates are given by

$$\begin{aligned} u &= B\cos(kx - \omega^+ t)\exp(-ky + \omega^- t), \\ v &= -B\sin(kx - \omega^+ t)\exp(-ky + \omega^- t), \\ w &= 0, \end{aligned} \tag{8}$$

where B, k, ω^+ and ω^- are arbitrary constants. Here the complex wave vector is $\mathbf{K} = k(\mathbf{i} + \iota\mathbf{j})$, and the complex angular frequency is $\omega = \omega^+ + \iota\omega^-$. This wave propagates along the x-axis with phase speed (ω^+/k). The planes of constant amplitude are orthogonal to the y-axis and the amplitude is propagated unchanged with speed (ω^-/k). For a second grade fluid, the velocity field (8) is an exact solution to the equation of motion (6) with the pressure field

$$p = p_0 + \frac{\rho}{k}(\omega^+ u + \omega^- v) + \{(4\alpha_2 + 6\alpha_1)k^2 - \frac{\rho}{2}\}(u^2 + v^2). \tag{9}$$

(2) Unidirectional Motion (Navier-Stokes Fluid)

Here the components of the velocity field $\mathbf{v}$ are

$$u = v = 0 \quad , \quad w = A\cos\xi\exp\eta, \tag{10}$$

with ξ and η defined by

$$\begin{aligned} \xi &= kmx\cos\phi - ky\sin\phi - \Omega t\sin 2\phi, \\ \eta &= -kmx\sin\phi - ky\cos\phi - \Omega t\cos 2\phi. \end{aligned} \tag{11}$$

Here A, Ω, ϕ and m are arbitrary constants, with $m^2 \neq 1$, and k is given by

$$k^2 = \frac{\rho\Omega}{\mu(m^2 - 1)}. \tag{12}$$

Here the complex wave vector is $\mathbf{K} = k(\exp\iota\phi)(m\mathbf{i} + \iota\mathbf{j})$, and the complex angular frequency is $\omega = -\iota\Omega\exp\iota 2\phi$. This wave is an exact solutions to the Navier-Stokes equations with a constant pressure field

$$p = p_0. \tag{13}$$

3 Energy flux

The kinetic energy density T, and the energy flux vector $\mathbf{R}$ are given by

$$T = \frac{1}{2}\rho v_i v_i \quad , \quad R_j = -v_i t_{ij}. \tag{14}$$

From the equations of motion (5) we have the balance of energy

$$\frac{dT}{dt} + \text{div }\mathbf{R} + t_{ij}v_{i,j} = 0. \tag{15}$$

Using, the constitutive equation (1), it may be shown that

$$t_{ij}v_{i,j} = \frac{dW}{dt} + D, \tag{16}$$

where W and D are called the "stored energy" and the "dissipation" respectively, and are given by

$$W = -(\alpha_1/4)\text{tr}(\mathbf{A}_1^2), \tag{17}$$

$$D = (\mu/2)\text{tr}(\mathbf{A}_1^2) + ((\alpha_2 - \alpha_1)/2)\text{tr}(\mathbf{A}_1^3) + \alpha_1\text{tr}(\mathbf{A}_1\mathbf{A}_2). \tag{18}$$

Hence, the balance of energy (15) may be written as

$$\frac{dE}{dt} + \text{div }\mathbf{R} + D = 0, \tag{19}$$

where the energy density $E = T + W$ is the sum of the kinetic energy and stored energy. Of course, for a Navier-Stokes fluid, E reduces to the kinetic energy density, and D to the usual expression for the dissipation.

(1) Circularly Polarized Wave (Second Grade Fluid)

For the motion (8), we find

$$\begin{aligned} R_1 &= p_0 u + (\rho/k)u(\omega^+ u + \omega^- v) - 2k\alpha_1\omega^+(u^2+v^2) \\ &\quad -(1/2)(\rho - 8k^2\alpha_1)(u^2+v^2)u, \\ R_2 &= p_0 v + (\rho/k)v(\omega^+ u + \omega^- v) + 2k(\mu + \omega^-\alpha_1)(u^2+v^2) \\ &\quad -(\rho/2)(u^2+v^2)v, \\ R_3 &= 0, \end{aligned} \tag{20}$$

and

$$E = T + W = (1/2)(\rho - 4k^2\alpha_1)(u^2+v^2), \tag{21}$$

$$D = 4k^2\{\mu + 2\alpha_1(\omega^- - kv)\}(u^2+v^2). \tag{22}$$

Accordingly, following the derivation outlined in [1], we compute the means $\tilde{\mathbf{R}}, \tilde{E}, \tilde{D}$ taken over a cycle at fixed amplitude. These are of the form

$$(\tilde{\mathbf{R}}, \tilde{E}, \tilde{D}) = (\hat{\mathbf{R}}, \hat{E}, \hat{D}) \exp -2(\mathbf{K}^-.x - \omega^- t), \tag{23}$$

with $\hat{\mathbf{R}}, \hat{E}, \hat{D}$ given by

$$\begin{aligned} \hat{R}_1 &= (\omega^+/2k)(\rho - 4k^2\alpha_1)B^2, \hat{R}_2 = (1/2k)\{\rho\omega^- + 4k^2(\mu + \alpha_1\omega^-)\}B^2, \hat{R}_3 = 0, \\ \hat{E} &= \hat{T} + \hat{W} = (1/2)(\rho - 4k^2\alpha_1)B^2, \\ \hat{D} &= 4k^2(\mu + 2\alpha_1\omega^-)B^2. \end{aligned} \tag{24}$$

These are respectively called the "weighted mean" energy flux, energy density, and dissipation.
We note that

$$\hat{\mathbf{R}}.\mathbf{K}^+ = k\hat{R}_1 = \omega^+ \hat{E}, \quad (25)$$
$$\hat{\mathbf{R}}.\mathbf{K}^- = k\hat{R}_2 = \omega^- \hat{E} + (1/2)\hat{D}. \quad (26)$$

These are the same moduli independent relations as those obtained for small amplitude waves in linearly viscous fluids [1]. here, however, the energy density is not only the kinetic energy density since it includes a stored energy term.

(2) Unidirectional Motion (Navier-Stokes Fluid)

Here $\alpha_1 = \alpha_2 = 0$, hence $W = 0$, and the energy density reduces to the kinetic energy density : $E = T$. The dissipation reduces to $D = (\mu/2)\mathrm{tr}(\mathbf{A}_1^2) = t_{ij}v_{i,j}$. For the motion given by (10) (11), we find

$$\begin{aligned} R_1 &= A^2 k\mu m \sin(\xi + \phi)\cos\xi \exp 2\eta, \\ R_2 &= A^2 k\mu \cos(\xi + \phi)\cos\xi \exp 2\eta, \\ R_3 &= p_0 w, \end{aligned} \quad (27)$$

and

$$E = T = (1/2)\rho w^2, \quad (28)$$
$$D = A^2 k^2 \mu \{m^2 \sin^2(\xi + \phi) + \cos^2(\xi + \phi)\}\exp 2\eta. \quad (29)$$

Again, integrating over a cycle at constant amplitude, we find that the means $\tilde{\mathbf{R}}, \tilde{E}, \tilde{D}$ have the form (23) with $\hat{\mathbf{R}}, \hat{E}, \hat{D}$ given by

$$\begin{aligned} \hat{R}_1 &= (1/2)A^2 k\mu m \sin\phi, \quad \hat{R}_2 = (1/2)A^2 k\mu \cos\phi, \quad \hat{R}_3 = 0, \\ \hat{E} &= (1/4)\rho A^2 \quad , \quad \hat{D} = (1/2)A^2 k^2 \mu (m^2 + 1). \end{aligned} \quad (30)$$

We note that

$$\begin{aligned} \hat{\mathbf{R}}.\mathbf{K}^+ &= k(\hat{R}_1 m \cos\phi - \hat{R}_2 \sin\phi) = \omega^+ \hat{E}, \\ \hat{\mathbf{R}}.\mathbf{K}^- &= k(\hat{R}_1 m \sin\phi + \hat{R}_2 \cos\phi) = \omega^- \hat{E} + (1/2)\hat{D}. \end{aligned} \quad (31)$$

Again, these are the same moduli independent relations as those obtained for small-amplitude waves in linearly viscous fluids [1].

Finally, we note that for ω real ($\omega^- = 0$), these relations become

$$\hat{\mathbf{R}}.\mathbf{S}^+ = \hat{E} \quad , \quad \hat{\mathbf{R}}.\mathbf{S}^- = (1/2\omega)\hat{D} \quad (32)$$

where $\mathbf{S} = \mathbf{S}^+ + \iota\mathbf{S}^-$ is the complex slowness defined by $\mathbf{K} = \omega\mathbf{S}$. The first of these was derived by Hayes [3] for linear conservative systems and the second by Buchen [4] for linear viscoelastic waves in isotropic media.

References

[1] Ph.BOULANGER and M.HAYES, Inhomogeneous Plane Waves in Viscous Fluids. Continuum Mech.Thermodyn.**2** (1990) 1-16.

[2] Ph.BOULANGER, M.HAYES, and K.R.RAJAGOPAL, Some unsteady exact Solutions in the Navier-Stokes and Second Grade Fluid Theories. Stability and Applied Analysis of Continuous Media (in print).

[3] M.HAYES, Energy Flux for trains of Inhomogeneous Plane Waves. Proc.R.Soc.London **A 370** (1980) 417-429.

[4] P.BUCHEN, Waves in Linear Viscoelastic Media. Geophys.J.R.Astr.Soc. **23** (1971) 531-542.

Philippe Boulanger
Département de Mathématique
Université Libre de Bruxelles
Campus Plaine, C.P.218/1
1050 Bruxelles, BELGIUM

Michael Hayes
Department of Mathematical Physics
University College
Belfield
Dublin 4, IRELAND

G CAVIGLIA

Wave propagation in viscoelastic media

ABSTRACT

Time-harmonic wave propagation in linear viscoelastic media is considered. It is shown that plane waves are only allowed in the form of inhomogeneous waves, which attenuate while propagation occurs, as follows from thermodymamic considerations. The behaviour of inhomogeneous waves at a plane interface is then analyzed. The corresponding results are applied to a discussion of scattering of a plane wave from a viscoleastic obstacle immersed in a fluid, in the framework of the Kirchhoff approximation.

1. INTRODUCTION

In many areas of scientific research such as underwater acoustics, seismology, and ultrasound analysis of living tissues, wave phenomena are modelled within the framework of linear viscoelasticity because this allows taking into account the dissipative properties of the materials involved. Through Fourier analysis the problem is then reduced to the study of time-harmonic wave propagation within a viscoelastic solid.

The first aim of this contribution is to to show that plane waves are only allowed in the form of two families of elliptically polarized waves, that are referred to as inhomogeneous waves. Such waves are most conveniently described through the introduction of a complex wavenumber vector whose real and imaginary part identify planes of constant phase and amplitude, respectively [1,2]. The second aim is to investigate the influence of thermodynamic restrictions on the behaviour of time-harmonic waves in viscoelastic solids. As to inhomogeneous waves, it is found in particular that the length of the axes of the polarization ellipse decreases exponentially in the direction of propagation of the phase [1,3,4]. The third aim is to discuss scattering of an incident longitudinal plane wave from a viscoelastic solid immersed in a fluid. Following a Kirchhoff-type approximation [5-7] an integral representation for the transmitted field is proposed and the corresponding far field is determined explicitly [7].

2. VISCOELASTIC SOLIDS

Consider a linear homogeneous isotropic viscoelastic body in a stress free reference configuration. The material properties are described be letting the Cauchy stress $\mathbf{T}$, at time t, be given by

$$\mathbf{T} = 2\mu_0 \mathbf{E} + \lambda_0 (\mathrm{tr}\,\mathbf{E})\,\mathbf{I} + \int_0^\infty [2\,\mu'(s)\,\mathbf{E}(t-s) + \lambda'(s)(\mathrm{tr}\,\mathbf{E})(t-s)\,\mathbf{I}]ds \tag{1}$$

where $\mathbf{E} = \mathrm{sym}\partial\mathbf{u}/\partial\mathbf{x}$ is the infinitesimal strain tensor; $\mathbf{u}$ is the displacement vector; $\mathbf{I}$ is the unit tensor; μ_0, and μ' are the initial value and the derivative of the relaxation modulus in shear; $\lambda + 2\mu/3$ is the relaxation modulus in dilatation: the initial value and the derivative of λ are denoted by λ_0 and λ', respectively.

The second law of thermodynamics is imposed by requiring that when $\mathbf{E}$ is subject to a cyclic history the energy dissipated along each cycle is non-negative [1,3]. In correspondence with a stress tensor of the form (1) this restriction leads to the inequalities

$$\int_0^\infty \mu'(s)\sin(\omega s)ds \leq 0, \qquad \int_0^\infty (\frac{2}{3}\mu' + \lambda')(s)\sin(\omega s)ds \leq 0, \tag{2}$$

holding for any positive angular frequency ω. We have also $\mu_0 > 0$ and $2\mu_0 + 3\lambda_0 > 0$. Finally the equations of motion are taken in the form

$$\rho_0 \ddot{\mathbf{u}} = \nabla \cdot \mathbf{T} \tag{3}$$

where ρ_0 is the (constant) mass density and a superposed dot denotes the time derivative.

3. TIME-HARMONIC WAVES

Suppose that the displacement vector is a harmonic function of time, that is $\mathbf{u}(\mathbf{x}, t; \omega) = \mathbf{U}(\mathbf{x}; \omega)\exp(-i\omega t)$, where ω denotes the constant angular frequency. For convenience the common factor $\exp(-i\omega t)$ is henceforth omitted. Choose for $\mathbf{U}$ the Helmholtz representation

$$\mathbf{U} = \nabla\Phi + \nabla \times \mathbf{\Psi}, \tag{4}$$

where Φ and $\mathbf{\Psi}$ denote the scalar and the vector potential, respectively. Substitution of (4) into (1) and comparison with (3) leads to a formulation of the equations of motion in terms of potentials as

$$\triangle\Phi + \frac{\rho_0 \omega^2}{2\hat{\mu} + \hat{\lambda}}\Phi = 0, \qquad \triangle\mathbf{\Psi} + \frac{\rho_0\omega^2}{\hat{\mu}}\mathbf{\Psi} = 0, \tag{5}$$

where

$$\hat{\lambda} = \lambda_0 + \int_0^\infty \lambda'(s)\exp(i\omega s)ds, \qquad \hat{\mu} = \mu_0 + \int_0^\infty \mu'(s)\exp(i\omega s)ds.$$

As a consequence of the thermodynamic restrictions (2) it is found that the the imaginary parts of the complex coefficients entering the Helmholtz equations (5) are non-negative, that is,

$$\Im\left(\frac{\rho_0\omega^2}{2\hat{\mu}+\hat{\lambda}}\right) \geq 0 \quad \text{and} \quad \Im\left(\frac{\rho_0\omega^2}{\hat{\mu}}\right) \geq 0, \tag{6}$$

where $\Im$ denotes the imaginary part.

Time-harmonic wave propagation in elastic bodies and in viscous or inviscid fluids is governed by equations of the form (5), where the scalar coefficients are suitably reinterpreted [1]. As to inviscid fluids we let the pressure p and the density ρ be related by $p = p(\rho)$; then we find that the vector potential vanishes identically, whereas the scalar potential is determined as a solution to the linearized equation of motion

$$\triangle\Phi + \kappa\Phi = 0, \tag{7}$$

where $\kappa = \omega^2/(dp/d\rho)$.

4. INHOMOGENEOUS WAVES

Inhomogeneous longitudinal and transverse waves are characterized as solutions to the scalar and the vector Helmholtz equations (5) that are represented in the form of complex exponentials. Specifically, for inhomogeneous longitudinal waves we find $\Phi = A\exp(i\mathbf{k}_L \cdot \mathbf{x})$ where A is constant and the (complex) wavenumber vector $\mathbf{k}_L$ is given by

$$\mathbf{k}_L \cdot \mathbf{k}_L = \frac{\rho_0\omega^2}{2\hat{\mu}+\hat{\lambda}} =: \kappa_L; \quad \mathbf{U} = i\mathbf{k}_L\Phi. \tag{8}$$

Similarly, inhomogeneous transverse waves are given by $\mathbf{\Psi} = \mathbf{q}\exp(i\mathbf{k}_T \cdot \mathbf{x})$ with $\mathbf{q}$ constant and $\mathbf{k}_T$ satisfying

$$\mathbf{k}_T \cdot \mathbf{k}_T = \frac{\rho_0\omega^2}{\hat{\mu}} =: \kappa_T; \quad \mathbf{U} = i\mathbf{k}_T \times \mathbf{\Psi}. \tag{9}$$

To evidence those properties of inhomogeneous waves that follow from the complex-valued wavenumber vectors it is convenient to represent $\mathbf{k}$ through its real and imaginary parts, say $\mathbf{k}_1$ and $\mathbf{k}_2$. Thus, on setting $\mathbf{k} = \mathbf{k}_1 + i\mathbf{k}_2$ we find that dependence on position is given by

$$\exp(i\mathbf{k} \cdot \mathbf{x}) = \exp(-\mathbf{k}_2 \cdot \mathbf{x}) \exp(i\mathbf{k}_1 \cdot \mathbf{x}) \tag{10}$$

whence it follows that the condition $\mathbf{k}_1 \cdot \mathbf{x} = c_1$ yields planes of constant phase while $\mathbf{k}_2 \cdot \mathbf{x} = c_2$ yields planes of constant amplitude.

The amplitude of inhomogeneous waves decreases exponentially in the direction of $\mathbf{k}_1$. The result follows from the identity

$$\mathbf{u}(\mathbf{x} + 2\pi n \mathbf{k}_1/k_1^2, t) = \exp(-\pi n b/k_1^2)\mathbf{u}(\mathbf{x}, t), \qquad b := \Im(\mathbf{k} \cdot \mathbf{k}), \tag{11}$$

holding for any positive integer n. In fact we have $b \geq 0$ as a consequence of the formulation (6) of the second law of thermodynamics, and thus (11) implies that the amplitude decays while phase propagates, the attenuation coefficient being related to the material parameters of the solid through (8) and (9). Of course, no decay occurs when $\mathbf{k}_1$ and $\mathbf{k}_2$ are orthogonal, which corresponds to the case when the coefficients entering the Helmholtz equations (5) are real, that is to the absence of dissipation. As to polarization, on letting $\mathbf{U}(\mathbf{x}; \omega) = (\mathbf{p}_1 + i\mathbf{p}_2) \exp(i\mathbf{k} \cdot \mathbf{x})$, with constant $\mathbf{p}_1$ and $\mathbf{p}_2$, it is found that

$$\mathbf{u}(\mathbf{x}, t) = \exp(-\mathbf{k}_2 \cdot \mathbf{x})[\mathbf{P}(\mathbf{x}) \cos \omega t + \mathbf{Q}(\mathbf{x}) \sin \omega t]$$

where $\mathbf{P}$ and $\mathbf{Q}$ are defined by

$$\mathbf{P}(\mathbf{x}) + i\mathbf{Q}(\mathbf{x}) = (\mathbf{p}_1 + i\mathbf{p}_2) \exp(i\mathbf{k}_1 \cdot \mathbf{x}),$$

thus showing that $\mathbf{u}$ is given by an elliptically-polarized wave [1,2]. The polarization ellipse depends on $\mathbf{x}$. It is also worth remarking that in general a longitudinal inhomogeneous wave corresponds to an elliptically polarized wave, unlike the case of linear elasticity.

5. BEHAVIOUR AT AN INTERFACE

Let S denote the boundary between two media with different material properties; it is assumed that S is sufficiently regular. As a result of the discontinuity at the boundary an incident wave is partially reflected and partially transmitted at S. The characteristics of the waves originated at S are determined through the requirements of continuity of the displacement $\mathbf{u}$ and of the traction $\mathbf{Tn}$, where $\mathbf{n}$ denotes the normal to the surface.

To simplify the analysis consider the boundary between a viscoelastic solid and an inviscid fluid, and a longitudinal wave travelling within the fluid and hitting the interface. Under these conditions, besides continuity of traction, we only require continuity of the normal component of $\mathbf{u}$. It is well known that, in the case of a plane boundary, a reflected longitudinal wave, and two transmitted waves, a longitudinal and a transverse one, originate at $\mathcal{S}$. The corresponding wavenumber vectors are determined through Snell's law, which is a consequence of the continuity requirements, and through conditions (8) and (9) [1]. The related amplitudes follow as a further consequence of the continuity conditions [1]. In particular, the reflection coefficient $\mathcal{R}$ is given by

$$\mathcal{R} = \frac{4\sqrt{\kappa - \xi^2}[\xi^2\sqrt{\kappa_L - \xi^2}\sqrt{\kappa_T - \xi^2} + (\xi^2 - \frac{1}{2}\kappa_T)^2] - \upsilon\kappa_T^2\sqrt{\kappa_L - \xi^2}}{4\sqrt{\kappa - \xi^2}[\xi^2\sqrt{\kappa_L - \xi^2}\sqrt{\kappa_T - \xi^2} + (\xi^2 - \frac{1}{2}\kappa_T)^2] + \upsilon\kappa_T^2\sqrt{\kappa_L - \xi^2}}, \qquad (12)$$

where $\upsilon = \rho/\rho_0$; ξ represents the common value of the projections onto $\mathcal{S}$ of the wavenumber vectors belonging to incident, reflected and transmitted waves; the roots $\sqrt{\kappa_L - \xi^2}$ and $\sqrt{\kappa_T - \xi^2}$ are chosen with positive real part.

6. SCATTERING PROBLEMS AND THE KIRCHHOFF APPROXIMATION

In a typical scattering problem it is assumed that the viscoelastic body is immersed in the fluid. For convenience we let the region occupied by the fluid be bounded and connected, and we let $\mathcal{S}$ denote its boundary. The problem is to find the field at infinity that results from the scattering of the incident plane wave.

The scattered field Φ^s obeys Helmholtz's equation (5a) in the domain exterior to $\mathcal{S}$; it is completely determined by its boundary values at $\mathcal{S}$ and by suitable conditions at infinity, stating that it behaves like a radiation field [8]. Use of the Green's identity and of the Sommerfeld radiation condition provides an integral representation for the scattered field as

$$-4\pi\Phi^s(\mathbf{x}_0) = \int_{\mathcal{S}} \left[\frac{1}{r}\exp(ikr)\frac{\partial\Phi^s}{\partial n} - \frac{\partial}{\partial n}\left(\frac{1}{r}\exp(ikr)\right)\Phi^s\right] da; \qquad (13)$$

here $k = \sqrt{\kappa}$; $r = |\mathbf{r}|$ with $\mathbf{r} = \mathbf{x} - \mathbf{x}_0$ and $\mathbf{x} \in \mathcal{S}$; $\mathbf{n} = \mathbf{n}(\mathbf{x})$ is the outward normal to $\mathcal{S}$.

The representation (13) determines the field Φ^s at any point $\mathbf{x}_0$ outside $\mathcal{S}$ in terms of the value of Φ^s and of its normal derivative at the obstacle boundary. To obtain such values we follow the idea of Kirchhoff approximation in the sense that any point of the reflecting surface which is "illuminated" by the incident wave is regarded as though the surface were locally flat [5-7]. Therefore the wave incident at that point is reflected according to Snell's law relative to the tangent plane while the amplitude is determined by the reflection coefficient $\mathcal{R}$. The scattered field is assumed to be zero on that part of $\mathcal{S}$ which is not "illuminated" by the incident wave.

To be specific, denote by $\mathbf{k}^i$ the wavenumber vector of the incident longitudinal plane wave. Since the wave is travelling within the fluid it is assumed that $\mathbf{k}^i$ is real. We regard $\mathcal{S}^- = \{\mathbf{x} \in \mathcal{S}\ \mathbf{n} \cdot \mathbf{k}^i < 0\}$ as the illuminated region: in the approximation adopted the integral (13) is thus reduced to an integral over $\mathcal{S}^-$.

As to the determination of the value of Φ^s and $\partial\Phi^s/\partial n$ at any point $\mathbf{x}$ of $\mathcal{S}^-$, we observe that $\mathbf{k}^i = -\sqrt{\kappa - \xi^2}\mathbf{n} + \xi\mathbf{e}$ where $\mathbf{e}$ is the unit vector of the projection of $\mathbf{k}^i$ on the tangent plane at $\mathbf{x}$ while $\xi = \xi(\mathbf{x})$ is the common value of the projections of the wavenumber vectors of incident, reflected, and transmitted waves. Then the specularly reflected wavenumber vector reads $\mathbf{k}^r = \sqrt{\kappa - \xi^2}\,\mathbf{n} + \xi\mathbf{e}$. Since the scattered field at $\mathcal{S}^-$ is identified with the reflected wave, we find

$$\Phi^s(\mathbf{x}) = \Phi_0 \mathcal{R} \exp(i\mathbf{k}^r \cdot \mathbf{x}), \qquad \frac{\partial \Phi^s}{\partial n}(\mathbf{x}) = -i(\mathbf{k}^i \cdot \mathbf{n})\Phi^s(\mathbf{x}), \tag{14}$$

for an incident wave of unit amplitude. In conclusion the integral representation (13) may be approximated by

$$-4\pi\Phi^s(\mathbf{x}_0) = \int_{\mathcal{S}^-} \frac{\exp(ikr)}{r}\left[-ik\cos(\mathbf{n}, \mathbf{k}^i) + (\frac{1}{r} - ik)\cos(\mathbf{n}, \mathbf{r})\right] \mathcal{R}\exp(i\mathbf{k}^i \cdot \mathbf{x})\, da \tag{15}$$

The result (15) provides the general expression for the field Φ^s at any point $\mathbf{x}_0$ exterior to $\mathcal{S}$. In particular the reflection coefficient $\mathcal{R}$, which depends on position through ξ, accounts for the material properties of the obstacle.

Especially in connection with the inverse scattering, it is of interest to determine the far field, namely the field at large distances with respect to the dimensions of the obstacle. To do this we approximate (15) by disregarding the terms $O(|\mathbf{x}|/r_0)$. Denoting by ϑ and φ the polar angles of the point $\mathbf{x}_0$ it follows that there exists a function $F(\vartheta, \varphi)$ such that

$$\Phi^s(\mathbf{x}_0) = \frac{\exp(ikr_0)}{r_0} F(\vartheta, \varphi) + O(\frac{1}{r_0^2}),$$

with $r_0 = |\mathbf{x}_0|$. The scalar function F is called far field. The representation (15) corresponds to

$$F = \frac{ik}{4\pi}\int_{\mathcal{S}^-} \exp\left[ik\left(\mathbf{n}^i - \frac{\mathbf{x}_0}{r_0}\right)\cdot \mathbf{x}\right]\left(\mathbf{n}^i - \frac{\mathbf{x}_0}{r_0}\right)\cdot \mathbf{n}\,\mathcal{R}\, da. \tag{16}$$

REFERENCES

1. G. CAVIGLIA, A. MORRO, and E. PAGANI, Inhomogeneous waves in viscoelastic media. Wave Motion **12** (1990) 143-159.
2. M. HAYES, Inhomogeneous plane waves. Arch. Ration. Mech. Anal. **85** (1984) 41-79.
3. G. CAVIGLIA, A. MORRO, and E. PAGANI, Reflection and refraction at elastic-viscoelastic interfaces. N. Cim. **12 C** (1989) 399-413.
4. G. CAVIGLIA, A. MORRO, and E. PAGANI, Time-harmonic waves in viscoelastic media. Mec. Res. Comm. **16** (1988) 53-60.
5. J. A. OGILVY, Wave scattering from rough surfaces, Rep. Prog. Phys. **50** (1987) 1553-1608.
6. N. BLEISTEIN, Large wave number aperture-limited Fourier inversion and inverse scattering, Wave Motion **11** (1989) 113-136.
7. G. CAVIGLIA and A. MORRO, Scattering problems for acoustic waves. Applied and Industrial Mathematics, R. SPIGLER ed. (Kluwer, Amsterdam, 1991).
8. D. COLTON, and R. KRESS, Integral Equation Methods in Scattering Theory (Wiley, New York, 1983).

Giacomo Caviglia
Dipartimento di Matematica
Università di Genova
Via Alberti 4, 16132 Genova, Italy

F CHARRU, J FABRE AND M MICHEL

Nonlinear waves at the interface of two viscous fluids

1. INTRODUCTION

Using a perturbation method, the equation describing the non linear behaviour of long waves at the interface between two viscous, incompressible and immiscible fluids, is derived for plane Couette-Poiseuille flow. This equation contains the Kuramoto-Sivashinsky equation and the importance of the extra-terms is discussed. The normal form of the ordinary differential equation corresponding to waves travelling without deformation is studied. The existence of solitary waves, shocks and doubly periodic waves is shown, and their characteristic wavelengths are computed.

2. DERIVATION OF THE WAVE EQUATION

We are interested in the non linear behaviour of the interface between two viscous, immiscible and incompressible fluids, flowing between two parallel plates. The fluids are driven either by the motion of the upper plate (Couette flow) or by pressure gradient (Poiseuille flow) or by their combination (Couette-Poiseuille flow, fig.1). The flow is assumed isothermal and two-dimensional. The linear stability analysis [1][2] shows that these flows can be unstable for disturbances of long wavelength λ, short waves being stabilized by interfacial tension, and that instability persists at arbitrarily small values of

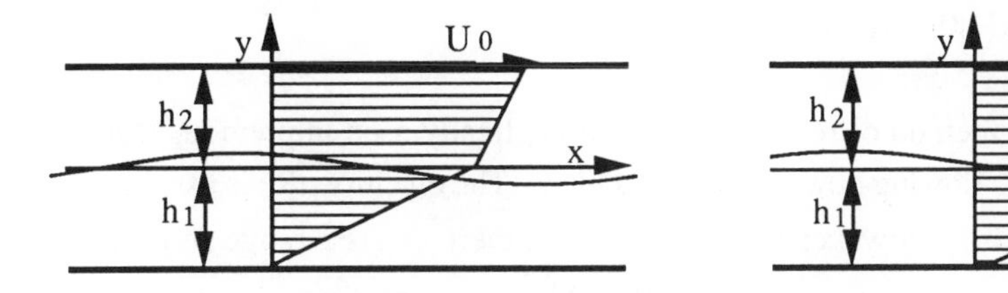

Fig. 1 - The flow configuration : Couette flow (on the left) and Poiseuille flow (on the right).

the Reynolds number. Thus one should focus on the non linear behaviour of long waves.

The physical quantities are expected to vary significantly over the wavelength λ in the x-direction and over the layer thickness in the y-direction. Thus following the methods used for non linear stability of liquid films flowing down an inclined plane, the x-coordinates will be normalized by λ and the y-coordinate with the lower layer thickness h_1. The time scale λ/U_I, where U_I is the velocity of the interface will be chosen.

The two length scales h_1 and λ introduce the "long wave parameter" $\mu = h_1/\lambda$. The problem involves six dimensionless independent parameters : the ratios d, m, r of the film thicknesses, the viscosities, and the densities, plus a Reynolds number, a Froude number and a Weber number, defined as follows :

$$d = \frac{h_2}{h_1} \qquad m = \frac{\mu_2}{\mu_1} \qquad r = \frac{\rho_2}{\rho_1}$$

$$Re = \frac{h_1 U_I}{\nu_1} \qquad Fr = \frac{U_I^2}{(1-r)gh_1} \qquad We = \frac{\rho_1 h_1 U_I^2}{\sigma} \qquad (1)$$

The streamfunction Ψ_k is decomposed into the primary flow streamfunction and the perturbation ψ_k. The normal-stress condition at the interface $\eta(x, t)$ involves the pressure perturbations p_1 and p_2 which are linked to the streamfunctions by their gradients in the Navier-Stokes equations. Thus, to eliminate the pressure, the normal-stress condition is differentiated with respect to the curvilinear coordinate of the interface, and the pressure gradients $p_{k,x}$ and $p_{k,y}$ (k=1,2) are replaced by their expressions from the Navier-Stokes equations. It must be pointed out that the interfacial tension is assumed to be high enough that the Weber number is $O(\mu^2)$. Although restrictive, this assumption has the advantage of taking into account the stabilizing effect of interfacial tension at the μ-order.

We seek the streamfunction solutions as expansions in powers of the small parameter μ :

$$\psi(x, y, t) = \sum_{i=0,n} \mu^i \, \psi^{(i)}(x, y, t) \qquad (2)$$

and we shall adopt the method of regular perturbation already used in previous studies [3][4][5] for liquid films flowing down an inclined plane. The quantity μRe is assumed to be small compared with 1. However large Re may be, there exists a range of μ small enough for the perturbation procedure to be valid. The wave amplitude is not *a priori* small. We now have to solve, for each order, two fourth-order differential equations in $\psi_k^{(i)}$ (one for each fluid), homogeneous at the zeroth-order, non-homogeneous at the upper orders, the streamfunctions $\psi_k^{(i)}$ having to satisfy the boundary conditions. The streamfunctions are thus determined as functions of y and of the unknown interface displacement $\eta(x,t)$. Details of the computations are available in [6][7]. The wave evolution equation is then given by the kinematic condition at the interface :

$$\eta_{,t} + [\Psi_1(\eta)]_{,x} = 0 \qquad \text{at } y = \eta(x, t) \quad (3)$$

3. SMALL AMPLITUDE WAVES

The expressions derived for the streamfunctions $\psi_k^{(1)}$ are too cumbersome to proceed further. In order to work with a more tractable interface equation, the study will be restricted to small amplitude waves. $\psi_k^{(1)}$ which does not contain any term in η_{xx}, is linear in $\eta_{,x}$ and $\eta_{,xxx}$; their coefficients, functions of η at the interface, will be developed in Taylor series for $\eta = 0$. The analytical expressions for the coefficients of these series have been calculated with the symbolic manipulation code "Mathematica". On retaining h_1 as the only length scale, the small parameter μ disappears from the equations and the interface equation, valid up to $O(\eta^3)$ and $O(\mu^2)$, becomes :

$$\eta_{,t} + \left\{ c_0 \eta + V \eta^2 + \mathrm{Re}\,(R + T\eta)\, \eta_{,x} + \frac{\mathrm{Re}}{\mathrm{We}} (S + U\eta)\, \eta_{,xxx} \right\}_{,x} = 0 \quad (4)$$

$$S = S(m,d) \quad ; \quad R = R_0(m, d, r) - \frac{S(m,d)}{Fr}$$

$$U = U(m,d) \quad ; \quad T = T_0(m, d, r) - \frac{U(m,d)}{Fr}$$

The analytical expressions for the coefficients of eq. (4) have been verified in handling the properties of symmetry. The method consists in searching the transformations which leave eq. (4), written in dimensional form, unchanged. The most interesting transformation consists in permuting the subscripts identifying the two fluids. Thus, the following relation must be satisfied by the coefficient R_0 for Couette flow :

$$r\, d^4 R_0(1/m, 1/d, 1/r) = m^3 R_0(m, d, r) \quad (5)$$

with similar relations for the other coefficients. We have checked with "Mathematica" that the coefficients of the present study satisfy these relationships.

The results of linear stability obtained for the first time by Yih [1], and then by Hooper & Grimshaw [2] in the presence of interfacial tension are recovered in the present study. The dispersion relation for infinitesimal disturbances of real wavenumber k and complex frequency ω gives :

$$\omega_r = c_0 k \quad \text{and} \quad \omega_i = \text{Re}\, k^2 \left(R - \frac{S}{\text{We}} k^2\right) \tag{6}$$

The first equation gives the velocity c_0 of infinitesimal disturbances. The second concerns the growth rate : for $R < 0$, the flow is stable whatever the wavenumber. For $R > 0$, there exists a wavenumber cutoff below which the long wavelength disturbances destabilize the flow. Nevertheless no critical Reynolds number do exist, contrary to the case of free falling films.

Moreover, eq.(4) contains the Kuramoto-Sivashinsky equation found by Hooper & Grimshaw from multiple scale perturbation method. However some discrepancies have been pointed out with this study. For typical amplitude η of O(0.1), R_0 and $T_0\eta$ are of the same order of magnitude : it confirms the importance of the second term of the Taylor expansion, which probably significantly modify the typically chaotic solutions of the Kuramoto-Sivashinsky equation.

The $O(k^2)$ growth rate of long waves is thus small so that periodic, solitary or shock waves are expected in the vicinity of R=0. Let us now discuss the existence and behaviour of such non linear waves.

4. WAVES TRAVELING WITHOUT DEFORMATION

The particular solutions of eq. (4) for the class of waves travelling with velocity c without deformation will now be studied for the small values of the wave velocity $\varepsilon = c - c_0$. In the frame moving at the wave velocity, the interface height depends only on the new space coordinate $\xi = x - ct$, so that eq. (4) may be integrated once. The constant of integration must be equal to zero in order that the flat interface ($\eta=0$) be a solution. Eq.(4) is scaled according to :

$$\eta \rightarrow \text{Re}^2\, \text{We} \frac{T^3}{S\, V^2}\, \eta \qquad \xi \rightarrow \frac{V}{\text{Re}\, T}\, \xi \qquad \varepsilon \rightarrow \text{Re}^2\, \text{We} \frac{T^3}{S\, V^3}\, \varepsilon \tag{7}$$

and then written as a system of ordinary differential equations. In the phase space spanned by η, $\eta_{,\xi}$ and $\eta_{,\xi\xi}$ where $\xi = x - ct$ is the moving coordinate, the flow defined by this O.D.E. admits two fixed points $\eta=0$ and $\eta=\varepsilon$. It corresponds to a conservative system : the trace of the Jacobian matrix is zero everywhere in the phase space. Dissipation appears only at next order in the perturbation scheme. The study of the linear stability of the two fixed points is similar to that of Pumir *et al.* [8] for free falling films. Heteroclinic and homoclinic curves can be expected in the vicinity of $\varepsilon=0$. Both a simple bifurcation and a Hopf bifurcation emanate from the singularity $\varepsilon=0$, simultaneously with the coalescence of the two fixed points. Such a degeneracy has been studied for the Kuramoto-Sivashinsky equation [9] and for a more general equation [10]. For settling the existence of homoclinic or heteroclinic trajectories, the linear study is not satisfactory. A more extended analysis of the flow behaviour may be obtained from the study of the normal form of the O.D.E.

Putting $R^* = Re^2\, We \dfrac{T^2}{S\, V^2} R \qquad U^* = \dfrac{1}{Re^2\, We} \dfrac{V^2}{T^3} U$

the normal form of the O.D.E., expressed in cylindrical coordinates, is given by the following equations :

$$R^*\, z_{,\xi} = \frac{\varepsilon^{*2}}{4} - z^2 - 2r^2$$

$$R^*\, r_{,\xi} = zr \tag{8}$$

$$R^{*1/2}\, \theta_{,\xi} = R^* + \frac{1-R^*U^*}{2}\left(z + \frac{\varepsilon^*}{2}\right)$$

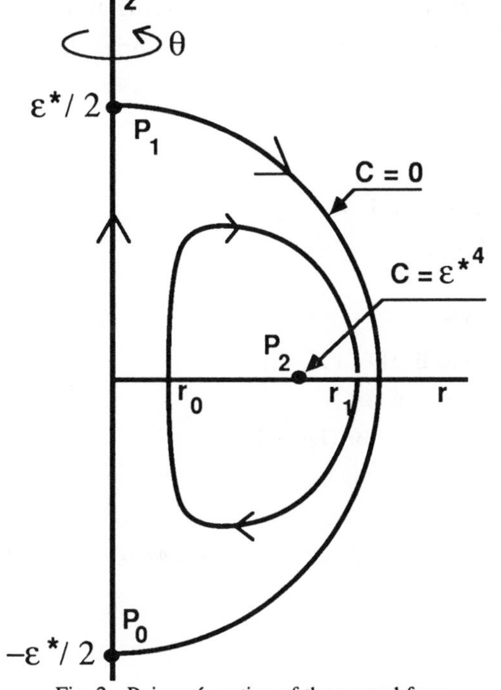

Fig. 2 - Poincaré section of the normal form.

This form differs from the normal form of the K-S equation [9][10] by the extra z-term in the angular velocity equation. The phase portrait of this normal form reveals two heteroclinic trajectories between the two fixed points $\mathbf{P}_0$ and $\mathbf{P}_1$, and a family of axisymmetrical tori surrounding a circular orbit (Fig.2). These tori are parametrized by the value C of a first integral independant of the azimuthal mode θ :

$$H(r, z) = 64\, r^2 \left(\frac{\varepsilon^*}{4} - z^2 - r^2\right) \tag{9}$$

where C is a constant specific to each trajectory, varying from 0 (heteroclinic orbits) to ε^{*4} (periodic orbit). The first wavelength Λ associated to the closed curve surrounding $\mathbf{P}_2$ may be calculated. The result is :

$$\Lambda(C)= 2R^* \int_{r_0}^{r_1} \frac{dr}{\left[(r^2 - r_0^2)\,(r_1^2 - r^2)\right]^{1/2}} \quad \text{where } r_{0,1}^2 = \frac{1}{8}\left[\varepsilon^{*2} \pm (\varepsilon^{*4} - C)^{1/2}\right] \tag{10}$$

Putting $r = r_1\, [1 - (1 - r_0^2 / r_1^2)\, \sin^2\phi]^{1/2}$, $\Lambda(C)$ has the canonical form of a Legendre elliptic integral of the first kind. The asymptotic values of $\Lambda(C)$ in the neighbourhood of the periodic orbit $\mathbf{P}_2$ and the two heteroclinic orbits $\mathbf{P}_0\mathbf{P}_1$, $\mathbf{P}_1\mathbf{P}_0$ can be calculated. The wavelength corresponding to the rotation around the z-axis can be calculated in the case of periodic waves for which the ratio of the two wavelenghts is rational. Finally, amplitude-celerity-wavelength correlations can be deduced, depending on the first integral value [7]. Higher order terms in the normal form break this symmetry, preserving torii and eventually giving rise to a homoclinic orbit and to chaotic trajectories in the neighbourhood of this homoclinic orbit [11].

1. C. S. YIH, *J. Fluid Mech.*, **27** (1967) 337-352.
2. A. P. HOOPER, R. GRIMSHAW, *Phys. Fluids* , **28** (1985) 37-45.
3. D. J. BENNEY, *J. Math. & Phys.*, **45** (1966), 150-155.
4. S. P. LIN, *J. Fluid Mech.*, **63** (1974), 417-429.
5. NAKAYA, C. *Phys. Fluids* , **18** (1975), 1407-1412.
6. F. CHARRU, J. FABRE, *C. R. Acad. Sci., Paris,* Série II, **310** (1991) 1711-1717.
7. F. CHARRU, *Thèse Institut Polytechnique de Toulouse* (1991).
8. A. PUMIR, Y. POMEAU, P. MANNEVILLE, *J. Fluid Mech.*, **135** (1983), 27-50.
9. H. C. CHANG, *Phys. Fluids* , **29** (1986), 3142-3147.
10. H. C. CHANG, *Chem. Engng. Sci.*, **42** (1987), 515-533.
11. J. GUCKENHEIMER, P. J. HOLMES, *Nonlinear Oscillations, Dynamical Systems and Bifurcations of Vector Fields.* Springer, 1983, New York.

F. Charru, J. Fabre & M. Michel
Institut de Mécanique des Fluides, C.N.R.S. / U.R.A. 005,
Avenue C. Soula, 31400 Toulouse, France.

T B MOODIE AND HE YUANPING

Weakly nonlinear hyperbolic waves with interior shocks

ABSTRACT

Weakly nonlinear hyperbolic waves arising from the action of small-amplitude, high-frequency boundary disturbances are studied. By directly introducing a nonlinear phase variable and specifying a class of 'single-wave mode' boundary disturbances, we are able to find an asymptotic solution and perform the requisite shock calculations. Our result suggests a systematic procedure for constructing weakly nonlinear hyperbolic waves with interior shocks and locating the shock initiation in space or time.

1. INTRODUCTION

Over the past decade, study of asymptotic methods for the analysis of weakly nonlinear hyperbolic waves generated by small-amplitude, high-frequency disturbances, has produced many new and important results. The readers may refer to Hunter and Keller [1], Majda and Rosales [2], and Hunter *et al.* [3] for some of these results. Here we develop a single-wave mode theory for the study of weakly nonlinear hyperbolic waves generated by such disturbances in one space dimension. We shall deal with quasi-linear, totally hyperbolic systems of the form

$$\boldsymbol{u}_{,t} + \boldsymbol{A}(\boldsymbol{u}, x)\boldsymbol{u}_{,x} = \boldsymbol{b}(\boldsymbol{u}, x), \tag{1}$$

where $x \geq 0$ is the space variable, $t \geq 0$ is time, $\boldsymbol{u}$ is the vector of n state variables, and $\boldsymbol{A}(\boldsymbol{u}, x)$ and $\boldsymbol{b}(\boldsymbol{u}, x)$ are smooth matrix and vector functions.

This paper is organized as follows. In §2 some preliminaries and notation are explained and the *signalling problem* defined. In §3 we introduce the nonlinear phase variable, carry out the *eikonal* transformation and obtain, in an asymptotic representation of the solution, the $O(1), O(\varepsilon)$, and $O(\varepsilon^k)$ problems. Solutions to these problems are presented in §4 where conditions of shock occurence and means of calculating their precise location are given. In the final section we show how our method can be employed to recover the recent results of Moodie and Swaters [4] for a problem involving inhomogeneous hyperelastic fluid-filled tubes. Full details and proofs for most results reported here will be submitted for publication elsewhere [5].

2. THE SIGNALLING PROBLEM

Throughout this report, (1) is totally hyperbolic, i.e., $\boldsymbol{A}$ has n distinct real eigenvalues $\{\lambda_i(\boldsymbol{u}, x)\}_{i=1}^n$. We further assume

$$\lambda_1(\boldsymbol{u}, x) < \lambda_2(\boldsymbol{u}, x) < \cdots < \lambda_p(\boldsymbol{u}, x) \leq 0 < \lambda_{p+1}(\boldsymbol{u}, x) < \cdots < \lambda_n(\boldsymbol{u}, x), \qquad (2)$$

when positive and negative eigenvalues are distinguished. We denote $\boldsymbol{\ell}^{(i)}(\boldsymbol{u}, x)$, $\boldsymbol{r}^{(i)}(\boldsymbol{u}, x)$ as the left and right eigenvectors associated with $\lambda_i(\boldsymbol{u}, x)$. They satisfy

$$\boldsymbol{\ell}^{(i)}\boldsymbol{A} = \lambda_i \boldsymbol{\ell}^{(i)}, \quad \boldsymbol{A}\boldsymbol{r}^{(i)} = \lambda_i \boldsymbol{r}^{(i)}, \quad i = 1, 2, \ldots, n. \qquad (3)$$

and the orthonormality condition

$$\boldsymbol{\ell}^{(i)}\boldsymbol{r}^{(j)} = \delta_{ij}, \quad i, j = 1, 2, \ldots, n, \qquad (4)$$

where δ_{ij} is the Kronecker function. In addition, we shall call $\boldsymbol{r}_0^{(i)}(x) \overset{\Delta}{=} \boldsymbol{r}^{(i)}(\boldsymbol{0}, x)$ a *wave mode* of the system (1).

Let $\boldsymbol{u} = \boldsymbol{u}_0(x, t)$ be a solution to (1). We call the i-th characteristic field *locally linearly degenerate* about $\boldsymbol{u} = \boldsymbol{u}_0$ if

$$(\text{grad}_{\boldsymbol{u}} \lambda_i) \boldsymbol{r}^{(i)}|_{\boldsymbol{u}=\boldsymbol{u}_0} \equiv 0. \qquad (5)$$

Now assuming a small-amplitude, high-frequency disturbance is excited on the boundary $x = 0$, we may expect a weakly nonlinear wave to propagate into the steady state region $(\boldsymbol{u} \equiv \mathbf{0})$. This signalling problem is specified by taking (1) in conjunction with

$$\boldsymbol{u} \equiv \mathbf{0}, \qquad x \geq 0, \quad t = 0, \tag{6}$$

$$\boldsymbol{u} = \epsilon \boldsymbol{g}_\epsilon(t/\delta), \quad x = 0, \quad t \geq 0, \tag{7}$$

where ϵ and δ are two small but not independent positive parameters introduced to describe the small-amplitude and high-frequency feature. We choose $\delta = \epsilon^m$, $m \geq 1$, where $\boldsymbol{g}_\epsilon(\cdot)$ is a smooth vector function with compact support satisfying $\boldsymbol{g}_\epsilon(0) = \boldsymbol{g}'_\epsilon(0) = \mathbf{0}$. In addition, $\boldsymbol{g}_\epsilon(\cdot)$ is assumed Taylor expandable about the small parameter ϵ. As it is known, to ensure well-posedness of the signalling problem, $\boldsymbol{g}_\epsilon(\cdot)$ cannot be given arbitrarily. In this report $\boldsymbol{g}_\epsilon(\cdot)$ will be identified in the course of solution.

If $\boldsymbol{g}_\epsilon(\cdot)$ is a wave mode, i.e., $\boldsymbol{g}_\epsilon(\cdot)$ is parallel to $\boldsymbol{r}_0^{(i)}(0)$ for some i, we shall call $\boldsymbol{g}_\epsilon(\cdot)$ a 'single-wave mode'. Since this does not happen in general we adopt the convention that $\boldsymbol{g}_\epsilon(\cdot)$ is a single-wave mode if its leading term $\boldsymbol{g}^{(0)}(\cdot)$ is. In order that our mathematical formulation agree with an acutal physical implementation of the boundary condition we impose the additional condition on (7) that

$$u_1(0, t) = \epsilon g_1^{(0)}(t/\delta). \tag{8}$$

3. EIKONAL TRANSFORM AND TRANSPORT EQUATIONS

Pick a positive eigenvalue from among $\lambda_{p+1}, \ldots, \lambda_n$ and denote it by $\lambda = \lambda(\boldsymbol{u}, x)$, and denote by $\boldsymbol{\ell} = \boldsymbol{\ell}(\boldsymbol{u}, x)$, $\boldsymbol{r} = \boldsymbol{r}(\boldsymbol{u}, x)$ the associated left and right eigenvectors, respectively. Now suppose $t = T(x)$, defined by

$$dt/dx = 1/\lambda_0(x), \quad t(0) = 0, \tag{9}$$

describes the leading wavefront, i.e., the time for the first disturbance to arrive at position x.

Introduce the nonlinear phase variable $\theta = \theta(x,t)$ associated with $\lambda = \lambda(\boldsymbol{u}, x)$ and define it as the solution of

$$\theta_{,t} + \lambda(\boldsymbol{u}, x)\theta_{,x} = 0, \tag{10}$$

$$\theta|_{x=0} = t/\epsilon^m, \quad t \geq 0. \tag{11}$$

The existence and smoothness of $\theta = \theta(x,t)$ is guaranteed by the existence and smoothness of $\boldsymbol{u} = \boldsymbol{u}(x,t)$. The inverse function of $\theta = \theta(x,t)$ gives the arrival time formula, i.e., the time for wave of phase θ to arrive at position x, which we write as

$$t = T(x, \theta; \epsilon). \tag{12}$$

Regard θ as an independent variable and transform the signalling problem (1), (7), (8) into (x, θ) coordinates obtaining

$$\theta_{,t}\left(\boldsymbol{I} - \frac{\boldsymbol{A}(\boldsymbol{U}, x)}{\lambda(\boldsymbol{U}, x)}\right)\boldsymbol{U}_{,\theta} = \boldsymbol{b}(\boldsymbol{U}, x) - \boldsymbol{A}(\boldsymbol{U}, x)\boldsymbol{U}_{x}, \tag{13}$$

$$\boldsymbol{U} \equiv \boldsymbol{0}, \quad \theta = 0, \quad x \geq 0, \tag{14}$$

$$\boldsymbol{U} = \epsilon \boldsymbol{g}_\epsilon(\theta), \quad x = 0, \quad \theta \geq 0, \tag{15}$$

where we write $\boldsymbol{u}(x,t) = \boldsymbol{U}(x, \theta; \epsilon)$ and ahead of the leading wavefront $\boldsymbol{u} \equiv \boldsymbol{0}$ is noted. For future shock calculations observe from (12) that

$$\theta_{,x} = T_{,\theta}^{-1}, \quad \theta_{,x} = -T_{,x}/T_{,\theta}, \tag{16}$$

which when substituted into (10), (11) followed by integration over x provides the exact nonlinear arrival time formula

$$t = \epsilon^m \theta + \int_0^x \frac{ds}{\lambda(\boldsymbol{U}(s, \theta; \epsilon), s)}. \tag{17}$$

The validity of the eikonal transform depends upon $\theta_{,t}$ and $T_{,\theta}$ being nonzero. This is easy to check near $x = 0$. An important feature of nonlinear hyperbolic systems is that the solution, however smooth initially, may develop shocks. A criterion for detecting such shocks is $T_{,\theta} = 0$ which in turn signals the breakdown of the eikonal transform.

Construct the asymptotic solution of (13)–(15) in the form

$$\boldsymbol{U}(x,\theta;\epsilon) = \epsilon \sum_{k=0}^{\infty} \frac{\epsilon^k}{k!} \boldsymbol{U}^{(k)}(x,\theta), \tag{18}$$

where $\boldsymbol{U}^{(k)}, \boldsymbol{U}_{,x}^{(k)}, \boldsymbol{U}_{,\theta}^{(k)} = O(1)$ as $\epsilon \to 0$, and proceed by rewriting (13) as

$$(\boldsymbol{I} - \boldsymbol{A}/\lambda)\boldsymbol{U}_{,\theta} = (\boldsymbol{b} - \boldsymbol{A}\boldsymbol{U}_{,x})T_{,\theta}, \tag{19}$$

and applying $\boldsymbol{\ell}$ to both sides to obtain

$$\boldsymbol{\ell b} - \boldsymbol{\ell A U}_{,x} = 0. \tag{20}$$

In addition, (14) and (8) are interpreted as

$$\boldsymbol{U}^{(k)}(x,\theta)|_{\theta=0} = \boldsymbol{0}, \quad k = 0,1,\ldots, \tag{21}$$

and

$$U_1^{(k)}(x,\theta)|_{x=0} = 0, \quad k = 1,2,\ldots. \tag{22}$$

Based upon (19) – (22) the solution can now be constructed. We expand all functions as Taylor series about $\boldsymbol{U} = \boldsymbol{0}$ noting in particular that

$$\lambda^{-1}(\boldsymbol{U},x) = \lambda_0^{-1}(x) + \sum_{k=1}^{\infty} \Lambda^{(k)}(\boldsymbol{U},\ldots,\boldsymbol{U})/k!, \tag{23}$$

$$(\boldsymbol{\ell b})(\boldsymbol{U},x) = \sum_{k=1}^{\infty} c^{(k)}(\boldsymbol{U},\ldots,\boldsymbol{U})/k! \tag{24}$$

where $\Lambda^{(k)}(\boldsymbol{U},\ldots,\boldsymbol{U})$ and $c^{(k)}(\boldsymbol{U},\ldots,\boldsymbol{U})$ are k-linear forms. Inserting (18) into each of these Taylor series give

$$\lambda^{-1}(\boldsymbol{U},x) = \lambda_0^{-1}(x) + \sum_{j=1}^{\infty} \epsilon^j \Lambda_j(x,\theta)/j!, \tag{25}$$

$$\boldsymbol{A}(\boldsymbol{U},x) = \sum_{j=0}^{\infty} \epsilon^j \boldsymbol{A}_j(x,\theta)/j!, \tag{26}$$

$$\boldsymbol{b}(\boldsymbol{U},x) = \sum_{j=1}^{\infty} \epsilon^j \boldsymbol{b}_j(x,\theta)/j!, \tag{27}$$

$$(\lambda\, \boldsymbol{\ell})(\boldsymbol{U},x) = \sum_{j=0}^{\infty} \epsilon^j \boldsymbol{a}_j(x,\theta)/j!, \tag{28}$$

$$(\boldsymbol{\ell}\,\boldsymbol{b})(\boldsymbol{U},x) = \sum_{j=1}^{\infty} \epsilon^j c_j(x,\theta)/j!, \tag{29}$$

$$(\boldsymbol{A}/\lambda)(\boldsymbol{U},x) = \sum_{j=0}^{\infty} \epsilon^j \boldsymbol{D}_j(x,\theta)/j!. \tag{30}$$

Substitution of (25) into (17) gives the arrival time formula

$$\begin{aligned} t &= \int_0^x \frac{ds}{\lambda_0(s)} + \sum_{j=1}^{m-1} \frac{\epsilon^j}{j!} \int_0^x \Lambda_j(s,\theta)ds \\ &+\epsilon^m\{\theta + \frac{1}{m!} \int_0^x \Lambda_m(s,\theta)ds\} + \sum_{j=m+1}^{\infty} \frac{\epsilon^j}{j!} \int_0^x \Lambda_j(s,\theta)ds \\ &\triangleq \sum_{j=0}^{\infty} \frac{\epsilon^j}{j!} T^{(j)}(x,\theta). \end{aligned} \tag{31}$$

Employing (25) – (31) in (19), (20), equating like powers of ϵ we obtain

$$O(1) \quad \textit{problem:} \quad (\boldsymbol{I} - \boldsymbol{A}_0/\lambda_0)\boldsymbol{U}_{,\theta}^{(0)} = \boldsymbol{0} \tag{32}$$

$$c^{(1)}(\boldsymbol{U}^{(0)}) - \boldsymbol{a}_0 \boldsymbol{U}_{,x}^{(0)} = 0. \tag{33}$$

$$O(\epsilon^k) \quad \textit{problem:} \quad (\boldsymbol{I} - \boldsymbol{A}_0/\lambda_0)\boldsymbol{U}_{,\theta}^{(k)} = \boldsymbol{M}_k(x,\theta), \tag{34}$$

$$c^{(1)}(\boldsymbol{U}^{(k)}) - \boldsymbol{a}_0 \boldsymbol{U}_{,x}^{(k)} = N_k(x,\theta), \tag{35}$$

where $\boldsymbol{M}_k$ and N_k are complicated expressions given in [5].

4. WEAKLY NONLINEAR WAVES AND SHOCK CALCULATIONS

The solution to the $O(1)$ problem has the form

$$\boldsymbol{U}^{(0)}(x,\theta) = \sigma_0(\theta)\, \boldsymbol{r}_0(x). \tag{36}$$

where $\sigma_0(\theta)$ is an arbitrary smooth scalar function with compact support and $\sigma_0(0) = \sigma_0'(0) = 0$. Also to be noted is that $\boldsymbol{r}_0(x) \triangleq \boldsymbol{r}(\boldsymbol{0}, x)$ and it is 'normalized' in the sense that

$$\Gamma_0(x) \triangleq (c^{(1)}(\boldsymbol{r}_0) - \boldsymbol{a}_0\boldsymbol{r}_0')/\boldsymbol{a}_0\boldsymbol{r}_0 \equiv 0. \tag{37}$$

Observe that

$$\boldsymbol{g}^{(0)}(\cdot) = \sigma_0(\cdot)\boldsymbol{r}_0(0), \tag{38}$$

i.e., $\boldsymbol{g}_\epsilon(\cdot)$ is a single-wave mode as mentioned before.

Rewriting (34), (35), and using (36) we find that the $O(\epsilon)$ problem takes the form

$$(\boldsymbol{I} - \boldsymbol{A}_0/\lambda_0)\boldsymbol{U}_{,\theta}^{(1)} = \boldsymbol{p}_1(x)\sigma_0(\theta)\sigma_0'(\theta), \tag{39}$$

$$c^{(1)}(\boldsymbol{U}^{(1)}) - \boldsymbol{a}_0\boldsymbol{U}_{,x}^{(1)} = q_1(x)\sigma_0^2(\theta), \tag{40}$$

whose solution can be proved to be

$$\boldsymbol{U}^{(1)} = \sigma_0^2(\theta)\,\boldsymbol{r}_1(x)/2, \tag{41}$$

where $\boldsymbol{r}_1(x)$ is a particular solution of the linear system

$$(\boldsymbol{I} - \boldsymbol{A}_0/\lambda_0)\,\boldsymbol{r}_1 = \boldsymbol{p}_1(x). \tag{42}$$

$\boldsymbol{r}_1(x)$ is 'normalized' in the sense that its first component vanishes at $x = 0$ (we assume the first component of $\boldsymbol{r}_0(0)$ is nonzero), and

$$K_1(x) \triangleq (c^{(1)}(\boldsymbol{r}_1) - \boldsymbol{a}_0\,\boldsymbol{r}_1 - 2q_1)/\boldsymbol{a}_0\,\boldsymbol{r}_0 \equiv 0. \tag{43}$$

The $O(\epsilon^k)$ problem has the form

$$(\boldsymbol{I} - \boldsymbol{A}_0/\lambda_0)\boldsymbol{U}_{,\theta}^{(k)} = \boldsymbol{p}_k(x)\sigma_0^k(\theta)\sigma_0'(\theta), \tag{44}$$

$$c^{(1)}(\boldsymbol{U}^{(k)}) - \boldsymbol{a}_0\boldsymbol{U}_{,x}^{(k)} = q_k(x)\sigma_0^{k+1}(\theta), \tag{45}$$

when $k < m$, with solution

$$\boldsymbol{U}^{(k)}(x,\theta) = \sigma_0^{k+1}(\theta)\, \boldsymbol{r}_k(x)/k+1, \tag{46}$$

where $\boldsymbol{r}_k(x)$ is a particular solution of the linear system

$$(\boldsymbol{I} - \boldsymbol{A}_0/\lambda_0)\, \boldsymbol{r}_k = \boldsymbol{p}_k(x) \tag{47}$$

'normalized' in the sense that its first component disappears as $x = 0$ and

$$K_k(x) \triangleq \left(c^{(1)}(\boldsymbol{r}_k) - \boldsymbol{a}_0\, \boldsymbol{r}_k' - (k+1)q_k\right)/\boldsymbol{a}_0\, \boldsymbol{r}_0 \equiv 0. \tag{48}$$

In summary, the asymptotic solution takes the form

$$\boldsymbol{U}(x,\theta;\epsilon) = \sum_{k=1}^{m} \frac{\epsilon^k}{k!}\, \sigma_0^k(\theta)\, \boldsymbol{r}_{k-1}(x) + O(\epsilon^{k+1}). \tag{49}$$

We now present the method for calculating shock initiation. Using the above results we rewrite the arrival time formula as

$$t = \int_0^x \frac{ds}{\lambda_0(s)} + \sum_{k=1}^{m-1} \frac{\epsilon^k}{k!}\, \sigma_0^k(\theta) \int_0^x [\Lambda_k]_{\sigma_0\equiv 1} ds$$

$$+\epsilon^m \{\theta + \frac{1}{m!}\, \sigma_0^m(\theta) \int_0^x [\Lambda_m]_{\sigma_0\equiv 1} ds\} + O(\epsilon^{m+1}), \tag{50}$$

where $[\Lambda_k]_{\sigma_0\equiv 1}$ is a function of x only. Then if there exists an integer $q \geq 1$ such that

$$[\Lambda_k]_{\sigma_0\equiv 1} \begin{cases} \equiv 0, & k = 1,2,\ldots,q-1, \\ \not\equiv 0, & k = q, \end{cases} \tag{51}$$

then when $m = q$ is chosen, a shock will form behind the leading wavefront if there exist (x,θ), $x > 0$, $\theta > 0$ such that $t_{,\theta} = 0$ or

$$1 + \frac{1}{(q-1)!}\, \sigma_0^{q-1}(\theta)\sigma_0'(\theta) \int_0^x [\Lambda_q]_{\sigma_0\equiv 1} ds = 0, \tag{52}$$

approximately. The shock initiates at (x_s, θ_s) in (x,θ) coordinates or (x_s, t_s) in (x,t) coordinates where

$$x_s = \min\{x > 0:\ 1 + \frac{1}{(q-1)!}\, \sigma_0^{q-1}(\theta)\sigma_0'(\theta) \int_0^x [\Lambda_q]_{\sigma_0\equiv 1} ds = 0\}, \tag{52}$$

and (x_s, θ_s) satisfies

$$t_s = \int_0^{x_s} \frac{ds}{\lambda_0(s)} + \epsilon^q \{\theta_s + \frac{1}{q!}\sigma_0^q(\theta_s)\int_0^{x_s}[\Lambda_q]_{\sigma_0\equiv 1} ds\} + O(\epsilon^{q+1}). \tag{54}$$

Since $[\Lambda_k]_{\sigma_0\equiv 1}$ $(k = 1, 2, \ldots, m)$ are independent of $\sigma_0(\theta)$ and m, the above result suggests a systematic procedure for constructing weakly non-linear waves with interior shocks and determining shock initiation distance and time.

5. EXAMPLE

Consider the propagation of weakly nonlinear waves in fluid-filled, hyperelastic, tethered tubes subjected to axial strain. The one dimensional model developed in [4] is

$$A_{,t} + (Au)_{,x} = 0, \tag{55}$$

$$u_{,t} + uu_{,x} + p_{,x} = 0, \tag{56}$$

where x, t are space and time, A is the cross sectional area, u the fluid velocity in the axial direction, p the transmural pressure, and the equations are nondimensional.

The consitutional relation gives

$$A = A_0(x) + \varphi_0(x)p + \varphi_1(x)p^2 + \varphi_2(x)p^3 + O(p^4), \tag{57}$$

where

$$\varphi_1(x) = 3\varphi_0^2(x)/2A_0(x), \quad \varphi_2(x) = [\tfrac{5}{2} - \beta(x)]\varphi_0^3(x)/A_0^2(x), \tag{58}$$

and φ_0, β, A_0 are all known functions involving the strain energy function W. In particular, $A_0(0) = 1$ and $\varphi_0(0) = 1/2$. The system admits a steady state solution $p = u = 0$.

Now, if the boundary is perturbed by

$$p|_{x=0} = \epsilon g(t/\delta), \quad t \geq 0, \tag{59}$$

we may consider the mixed initial and boundary problem prescribed by (59) and $p = u = 0,\ t = 0,\ x \geq 0.$

As proved in [5], a shock is not possible on the leading wavefront but it is possible to find a solution leading to interior shocks. We apply the theory developed here to verify this fact and compute shock initiation distance and time.

Writing (55), (56) in canonical form the positive (about $\boldsymbol{u} = \boldsymbol{0}$) eigenvalue and corresponding eigenvectors are found to be $\lambda = u + (AA,_p^{-1})^{1/2}$ and

$$\boldsymbol{\ell} = \tfrac{1}{2}(AA,_p^{-1})^{-1/2}, 1), \quad \boldsymbol{r} = \left((AA,_p^{-1})^{1/2}, 1\right)^T. \tag{60}$$

Taylor expand $(1/\lambda)$ about $\boldsymbol{u} \equiv \boldsymbol{0}$ obtaining

$$(1/\lambda) = \left(1/\lambda_0(x)\right) + \Lambda^{(1)}(\boldsymbol{u}) + \tfrac{1}{2}\Lambda^{(2)}(\boldsymbol{u},\boldsymbol{u}) + O(\|\boldsymbol{u}\|^3), \tag{61}$$

to find that $\Lambda^{(1)}(\boldsymbol{r}_0) \equiv 0$, i.e. the leading wavefront is locally linearly degenerate. Therefore, to construct the weakly nonlinear wave solution that will build up interior shocks, we must have $m \geq 2$. With $m = 2$, (49) and (50) provide

$$\boldsymbol{u} = \epsilon\sigma_0(\theta)\boldsymbol{r}_0(x) + \tfrac{1}{2}\epsilon^2\sigma_0^2(\theta)\boldsymbol{r}_1(x) + O(\epsilon^3), \tag{62}$$

$$t = \int_0^x \frac{ds}{\lambda_0(s)} + \epsilon^2\{\theta + \tfrac{1}{2}\sigma_0^2(\theta)\int_0^x [\Lambda_2]_{\sigma_0\equiv 1} ds\} + O(\epsilon^3), \tag{63}$$

where $[\Lambda_2]_{\sigma_0\equiv 1} = \Lambda^{(2)}(\boldsymbol{r}_0,\boldsymbol{r}_0) + \Lambda^{(1)}(\boldsymbol{r}_1)$.

Solving for $\boldsymbol{r}_0, \boldsymbol{r}_1$, and normalizing according to above scheme, gives

$$\boldsymbol{r}_0 = \left((A_0/\varphi_0)^{1/2}, 1\right)^T [A_0/A_0(0)]^{-3/4}[\varphi_0/\varphi_0(0)]^{1/4}, \tag{64}$$

$$\boldsymbol{r}_1 = \left(0, (A_0/\varphi_0)^{-1/2}\right)^T [A_0/A_0(0)]^{-3/4}[\varphi_0/\varphi_0(0)]^{1/2}. \tag{65}$$

Finally, comparing (59) and (62) and carrying out some calculations gives

$$\boldsymbol{u} = \begin{pmatrix} p \\ u \end{pmatrix} = \epsilon 2^{-1/4} g(\theta) A_0^{-3/4} \varphi_0^{1/4} \begin{pmatrix} (A_0/\varphi_0)^{1/2} \\ 1 \end{pmatrix}$$

$$+\epsilon^2 2^{-3/2} g^2(\theta) A_0^{-3/2} \varphi_0^{1/2} \begin{pmatrix} 0 \\ (A_0/\varphi_0)^{-1/2} \end{pmatrix} + O(\epsilon^3), \tag{66}$$

$$t = \int_0^x \varphi_0(\eta) A_0^{-1}(\eta) d\eta + \epsilon^2 \{\theta - (3/2^{3/2}) g^2(\theta)$$

$$\times \int_0^x \beta(\eta) \varphi_0^2(\eta) A_0^{-3}(\eta) d\eta\} + O(\epsilon^3). \tag{67}$$

The readers will find that (67) recovers the same result as in [4] (replacing ϵ by $\epsilon^{1/2}$).

REFERENCES

1. J. HUNTER AND J.B. KELLER, Weakly Nonlinear High Frequency Waves. Comm. Pure Appl. Math. **36** (1983) 547-569.
2. A. MAJDA AND R. ROSALES, Resonantly Interacting Weakly Nonlinear Hyperbolic Waves I: A Single Space Variable. Stud. Appl. Math. **71** (1984) 149-179.
3. J. HUNTER, A. MAJDA AND R. ROSALES, Resonantly Interacting Weakly Nonlinear Hyperbolic Waves II: Several Space Variables. Stud. Appl. Math. **75** (1986) 187-226.
4. T.B. MOODIE AND G.E. SWATERS, Nonlinear Waves and Shock Calculations for Hyperelastic Fluid-Filled Tubes. Quart. Appl. Math. **47** (1989) 705-722.
5. Y. HE AND T.B. MOODIE, The Signalling Problem in Nonlinear Hyperbolic Wave Theory, to be submitted.

T.B. Moodie and Yuanping He
Applied Mathematics Institute
Department of Mathematics
University of Alberta
Edmonton, Alberta
Canada T6G 2G1

D F PARKER AND A P MAYER

Dissipation of surface acoustic waves

ABSTRACT

Surface acoustic waves (SAWs) are guided waves with a linearized mode structure determined by solving a boundary value problem. In such problems, a change in the governing equations at any point of the cross-section influences the whole mode shape. Consequently, constitutive nonlinearity has a nonlocal influence.

Recent papers [1,2,3] derive the evolution equation applying to nonlinear SAWs travelling across the surface of either an elastic or an electro-elastic material. It is shown here that the method can be compactly adapted to include dissipative effects.

1. INTRODUCTION

Although the Euler equations of nonlinear elasticity are hyperbolic, surface acoustic waves (SAWs) have many features more representative of elliptic systems, since they are guided modes which usually travel subsonically. Consequently, when dissipative effects are included into a theory, their action is non-local and so they cannot always be modelled by introducing into the evolution equation terms involving higher derivatives. This paper reports how an appropriate equation is obtained by relatively straightforward modification of the procedure described in [1,2] for analysing nonlinear elastic and piezo-electric SAWs.

A compact representation of the complicated deformation fields within linearly elastic or piezo-electric SAWs is first given. Since linearized SAWs are nondispersive, all wavelengths travel at the same speed and may be superposed to allow nondistorting waves having arbitrary profile of surface elevation. The procedures of Lardner [3] and Parker et al [1,2] allow calculation of the quadratically nonlinear terms causing gradual evolution of the Fourier transform of the elevation. Since numerical calculation shows how energy cascades to all higher frequencies causing profile 'break-up', somewhat akin to shock formation in hyperbolic waves, it is pertinent to derive a dissipative theory in which this 'break-up' might be inhibited.

The derivation procedures in [1,2,4] do not require that perturbations to the deformation field are calculated. Here we show how the influence of heat conduction and viscosity may similarly be included without calculation of the perturbed displacements. We also indicate how the resulting evolution equation may be recast as an equation governing the elevation profile directly, but including nonlocal operators for both nonlinear and dissipative effects.

2. THE UNDERLYING LINEARIZED THEORY

Waves travelling along the surface $X_2 = 0$ of a homogeneous elastic half-space $X_2 < 0$ have components τ_{Lj} of Piola-Kirchhoff stress related to the components $F_{jL} \equiv \partial x_j/\partial X_L = x_{j,L}$ of deformation gradient by

$$\tau_{Lj} = \partial W/\partial F_{jL} \quad ,$$

where $W = W(\underset{\sim}{F})$ is the isentropic strain energy density, X_L ($L = 1,2,3$) denote Lagrangian coordinates and x_j ($j = 1,2,3$) are the current (Eulerian) coordinates. Elastic surface waves are solutions of the Euler (momentum) equations, traction-free boundary conditions and decay conditions

$$\tau_{Lj,L} = \rho \ddot{x}_j \qquad (X_2 < 0) \quad , \tag{1}$$

$$\tau_{2j} = 0 \qquad (X_2 = 0) \quad , \tag{2}$$

$$u_j \equiv \varepsilon^{-1}(x_j - X_L\delta_{Lj}) \to 0 \qquad (X_2 \to -\infty) \quad .$$

In linear theory, the constitutive law reduces to

$$\tau_{Lj} = \varepsilon c_{jLmM}\, u_{m,M} \quad , \tag{3}$$

where the tensor elements c_{jLmM} are linear elastic moduli which depend on the material anisotropy. The corresponding linearized SAWs are constructed by superposing <u>partial waves</u>, which are solutions of (1) having the form

$$\underset{\sim}{u} = \underset{\sim}{a} \exp ik(\theta + sX_2) \qquad (k > 0,\ \mathrm{Im}\, s < 0)$$

as

$$\underset{\sim}{u} = \sum_{p=1}^{3} B^{(p)}\underset{\sim}{a}^{(p)} \exp ik(\theta + s^{(p)}X_2) \qquad (4)$$

Here, $\theta = X_1 - ct$ is the phase variable, c is the speed of surface waves, k is the wavenumber, while $s^{(p)}$ $(p = 1,2,3)$ are the depth factors and $\underset{\sim}{a}^{(p)}$ $(p = 1,2,3)$ the associated wave polarizations determined as roots of the algebraic eigenvalue problem

$$\{c_{j1m1} - \rho c^2\delta_{jm} + s(c_{j1m2} + c_{j2m1}) + s^2 c_{j2m2}\}a_m = 0 \quad (5)$$

arising from equations (1). Although for each c, this gives six roots $s = s^{(p)}$ which are either real or else complex conjugate in pairs, only those roots with $\mathrm{Im}\, s^{(p)} < 0$ may contribute to displacements (4) in SAWs. Thus, the summation in (4) extends over $p = 1,2,3$ if c is subsonic, but cannot include more than two partial waves if c exceeds one or more characteristic speeds of plane waves travelling in the X_1-direction.

The coefficients $B^{(p)}$ are determined by substituting (4) into (2), so yielding the algebraic system

$$\sum_{p=1}^{3} (c_{j2m1} + s^{(p)}c_{j2m2})a_m^{(p)}B^{(p)} \equiv M_{jp}B^{(p)} = 0 \ .$$

Vanishing of the determinant ($\det M_{jp} = 0$) then fixes the speed c (independently of wavenumber k) and so determines the sets of values $s^{(p)}$, $\underset{\sim}{a}^{(p)}$ and $B^{(p)}$ apart from normalization. Consequently, the displacement field in a harmonic wave of wavelength $2\pi/k$ may be written as

$$\underset{\sim}{u} = C\,\underset{\sim}{A}(kX_2)e^{ik\theta} + \text{c.c.} \quad , \qquad (6)$$

where c.c. denotes the complex conjugate and where $\underset{\sim}{A}$ is

$$\underset{\sim}{A}(kX_2) = \sum_{p=1}^{3} B^{(p)}\underset{\sim}{a}^{(p)} \exp iks^{(p)}X_2 \ , \quad k > 0 \ . \qquad (7)$$

Fourier superposition then shows that the general linearized, non-dissipative disturbance travelling at speed c is

$$\underset{\sim}{u}(X_1,X_2,t) = \int_{-\infty}^{\infty} C(k)\underset{\sim}{A}(kX_2)e^{ik\theta}dk \quad , \qquad (8)$$

where, for $k < 0$, C and $\underset{\sim}{A}$ are given by the complex conjugate relations

$$C(k) = C^*(-k) \qquad , \qquad \underset{\sim}{A}(kX_2) = \underset{\sim}{A}^*(-kX_2) \quad .$$

Furthermore, the normalization $A_2(0) = 1$ gives

$$u_2(X_1, 0, t) = \int_{-\infty}^{\infty} C(k) e^{ik\theta} dk \equiv u(X_1 - ct)$$

and so allows $C(k)$ to be identified as the Fourier transform

$$C(k) = \frac{1}{2\pi} \int_{-\infty}^{\infty} u(\theta) e^{-ik\theta} d\theta \tag{9}$$

of the surface elevation profile $u_2(X_1, 0, t) = u(\theta)$, which is arbitrary.

As described in [2], the foregoing analysis is readily adapted to SAWs on piezoelectric materials, by using the notational device of Taylor and Crampin [5] in which the electric potential is regarded as a fourth component u_4 of $\underset{\sim}{u}$. The linearized constitutive laws then retain the form (3), where lower case indices range from 1 to 4, where τ_{L4} are the components of the (material) electric displacement with c_{4LmM} and c_{jL4M} being piezoelectric coefficients and c_{4L4M} being dielectric coefficients. The Gauss equation is $\tau_{L4,L} = 0$, so that (5) is changed only by allowing both j and m to range from 1 to 4 and by replacing δ_{jm} by $\eta_{jm} \equiv \delta_{jm} - \delta_{j4}\delta_{m4}$.

Various electric boundary conditions are relevant. The simplest, $\tau_{24} = 0$ on $X_2 = 0$, has the form (2), so that (4) and (7) are amended only by treating $\underset{\sim}{u}$, $\underset{\sim}{A}$ and $\underset{\sim}{a}^{(p)}$ as four-vectors and by including contributions with $p = 4$. Alternatively, the 'earthed' boundary condition, $u_4 = 0$ on $X_2 = 0$, amends the boundary matrix elements M_{4p}, so altering the wavespeed c and functional form $\underset{\sim}{A}(kX_2)$ while allowing the general disturbance to have the description (8) and (9). In [2], a treatment may be found of the case in which the surface $X_2 = 0$ is the interface between a piezoelectric material and free space. This also leads to (8).

In all these cases, elastic or piezoelectric, the linear theory of SAWs propagating in the X_1 direction predicts that <u>any</u> surface elevation may travel without distortion, as $u_2 = u(X_1 - ct)$. References [1-3] show how quadratic nonlinearity causes slow evolution in which all Fourier components have perfect phase matching and so interact strongly. This paper shows how a similar treatment can be used to include heat conduction and viscous diffusion and also to account for thermal boundary conditions, which were absent from a previous treatment [6].

3. NONLINEARITY AND DIFFUSION

Inclusion into the constitutive law (3) of nonlinear terms $\varepsilon^2 c_{jLmMnN} u_{m,M} u_{n,N} + \ldots$ alters the representation (8) by $O(\varepsilon)$ terms and also causes profile evolution on the long scale measured by the coordinate $X \equiv \varepsilon X_1$. The technique, described in [1-3] for accommodating this evolution by writing $C = C(k,X)$, allows also for inclusion into (3) of viscous stresses $\nu_{jLmM}(\varepsilon \dot{u}_{m,M})$, when all active components ν_{jLmM} of the viscosity tensor are $O(\varepsilon)$. Similarly (see [7]) thermoelastic coupling may be incorporated, provided that the relevant coupling coefficients also are $O(\varepsilon)$.

We replace (3) by the constitutive law

$$\tau_{Lj} = \varepsilon c_{jLmM} u_{m,M} + \varepsilon^2 c_{jLmMnN} u_{m,M} u_{n,N} + \varepsilon \nu_{jLmM} \dot{u}_{m,M} - \varepsilon K_{jL} \hat{T} \quad , \tag{10}$$

which, besides including nonlinear and viscous tems, includes also coupling to the temperature perturbation field $\varepsilon \hat{T}$. In equation (10), it is assumed that all terms neglected are $O(\varepsilon^2)$. Also we assume that it is sufficient to take $\hat{T}$ to be governed by the linearized thermoelastic equations

$$(\kappa_{LM} \hat{T}_{,M})_{,L} - c_\nu \dot{\hat{T}} = T_o K_{jL} \dot{u}_{j,L} \quad . \tag{11}$$

We leave the boundary condtions (2) (or their piezoelectric generalizations unchanged and treat the surface $X_2 = 0$ as either isothermal or insulated, so giving the linear boundary conditions

$$\hat{T} = 0 \quad (X_2 = 0) \qquad \text{or} \qquad K_{2M} \hat{T}_{,M} = 0 \quad (X_2 = 0) \tag{12a,b}$$

respectively. As $X_2 \to -\infty$, both $\hat{T}$ and u_j are to vanish. (It is known, within linear theory (see [8]), that attenuation at low frequencies is much greater for the isothermal boundary condition

(12a) than for the insulating condition (12b).

We seek solutions in the form

$$\underset{\sim}{u} = \underset{\sim}{U}(\theta, X_2, X) + \varepsilon\underset{\sim}{w}(\theta, X_2, X; \varepsilon)$$

$$\hat{T} = T(\theta, X_2, X) + \varepsilon\tilde{T}(\theta, X_2, X; \varepsilon)$$

where

$$\underset{\sim}{U}(\theta, X_2, X) \equiv \int_{-\infty}^{\infty} C(k,X)\underset{\sim}{A}(kX_2)e^{ik\theta}dk \quad . \tag{13}$$

Then, using $_{,1}$ and $_{,X}$ to denote $\partial/\partial\theta$ and $\partial/\partial X$ respectively and allowing Greek indices to range only over the values 1 and 2 , we find that the governing systems have the structure

$$\begin{aligned} &\rho c^2 w_{j,11} - c_{j\alpha m\beta}w_{m,\alpha\beta} = P_j(\theta, X_2; X, \varepsilon) && (X_2 < 0) \\ &c_{j2m\beta}w_{m,\beta} = -Q_j(\theta; X, \varepsilon) && (X_2 = 0) \quad (14) \\ &w_m \to 0 && (X_2 \to -\infty) \end{aligned}$$

and

$$\begin{aligned} &(\kappa_{\alpha\beta}T_{,\beta})_{,\alpha} - c_\nu\dot{T} + cT_o K_{j\beta}U_{j,1\beta} = 0(\varepsilon) && (X_2 < 0) \\ &T = 0 \qquad \text{or} \qquad \kappa_{2\beta}T_{,\beta} = 0 && (X_2 = 0) \quad (15) \\ &T \to 0 && (X_2 \to -\infty) \end{aligned}$$

respectively. Here, P_j and Q_j may be written out without approximation, but for our purposes the information

$$\begin{aligned} P_j = {} & (c_{j1m\beta} + c_{j\beta m1})U_{m,\beta X} + c_{j\alpha m\beta n\gamma}(U_{m,\beta}U_{n,\gamma})_{,\alpha} \\ & -c\varepsilon^{-1}\nu_{j\alpha m\beta}U_{m,\alpha\beta 1} - \varepsilon^{-1}K_{j\alpha}T_{,\alpha} + o(1) \quad , \end{aligned} \tag{16}$$

$$\begin{aligned} Q_j = {} & c_{j2m\beta n\gamma}U_{m,\beta}U_{n,\gamma} + c_{j2m1}U_{m,X} \\ & -c\varepsilon^{-1}\nu_{j2m\beta}U_{m,\beta 1} - \varepsilon^{-1}K_{j2}T + o(1) \end{aligned} \tag{17}$$

is sufficient.

Since, to this accuracy, $T(\theta, X_2, X)$ is related to $U(\theta, X_2, X)$ through the linear, inhomogeneous boundary-value problem (15) we can, to leading order, treat problem (14) as if it were an inhomogeneous problem for $\underset{\sim}{w}$ in which P_j and Q_j were specified functions. For such problems, solutions exist only if certain compatibility conditions are satisfied. For the system (14), the constraint has been derived and discussed in references [1] and [2]. It imposes a relationship between P_j and Q_j , which may be envisaged as uniformly travelling distributions of body force and surface traction in a linearly elastic problem.

The constraint is compactly represented (see [1]) as

$$\int_{-\infty}^{0} A_j^*(kX_2)\bar{P}_j(k,\ X_2;\ X,\ \varepsilon)dX_2 - A_{oj}^*\bar{Q}_j(k;\ X,\ \varepsilon) = 0\ . \quad (18)$$

Here $\bar{P}_j$ and $\bar{Q}_j$ denote Fourier transforms of P_j and Q_j with respect to θ, while $A_j^*(kX_2)$, the complex conjugate of $A_j(kX_2)$, is the solution of the homogeneous problem adjoint to the Fourier transform of problem (14). Also A_{oj}^* denotes $\lim_{X_2\to 0-} A_j^*(kX_2)$. Application of (18) to the terms given explicitly in (16) and (17) yields the evolution equation

$$iJ\frac{\partial C}{\partial X}(k,X) + \int_0^\infty K(\ell/k)(k-\ell)\ell C(k-\ell,X)C(\ell,X)d\ell$$
$$+ ic\{Mk^2 + \bar{\chi}(k)\}C(k,X) = 0 \quad , \quad (19)$$

where the constant J and kernel $K(\ell/k)$ are the same as in the non-dissipative theory [2], while

$$M \equiv \int_{-\infty}^{0} \varepsilon^{-1}\{\nu_{j1m1}A_m(\xi)A_j^*(\xi) + i\nu_{j2m1}[A_m(\xi)A_j^{*\prime}(\xi) - A_m^*(\xi)A_j^\prime(\xi)]$$
$$+ \nu_{j2m2}A_m^\prime(\xi)A_j^{*\prime}(\xi)\}d\xi$$

is the viscous coefficient which is purely real and

$$\bar{\chi}(k) = -\int_{-\infty}^{0} T_o\varepsilon^{-1}\{K_{j1}A_j^*(\xi) + iK_{j2}A_j^{*\prime}(\xi)\}S(\xi,k)d\xi$$

is the thermal dissipation kernel, in which $S(\xi,k)$ describes the thermal field which is coupled to $\underset{\sim}{U}$ through problem (15) and is the solution to the problem

$$\kappa_{22}\frac{d^2S}{d\xi^2} + i(\kappa_{12} + \kappa_{21})\frac{dS}{d\xi} - (\kappa_{11} - i\frac{cc_\nu}{k})S$$
$$= K_{j1}^\prime A_j(\xi) - iK_{j2}A_j^\prime(\xi) \qquad (X_2 < 0)$$
$$S = 0 \quad \text{or} \quad i\kappa_{21}S + \kappa_{22}\frac{dS}{d\xi} = 0 \qquad (X_2 = 0)$$
$$S \to 0 \qquad (\xi \to -\infty,\ k > 0).$$

Equation (19) shows that the thermoviscous contribution Mk^2 to the attenuation does correspond exactly to a second-derivative damping term, with coefficient M which takes account of the depth-dependence $\underset{\sim}{A}(\xi)$ of the displacement field in surface waves. The thermal field $S(\xi,k)$ depends explicitly on k, as well as on the scaled depth $\xi = kX_2$. Consequently the contribution $\bar{\chi}$ models attenuation of the hereditary integral type and also introduces dispersion. The procedure leading to the evolution equation (19) based on the compatibility condition (18) is closely related to the 'projection method' of Maradudin and Mayer [4] and the perturbation method of Tiersten and Sinha [9].

4. EVOLUTION EQUATION FOR THE ELEVATION

Equation (19) governs the evolution of the Fourier transform $C(k,X)$ of the surface elevation. For the isotropic, elastic, non-dissipative case (in which $M = 0$, $\bar{\chi} = 0$), Hunter [10] has shown how to reformulate the theory to yield directly an equation for the θ-derivative of the elevation

$$u(\theta,X) \equiv \int_{-\infty}^{\infty} C(k,X)e^{ik\theta}dk = U_2(X_1-ct,0,\varepsilon X_1) \quad . \qquad (20)$$

Clearly, the resulting equation is the inverse transform of (19). The difficult step, obtaining a neat representation for the inverse transform of the nonlinear term, may be taken directly from Hunter's description, since it is a property of the non-dissipative contributions to (19). It is then found that the inverse transform of (19) has the form

$$J\frac{\partial u_\theta}{\partial X} + \frac{\partial}{\partial\theta}\left\{\int_{-\infty}^{\infty}\int_{-\infty}^{\infty} G(\theta-\sigma,\theta-\tilde{\sigma})u_\theta(\sigma,X)u_\theta(\tilde{\sigma},X)d\sigma d\tilde{\sigma}\right\}$$
$$= cM\frac{\partial^2 u_\theta}{\partial\theta^2} - c\int_{-\infty}^{\infty}\chi(\theta-\sigma)u_\theta(\sigma,X)d\sigma \quad , \qquad (21)$$

where $G(\theta,\tilde{\theta})$ is a (symmetric) nonlocal interaction kernel related to the kernel $K(\ell/k)$ occurring in (19), while $\chi(\theta)$ is the inverse Fourier transform of $\bar{\chi}(k)$.

Equation (21) governs the surface slope, or equivalently vertical speed, u_θ . It shows that the viscous terms in (10) introduce a familiar second-derivative term, with coefficient M which fully takes into account the distribution of dissipative stresses at all depths $X_2 < 0$. It is perhaps surprising that this non-local dissipation should yield a local term in (21). The reason lies in the fact that the relevant term from (10) is a linear combination of second derivatives of $\underset{\sim}{U}(\theta,X_2,X)$. The effect of thermoelastic interaction is, however, different. The field $T(\theta,X_2,X)$ is governed by a boundary value problem (15), in which $\underset{\sim}{U}$ appears as a driving force and which is not homogeneous in the order of derivatives.

The non-local nature of the quadratic nonlinearity arises because the quadratic terms in $\underset{\sim}{P}(\tilde{\theta},X_2;\ X,\varepsilon)$ depend on $\underset{\sim}{U}(\tilde{\theta},X_2,X)$ which depends non-locally on the boundary values $u(\theta,X)$. Indeed $\underset{\sim}{U}(\tilde{\theta},X_2,X)$ has a representation, equivalent to the inverse Fourier transform of (13), of the form

$$\underset{\sim}{U}(\tilde{\theta},X_2,X) = \int_{-\infty}^{\infty} \underset{\sim}{g}(\tilde{\theta}-\sigma,X_2)u(\sigma,X)d\sigma \quad ,$$

where the kernel $\underset{\sim}{g}$ generalizes that in the Poisson integral formula representing a harmonic function in a half-plane in terms of its boundary behaviour. Consequently, the compatibility condition allowing problem (14) to be solved, relates boundary behaviour at $(\theta,0,X)$ to a weighted integral of $\underset{\sim}{P}(\tilde{\theta},X_2;\ X,\varepsilon)$, which itself is a double integral involving the behaviour of u at (σ,X) and $(\tilde{\sigma},X)$ for all $-\infty < \sigma,\ \tilde{\sigma} < \infty$ (see Fig.1). The special properties of the kernel G in (21) described in [10] follow from the 'scale invariance' of the nonlinear, non-dissipative problem, which applies to the anisotropic and electro-elastic cases also.

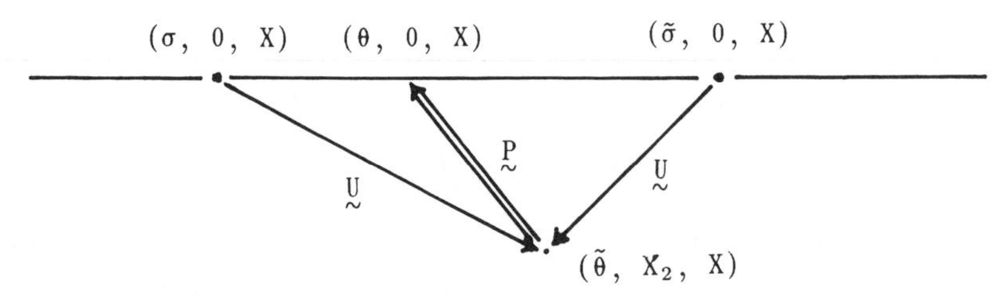

Fig.1. Diagram showing how $\underset{\sim}{P}(\tilde{\theta},X_2,X)$ is influenced by $u(\sigma,X)$ and $u(\tilde{\sigma},X)$ and itself influences the boundary behaviour at $(\theta,0,X)$.

REFERENCES

1. D.F. PARKER, Waveform Evolution for Nonlinear Surface Acoustic Waves, Int. J. Engng Sci. 26 (1988) 59-75.
2. D.F. PARKER and E.A. DAVID, Nonlinear Piezoelectric Surface Waves, Int. J. Engng Sci. 27 (1989) 565-581.
3. R.W. LARDNER, Nonlinear Surface Acoustic Waves on an Elastic Solid of General Anisotropy, J. Elast. 16 (1986) 63-73.
4. A.A. MARADUDIN and A.P. MAYER, Surface Acoustic Waves on Nonlinear Substrates, in: Nonlinear Waves in Solid State Physics (eds. A.D. Boardman, T. Twardowski and M. Bertolotti, Plenum New York 1991).
5. D.B. TAYLOR and S.G. CRAMPIN, Surface Waves in Anisotropic Media: Propagation in a Homogeneous Piezoelectric Halfspace, Proc. Roy. Soc. A364 (1978) 161-179.
6. R.W. LARDNER, Nonlinear Rayleigh Waves: Harmonic Generation Parametric Amplification, and Thermoviscous Damping, J.Appl. Phys.55 (1984) 3251-3260.
7. A.P. MAYER, Thermoelastic Attenuation of Surface Acoustic Waves, Int.J.Engng Sci. 28 (1990) 1073-1082.
8. R.J. ATKIN and P. CHADWICK, Surface Waves in a Heat-conducting Elastic Body, J. Thermal Stresses 4 (1981) 509-521, and references therein.
9. H.F. TIERSTEN and B.K. SINHA, A Perturbation Analysis of the Attenuation and Dispersion of Surface Waves, J.Appl.Phys.49 (1978) 87-95.
10. J.K. HUNTER, Nonlinear Surface Waves, in: Current Progress in Hyperbolic Systems: Riemann Problems and Computations (ed. W.B. Lindquist, Am.Math.Soc.1989) 185-202.

D.F. Parker & A.P. Mayer
Department of Mathematics
University of Edinburgh
Mayfield Road
Edinburgh, EH9 3JZ, UK

A STRUMIA

Relativistic M. H. D. with finite electrical conductivity

1 Equations and field variables

The field equations required to govern relativistic M.H.D. with finite electrical conductivity are the following

$$\nabla_\alpha v^\alpha = 0 \qquad (continuity\ equation) \tag{1.1}$$

$$\nabla_\alpha T^{\alpha\beta} = k^\beta \qquad (energy - momentum\ balance) \tag{1.2}$$

$$\nabla_\alpha F^{\alpha\beta} = -\frac{J^\beta}{\varepsilon} \qquad (Maxwell\ equations\ I) \tag{1.3}$$

$$\nabla_\alpha G^{\alpha\beta} = 0 \qquad (Maxwell\ equations\ II) \tag{1.4}$$

$$\nabla_\alpha A^{\alpha\beta} = a^\beta \tag{1.5}$$

where

$$v^\alpha = n u^\alpha, \qquad u_\alpha u^\alpha = 1 \tag{1.6}$$

is the particle flux, n being the number of particles per unit volume and u^α the particle fourvelocity normalized to unity.

$$\begin{aligned} T^{\alpha\beta} = {} & \left(e + p + \varepsilon E^2 + \mu H^2\right) u^\alpha u^\beta - \left[p + \tfrac{1}{2}\left(\varepsilon E^2 + \mu H^2\right)\right] g^{\alpha\beta} - \\ & -\varepsilon E^\alpha E^\beta - \mu H^\alpha H^\beta - \tfrac{1}{c}\left(u^\alpha S^\beta + u^\beta S^\alpha\right) \end{aligned} \tag{1.7}$$

is the energy-momentum tensor and $F^{\alpha\beta}, G^{\alpha\beta}$ are the Maxwell tensors for a medium of constant electric permutivity ε and magnetic permaeability μ; E^β, H^β are the proper electric and magnetic fields, S^α is the relativistic Poyinting vector. The thermodynamic variables e, p are the equilibrium internal energy and the equilibrium pressure. The energy flux

$$k^\beta = Q u^\beta \tag{1.8}$$

is a term arising from Joule effect, while J^β is the electric current density vector and is considered as an additional field variable. In the present approach we propose that non equilibrium entropy is a function also of the current density and the other elecromagnetic vectors extended thermodynamics), while modification of internal energy and pressure are not required. We introduce also the proper charge density of the fluid

$$\rho = J_\beta u^\beta \tag{1.9}$$

and the conduction current density

$$J_c^\beta = J^\beta - \rho u^\beta, \qquad J_c^\beta u_\beta = 0 \tag{1.10}$$

The eq. (1.5) includes one scalar equation, representing the *charge conservation law* and three more equations representing a *generalized Ohm's law*. [1] The field equations (1.1)–(1.5) consist of a system of *generalized conservation laws* in the form

$$\nabla_\alpha \boldsymbol{f}^\alpha = \boldsymbol{f} \tag{1.11}$$

compatible with the *supplementary* entropy balance

$$\nabla_\alpha h^\alpha = g \leq 0 \tag{1.12}$$

$$h^\alpha = -n\hat{S}u^\alpha \tag{1.13}$$

$$\hat{S} = S + \Sigma \tag{1.14}$$

being the non equilibrium entropy, S the equilibrium entropy and Σ a constitutive function characterizing the entropy increment in presence of electric charges and conduction. Compatibility between the system of field equations (1.11) and the entropy balance (1.12) is ensured by the conditions

$$\boldsymbol{U}' \cdot \delta \boldsymbol{f}^\alpha + \omega_I^\alpha \delta \Phi_I \equiv \delta h^\alpha, \qquad \boldsymbol{U}' \cdot \boldsymbol{f} = g \tag{1.15}$$

where $\boldsymbol{U}'$ is the *main field* and ω_I^α are the Lagrange multipliers relative to the constraints $\Phi_I(\boldsymbol{U}') = 0$ among the field variables one chooses to evaluate the variations. A special role is played by the *generator*

$$h'^\alpha = \boldsymbol{U}' \cdot \boldsymbol{f}^\alpha + \omega_I^\alpha \Phi_I - h^\alpha \tag{1.16}$$

In particular in relativistic fluid dynamics [2], relativistic extended thermodynamics of non degenerate gases [5] and in relativistic M.H.D. [3] thanks to the constraints among the field variables, h'^α results to be collinear with the velocity of the fluid particle u^α. In the present approach the equation (1.5) which completes the theory of a relativistic M.H.D. with finite electrical conductivity, is determined following the ideas of extended thermodynamcs [6] requiring that i) the tensor $A^{\alpha\beta}$ and the vector a^β lead to a vector equation compatible with charge conservation and consistent with a generalized Ohm's law; ii) the generator h'^α is parallel to the fluid velocity u^α; iii) the entropy principle is fulfilled. Convexity condition and positivity of h' are assumed to hold as selecting criteria among the constitutive functions Σ.

2 Results

The eq. (1.5) must include one scalar equation, representing *charge conservation law* for the scalar variable ρ and three more equations for the constrained vector J_c^β, representing a *generalized Ohm's law*. Then the charge conservation law must be obtained by (1.5) through a contraction with some *multiplier* unit vector τ_β. It follows that

$$A^{\alpha\beta} = J^\alpha \tau^\beta \tag{2.1}$$

$$a^\beta = -\frac{1}{\kappa}(J_c^\beta - \sigma E^\beta) \tag{2.2}$$

where κ plays the role of a *relaxation time*. A similar structure for the generalized Ohms law is usual in non relativistic plasma physics [7] [8]. τ^α is to be thought as a constitutive function, J^α being assumed as a field variable. When $\kappa \to 0$ (1.5) gives the usual Ohms law.

3 Main field

The assumption of collinearity between the generator and the fluid velocity leads to great simplification in order to determine the main field U'. We point out that in the present case only one scalar identity is enough to determine completely the main field, instead of the the four component vector condition (1.15). The structure of the main field then results to be

$$U' \equiv (\psi, w_\beta, \Lambda_\beta, \Omega_\beta, \chi_\beta) \tag{3.1}$$

$$\psi = \frac{e+p}{n\hat{T}} - \hat{S} - n\frac{\partial \Sigma}{\partial n} \tag{3.2}$$

$$w_\beta = -\tfrac{1}{\hat{T}}\left[u_\beta - \tfrac{1}{c\eta}\left(S_\beta + n\hat{T}\tfrac{\partial\Sigma}{\partial I_4}\zeta_\beta - n\hat{T}\tfrac{\partial\Sigma}{\partial I_5}I_4E_\beta - n\hat{T}\tfrac{\partial\Sigma}{\partial I_6}I_4H_\beta\right)\right] \tag{3.3}$$

$$\begin{aligned}\Lambda_\beta = \tfrac{1}{\hat{T}}\Big[&\left(\tfrac{2}{c^2\eta}I_2 - n\hat{T}\tfrac{\partial\Sigma}{\partial I_1} - \varepsilon\right)E_\beta - \left(n\hat{T}\tfrac{\partial\Sigma}{\partial I_3} + \tfrac{I_3}{c^2\eta}\right)H_\beta - \\ &-n\hat{T}\tfrac{\partial\Sigma}{\partial I_5}\left(\zeta_\beta + \tfrac{1}{c\eta}S_\beta\right) + \tfrac{1}{c\eta}n\hat{T}\tfrac{\partial\Sigma}{\partial I_4}\,\eta_{\beta\lambda\mu\nu}\zeta^\lambda H^\mu u^\nu\Big]\end{aligned} \tag{3.4}$$

$$\begin{aligned}\Omega_\beta = \tfrac{1}{\hat{T}}\Big[&\left(\tfrac{2}{c^2\eta}I_1 - n\hat{T}\tfrac{\partial\Sigma}{\partial I_2} - \mu\right)H_\beta - \left(n\hat{T}\tfrac{\partial\Sigma}{\partial I_3} + \tfrac{I_3}{c^2\eta}\right)E_\beta - \\ &-n\hat{T}\tfrac{\partial\Sigma}{\partial I_6}\left(\zeta_\beta + \tfrac{1}{c\eta}S_\beta\right) - \tfrac{1}{c\eta}n\hat{T}\tfrac{\partial\Sigma}{\partial I_4}\,\eta_{\beta\lambda\mu\nu}\zeta^\lambda E^\mu u^\nu\Big]\end{aligned} \tag{3.5}$$

$$\begin{aligned}\chi_\beta = -\tfrac{1}{\hat{T}}\Big\{&n\hat{T}\tfrac{\partial\Sigma}{\partial\rho}\tau_\beta + \tfrac{n\hat{T}}{\rho}\Big[+\tfrac{\partial\Sigma}{\partial I_4}(u_\beta - I_4\tau_\mu\tau^\mu\tau_\beta) + \\ &+\tfrac{\partial\Sigma}{\partial I_5}(E_\beta - I_5\tau_\mu\tau^\mu\tau_\beta) + \tfrac{\partial\Sigma}{\partial I_6}(H_\beta - I_6\tau_\mu\tau^\mu\tau_\beta)\Big]\Big\}\end{aligned} \tag{3.6}$$

$$\begin{aligned}&\omega_0 = \tfrac{1}{\hat{T}}\left(\eta - n\hat{T}\tfrac{\partial\Sigma}{\partial I_4}\tau_\alpha u^\alpha + \tfrac{1}{c}w_\alpha S^\alpha\right), \qquad \omega_1 = -n\tfrac{\partial\Sigma}{\partial I_5}I_4, \\ &\omega_2 = -n\tfrac{\partial\Sigma}{\partial I_6}I_4, \qquad \omega_3 = -n\tau_\mu\tau^\mu\left(\tfrac{\partial\Sigma}{\partial I_4}I_4 + \tfrac{\partial\Sigma}{\partial I_5}I_5 + \tfrac{\partial\Sigma}{\partial I_6}I_6 - \tfrac{\partial\Sigma}{\partial\rho}\rho\right)\end{aligned} \tag{3.7}$$

where the notations have been introduced

$$\eta = e + \frac{1}{2}\left(\varepsilon E^2 + \mu H^2\right), \qquad \hat{T} = \frac{T}{1+\frac{\partial\Sigma}{\partial S}} \tag{3.8}$$

$$\zeta_\beta = \tau_\beta - I_4 u_\beta, \qquad \zeta_\beta u^\beta = 0 \tag{3.9}$$

$$\begin{aligned}&I_1 = -\tfrac{1}{2}E^2, \quad I_2 = -\tfrac{1}{2}H^2, \quad I_3 = E_\alpha H^\alpha \\ &I_4 = \tau_\alpha u^\alpha, \quad I_5 = \tau_\alpha E^\alpha, \quad I_6 = \tau_\alpha H^\alpha\end{aligned} \tag{3.10}$$

4 Collinearity condition and generator

The colinearity condition provides the following information for the generator

$$h' = h'^{(0)} + h'^{(1)} \tag{4.1}$$

$$h'^{(0)} = \frac{1}{\hat{T}}\left[p + \frac{1}{2}\left(\varepsilon E^2 + \mu H^2\right) - \frac{3}{c^2\eta}S_\beta S^\beta\right] \geq 0 \tag{4.2}$$

$$h'^{(1)} = n\left\{n\frac{\partial\Sigma}{\partial n} - \rho\frac{\partial\Sigma}{\partial\rho}\tau_\beta\tau^\beta + \frac{\partial\Sigma}{\partial I_1}E^2 + \frac{\partial\Sigma}{\partial I_2}H^2 - 2\frac{\partial\Sigma}{\partial I_3}E_\beta H^\beta - \right.$$

$$\left. -\frac{3}{c}\frac{\partial\Sigma}{\partial I_4}S_\beta\zeta^\beta - \frac{\partial\Sigma}{\partial I_5}E_\beta\zeta^\beta - \frac{\partial\Sigma}{\partial I_6}H_\beta\zeta^\beta\right\} \tag{4.3}$$

and a constitutive relation between J_c^α and τ^α

$$J_c^\alpha = -\frac{\tau_\mu\tau^\mu}{n\hat{T}\frac{\partial\Sigma}{\partial\rho}}\left(J_S S^\alpha + J_\zeta\zeta^\alpha + J_E E^\alpha + J_H H^\alpha\right) \tag{4.4}$$

$$J_S = \frac{2\varepsilon I_1 + 2\mu I_2 - (\eta+\Pi)c^2\varepsilon\mu}{n\hat{T}\eta c^2\varepsilon\mu} -$$

$$-\frac{1}{c\varepsilon}\left(-\frac{\partial\Sigma}{\partial I_1} + \frac{\partial\Sigma}{\partial I_5}\frac{2I_2I_5 - I_3I_6}{4I_1I_2}\right) - \frac{1}{c\mu}\left(\frac{\partial\Sigma}{\partial I_2} + \frac{\partial\Sigma}{\partial I_6}\frac{2I_1I_6 - I_3I_5}{4I_1I_2}\right) \tag{4.5}$$

$$J_\zeta = \frac{\partial\Sigma}{\partial I_4}\frac{\Pi c^2\varepsilon\mu - 2\varepsilon I_1 - 2\mu I_2}{\eta c^2\varepsilon\mu} \tag{4.6}$$

$$J_E = \frac{1}{\eta}\left[\frac{\partial\Sigma}{\partial I_4}\varepsilon I_5 - \frac{\partial\Sigma}{\partial I_5}(p + \varepsilon I_1 - \mu I_2)I_4 - \frac{\partial\Sigma}{\partial I_6}\varepsilon I_3 I_4 - \right.$$

$$\left. -\frac{\partial\Sigma}{\partial I_5}\frac{4I_1I_2 - \eta c S_\beta\zeta^\beta}{2c^2\varepsilon I_1} + \frac{\partial\Sigma}{\partial I_6}\frac{4I_1I_2 - \eta c S_\beta\zeta^\beta}{4c^2\mu I_1 I_2}I_3 + \frac{\partial\Sigma}{\partial I_4}\frac{I_5}{c^2\mu}\right] \tag{4.7}$$

$$J_H = \frac{1}{\eta}\left[\frac{\partial\Sigma}{\partial I_4}\mu I_6 - \frac{\partial\Sigma}{\partial I_6}(p - \varepsilon I_1 + \mu I_2)I_4 - \frac{\partial\Sigma}{\partial I_5}\mu I_3 I_4 - \right.$$

$$\left. -\frac{\partial\Sigma}{\partial I_6}\frac{4I_1I_2 - \eta c S_\beta\zeta^\beta}{2c^2\mu I_2} + \frac{\partial\Sigma}{\partial I_5}\frac{4I_1I_2 - \eta c S_\beta\zeta^\beta}{4c^2\varepsilon I_1 I_2}I_3 + \frac{\partial\Sigma}{\partial I_4}\frac{I_6}{c^2\varepsilon}\right] \tag{4.8}$$

$$\Pi = p + \frac{1}{2}(\varepsilon E^2 + \mu H^2) \tag{4.9}$$

5 Entropy principle and Joule effect

The electromagnetic processes being reversible and heat flux being supposed equal zero the *entropy principle* must be fulfilled as an equality. Therefore we have

$$w_\beta k^\beta - \frac{1}{\varepsilon}\Lambda_\beta J_c^\beta + \chi_\beta a^\beta = 0 \tag{5.1}$$

and thanks to (1.8)

$$Q = \hat{T}\left(\chi_\beta a^\beta - \frac{1}{\varepsilon}\Lambda_\beta J_c^\beta\right) \tag{5.2}$$

which generalizes the production of energy Q responsible of Joule effect. Of course Q vanishes in absence of electrical conduction.

References

[1] D. Jou, J. Casas-Vasquez and G. Lebon, *Extended irreversible thermodynamics*, Rep. Progr. Phys. **51** (1988) 1105;

[2] T. Ruggeri and A. Strumia, *Main field and convesx covariant density for quasi-linear hyperbolic systems. Relativistic fluid dynalmics*, Ann. Inst. Henri Poincaré **34** (1981) 65;

[3] A. Strumia, *Wave propagation and symmetric hyperbolic systems of conservation laws with constrained field variables. II - Symmetric hyperbolic systems with constrained fields*, Il Nuovo Cimento **101B** (1988) 19;

[4] A. Strumia, *Some ramarks on conservative symmetric-hyperbolic systems governing relativistic theories*, Annales de l'Institut Henri Poincaré **38A** (1983) 113;

[5] T. Tuggeri, *Convexity and simmetrization in relativistic theories. Privileged time-like congruence and entropy*, preprint;

[6] I.S.Liu, I.Müller and T.Ruggeri, *Relativistic thermodynamics of gases*, Annals of Physics **169** (1986), 191-219;

[7] A.N. Krall and A.W Trivelpiece, *Principles of plasma physics*, New York, McGraw-Hill (1973);

[8] G.H. Wannier, *Statistical Physics*, New York, Wiley (1966).

Strumia Alberto
Dipartimento di Matematica dell'Università
C.I.R.A.M.
Via Saragozza 8 - 40123 Bologna (Italy)

Part II

Analytical and Numerical Methods for the Solution of Nonlinear Evolutionary Models

S CAORSI

Numerical approaches to nonlinear electromagnetic problems

ABSTRACT

Nonlinear electromagnetic problems are addressed, where nonlinearities are defined by dielectric characteristics dependent on the electric field vector inside a dielectric object.

Numerical methods are analyzed in order to solve the related nonlinear Maxwell equations. Time-domain techniques and integral equation approaches are considered.

1. INTRODUCTION

Nonlinear electromagnetic problems have generally been studied in relation to nonlinear optics and for applications to electronic components and systems [1].

In spite of the growing interest in such areas, only few recent papers are devoted to the study of nonlinear electromagnetic scattering.

For instance, the scattering from weakly nonlinear systems has been investigated by using the Volterra series approach [2], and an integral equation formalism for nonlinear nondissipative dielectrics, both in free space and in guided structures, has been proposed by the author [3] [4].

In the present work, the interest is also extended to nonlinear dissipative objects and to the numerical approaches that can be used. In particular, on the basis of a previous paper [4], the integral equation approach is summarized, and a numerical technique is developed in the time domain by using finite difference operators.

2. NONLINEAR ELECTROMAGNETIC PROBLEM

We consider a homogeneous spatial domain D bounded by a surface S. In this domain, we assume the presence of electromagnetic sources and of nonlinear dielectric objects (see fig.1).

The nonlinearity is described through the permittivity characteristics assumed to be dependent on the electric field vector:

$$\begin{aligned} \varepsilon_{NL} &= \varepsilon_0\left[\varepsilon_b + L_\varepsilon(\bar{E})\right] \\ \sigma_{NL} &= \sigma_b + L_\sigma(\bar{E}) \end{aligned} \qquad (1)$$

where ε_{NL} and σ_{NL} represent the dielectric permittivity and the electric conductivity, respectively; L_ε and L_σ stand for integral differential operators with scalar ranges.

The problem we want to face is the development of a numerical approach to determine the distributions of the electric and magnetic fields inside a dielectric object so defined.

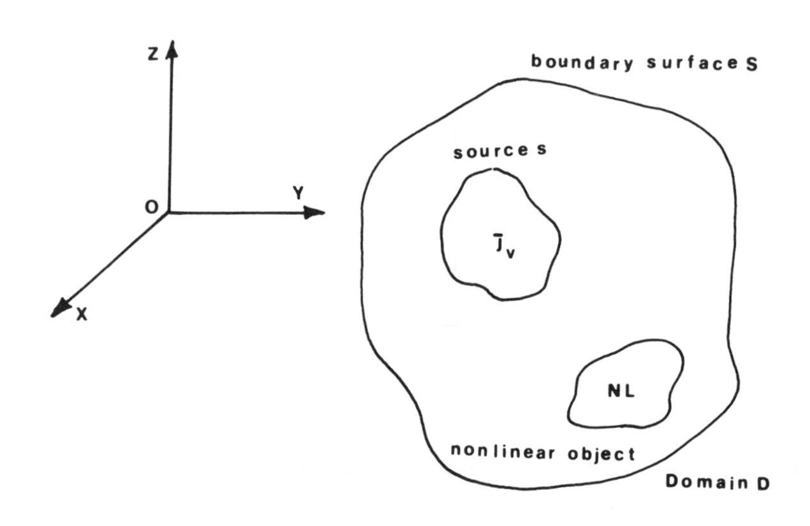

- FIGURE 1 Problem Geometry -

3. MATHEMATICAL FORMULATION AND NUMERICAL APPROACH

According to the nonlinear electromagnetic problem as defined in the previous section, the Maxwell equations can be written as follows:

$$\nabla x\bar{E} = -\mu \frac{\partial}{\partial t}\bar{H}$$
$$\nabla x\bar{H} = \sigma_{NL}\bar{E} + \frac{\partial}{\partial t}\varepsilon_{NL}\bar{E} \tag{2}$$

where σ_{NL} and ε_{NL} stand for the electric conductivity and the dielectric permittivity, respectively, both dependent on the electric field, as stated before.

Using relations (2) it is possible both to formulate, in the time domain, a numerical scheme based on the finite difference approach, [5], [6], and to deduce inhomogeneous wave equations for the harmonic components of the electric field vector. As a result, an integral equation approach can be developed in the frequency domain through the moment-method numerical solution [3], [4].

Finite difference approach

The classical FDTD (Finite Difference Time Domain) scheme based on the Yee cell [5] and on centered finite difference operators has been considered.

Consequently the following relation can be obtained:

$$\bar{H}(\bar{r},t+\frac{\Delta t}{2}) = 2\,\frac{\Delta t}{\mu}\,\langle\nabla x\bar{E}\rangle\Big|_{\bar{r},t} + \bar{H}(\bar{r},t-\frac{\Delta t}{2})$$
$$(\varepsilon_{NL}\bar{E})\Big|_{\bar{r},t+\Delta t} = (\varepsilon_{NL}\bar{E})\Big|_{\bar{r},t} - 2\Delta t(\sigma_{NL}\bar{E})\Big|_{\bar{r},t+\Delta t/2} + 2\Delta t\langle\nabla x\bar{H}\rangle\Big|_{\bar{r},t+\Delta t/2} \tag{3}$$

where <......> stands for the centered finite difference of the rotor operator.

Relations (3) can be defined as the finite-difference Maxwell equations.

From these relations, one can deduce that it is necessary both to define the initial conditions and to introduce two projection processes.

According to relations (3), once the above two requirements have been met, calculations of the distributions of the electric and magnetic fields can be performed in the time domain, step by step and alternately.

In other words, it is possible to develop two parallel numerical processes, one for each field vector, running in the time domain a half-step away from each other.

In this scheme, it is necessary to know the electric (or magnetic) field vector and the dielectric characteristics related to the previous half time step, and hence the two processes are linked to each other.

Concerning the initial conditions, if t = to indicates the starting time and the electric field vector is the first quantity to be calculated, one must specify:

$$\bar{E}(\bar{r},t_0),\ \bar{H}(\bar{r},t_0-\frac{\Delta t}{2}),\ \bar{E}(\bar{r},t_0-\Delta t)$$

Moreover, in accordance with relations (3), two projection processes are also required to predict the dielectric characteristics for the current time steps and the electric field vector for a time step that does not belong to the electric numerical running.

In order to show the devolopment of the numerical approach, rugh assumptions can be made about the more or less rapid variations in the fields; therefore, local linearized behaviours can be deduced from the length of the time-period. Thus one can write:

$$\bar{E}(\bar{r},\ t+\frac{\Delta t}{2}) = \frac{1}{2}\left[\bar{E}(\bar{r},\ t) + \bar{E}(\bar{r},\ t+\Delta t)\right] \tag{4}$$

$$\varepsilon_{NL}(\bar{r},\ t+\Delta t) = \varepsilon_{NL}\left[\bar{E}(\bar{r},\ t)\right] \tag{5}$$

$$\sigma_{NL}(\bar{r},\ t+\frac{\Delta t}{2}) = \sigma_{NL}\left[\bar{E}(\bar{r},\ t)\right] \tag{6}$$

Under these assumptions, the finite difference Maxwell equations (3) are not completely defined, and the numerical scheme can be implemented on the basis of the following relations:

$$\bar{H}(\bar{r},t+\frac{\Delta t}{2}) = 2\,\frac{\Delta t}{\mu}\,\langle\nabla x\bar{E}\rangle\Big|_{\bar{r},t} + \bar{H}(\bar{r},t-\frac{\Delta t}{2})$$

$$\bar{E}(\bar{r},t+\Delta t) = \alpha_{NL}(\bar{r},t)\,\bar{E}(\bar{r},t) + \frac{\Delta t}{\beta_{NL}(\bar{r},t)}\,\langle\nabla x\bar{H}\rangle \qquad (7)$$

where

$$\alpha_{NL}(r,t) = \frac{1-\dfrac{\Delta t}{2}\dfrac{\sigma_{NL}(\bar{r},t)}{\varepsilon_{NL}(\bar{r},t)}}{1+\dfrac{\Delta t}{2}\dfrac{\sigma_{NL}(\bar{r},t)}{\varepsilon_{NL}(\bar{r},t)}} \qquad (8)$$

$$\beta_{NL}(\bar{r},t) = 1+\frac{\Delta t}{2}\,\frac{\sigma_{NL}(\bar{r},t)}{\varepsilon_{NL}(\bar{r},t)} \qquad (9)$$

Integral equation approach

With regard to the nonlinear electromagnetic problem described in section 2, it is possible to indicate the total electric field at any point inside the region D as:

$$\bar{E}(\bar{r},t) = \bar{E}_{inc}(\bar{r},t) + \bar{E}_{scatt}(\bar{r},t)$$

where $\bar{E}_{inc}$ stands for the incident electric field vector produced by the source, and $\bar{E}_{scatt}$ stands for the electric field vector scattered by the nonlinear object.

By introducing the equivalent current density:

$$\bar{J}_{eq} = \sigma_b\bar{E} + L_\sigma(\bar{E})\,\bar{E} + \varepsilon_0\frac{\partial}{\partial t}\Big[(\varepsilon_b-1)\,\bar{E} + L_\varepsilon(\bar{E})\,\bar{E}\Big] \qquad (10)$$

the effect of a nonlinear dielectric object can be studied by means of the scattered electric field through the related inhomogeneous wave equation:

$$\nabla x\nabla x\bar{E}_{scatt}(\bar{r}) + \mu_0\varepsilon_0\frac{\partial}{\partial t}\,\bar{E}_{scatt}(\bar{r}) = -\mu_0\frac{\partial}{\partial t}\bar{J}_{eq}(\bar{r})$$

This relation represents a linear operator with a nonlinear source; such an operator can be formally solved in the frequency domain by using the well-known electric field integral equation [EFIE]:

$$\bar{E}_{scatt_n}(\bar{r})=\int_D k^2_{0n}\left[\tilde{\tau}_b\bar{E}_n+\bar{N}\bar{L}_n\right]\cdot\bar{\bar{G}}_n(r/r')dr' + \oint_S\left\{j\omega\mu_0\left[\hat{n}\times\bar{H}_{scatt_n}\right]\cdot\bar{\bar{G}}_n(r/r') + \left[\hat{n}\times\bar{E}_{scatt}\right]\cdot\nabla x\bar{\bar{G}}_n(r/r')\right\}dr' \qquad (12)$$

where $k^2_{0n}= n^2\omega^2\mu_0\varepsilon_0$

$$\tilde{\tau}_b= (\varepsilon_b-1) - j\frac{\sigma_b}{n\omega_0\varepsilon_0}$$

$$\bar{N}\bar{L}_n= \bar{Q}_n- j\frac{j}{n\omega_0\varepsilon_0}\bar{R}_n$$

and

$$L_\varepsilon(\bar{E})\bar{E} = \sum_n \bar{Q}_n e^{jn\omega_0 t}, L_\sigma(\bar{E})\bar{E} = \sum_n \bar{R}_n e^{jn\omega_0 t}$$

$\bar{\bar{G}}_n(r/r')dr'$ is the Dyadic Green function for the problem defined by specific boundary conditions on the surface S.

Relation (12) is the basis for the numerical scheme; once the operator $L_\sigma(\bar{E})\bar{E}$ and $L_\varepsilon(\bar{E})\bar{E}$ have been specified, it is possible to derive the dependences of $\bar{Q}_n$ and $\bar{R}_n$ on all the harmonic field components.

Therefore, through the approximation of a finite number of harmonics, relation (12) (written for each of them) can be transformed, into an algebraic nonlinear system by using the moment method. Solutions to this kind of algebraic system can be reached by means of suitable algorithms; then it is possible to obtain approximate distribution of the harmonic components of the electric field vector inside the nonlinear object.

REFERENCES

1. P.L.E. USLENGHI, Nonlinear electromagnetics (Academic Press, New York, 1980).

2. D.CENSOR, Scattering by weakly nonlinear object. SIAM, .App.Math. 43 (1983), 1400-1417.

3. S.CAORSI, M.PASTORINO, A theoretical analysis of the electromagnetic field for baunded non-linear media in free-space and rectangular waveguide. Proc. URSI, Int.Symp. on Electromagnetic Theory, Stockholm 1989, 198-200.

4. S.CAORSI, M.PASTORINO, Integral equation formulation of electromagnetic scattering by nonlinear dielectric objects. To appear in Electromagnetics (1991).

5. K.S.YEE, Numerical solution of initial boundary valueproblems involving Maxwell's equations in isotropic media. IEEE Trans. on AP, 14 (1966), 302-307.

6. D.M.CHOI, J.R. HOEFER, The Finite Difference Time Domain method and its applications to eigenvalue problems. IEEE Trans. on MTT, 34 (1986), 1464-1469.

Salvatore Caorsi

Department of Biophysical
and Electronic Engineering

University of Genoa

16145 Genova, ITALY

A DONATO AND A PALUMBO

Approximate invariant solutions to dissipative systems

ABSTRACT

We develop a procedure based on the concept of approximate symmetries [1] to try for approximate invariant solutions to dissipative systems.

INTRODUCTION

Wave propagation problems, even where non linear interaction are involved, are ruled by first order systems of partial differential equations of hyperbolic type that very often possess properties of symmetry investigated by classical group analysis.

These symmetries are not preserved when dissipative effects occurs. To take into account of dissipation one is led to consider more general equations where higher order space derivatives are involved multiplied by parameters directly related to the dissipative mechanism.

If these parameters are small and of the same order of magnitude it is possible to prevent the breaking of the nonlinear wave profile.

Following the main ideas of the procedure developed in [1], concerning approximate symmetries of differential equations with a small parameter, we show how to compute approximate invariant solutions of the whole system that at the zero order terms become invariant solutions of the associated hyperbolic system.

As an application we consider a non-heat conducting axi-symmetric fluid with small viscosity.

1. APPROXIMATE SYMMETRIES

We are interested to consider systems of the form:

$$\frac{\partial u}{\partial t} + \frac{\partial f(u)}{\partial x} + \epsilon\, K(x,t,u,u_x)\, \frac{\partial^2 u}{\partial x^2} = b(u,x,t) \tag{1.1}$$

where **u** is the unknown field having N-components, **f** and **b** are vectors and K an NxN matrix, all assumed to be known in terms of their arguments, t is the time and x the space variable.

Any solution of (1.1) will be of the form $u = u(x,t;\epsilon)$; we assume that for $\epsilon=0$ $u(x,t;0) = u_0(x,t)$ is a solution of the associated hyperbolic system:

$$L(u_0) = \frac{\partial u_0}{\partial t} + \frac{\partial f(u_0)}{\partial x} - b(u_0,x,t) = 0 \tag{1.2}$$

Systems like (1.2) usually describe non-linear wave phenomena in many physical contexts and, very often, they admit groups of simmetries determined by classical group analysis [2, 3].

Of course, a group of invariance for (1.2) is not, in general, admitted by (1.1). For instance it may occur that the system (1.2) is invariant with respect to a dilatation group of trasformations having the generators of the form:

$T = \gamma\, t$; $\qquad X = x$; $\qquad \eta_A = \alpha_A\, u_A$

provided that some restrictions on the functional form of **f** and **b** that has been deduced in [4], are satisfied. Recentely in [1] an approximate group analysis of differential equations with a small parameter has been carried on allowing to construct approximate symmetries that are stable under a small perturbation of the differential equations.

Let us now to consider an expansion of the form

$$u = u_0(x,t) + \epsilon\, u_1(x,t) + O(\epsilon^2) \tag{1.3}$$

where $u_0(x,t)$ is a non-constant solution of (1.2), then by substituting (1.3) in (1.1) to the first order in ϵ we get:

$$\frac{\partial u_1}{\partial t} + A(u_0)\frac{\partial u_1}{\partial x} = -(\nabla_u A)_0\, u_1 \frac{\partial u_0}{\partial x} - K(x,t,u_0,u_{0x})\frac{\partial^2 u_0}{\partial x^2} + \qquad (1.4)$$
$$+ (\nabla_u b)_0\, u_1$$

where

$$A(u_0) = (\nabla_u f)_{u_0}$$

that is a linear system to determine u_1.

If we denote by

$$\zeta \cdot \partial = X(u,x,t;\epsilon)\,\partial_x + T(u,x,t;\epsilon)\,\partial_t + \eta(u,x,t;\epsilon)\frac{\partial}{\partial u} \qquad (1.5)$$

any first order infinitesimal operator generating a group of trasformation for (1.1), the invariance conditions are:

$$\underset{1}{\zeta} \cdot \partial L(u) + \epsilon(\underset{1}{\zeta} \cdot \partial K)\frac{\partial^2 u}{\partial x^2} + \epsilon\, K\, \eta^{(xx)} = 0 \qquad (1.6)$$

where

$$\underset{1}{\zeta} \cdot \partial = \zeta \cdot \partial + \eta^{(t)}\frac{\partial}{\partial u_t} + \eta^{(x)}\frac{\partial}{\partial u_x} \qquad (1.7)$$

and $\eta^{(t)}$, $\eta^{(x)}$, $\eta^{(xx)}$ are, respectively, the first and the second extension computed with the usual technique for prolungation of the infinitesimal operators [3].

Because of the assumption (1.3) we have

$$\zeta = \zeta_0(u_0,x,t) + \epsilon\,(\zeta_1(u_0,x,t) + (\nabla_u \zeta)_0\, u_1) + O(\epsilon^2) \qquad (1.8)$$

where

$$\zeta\,|_{\epsilon=0} = \zeta_0 \quad \frac{d\zeta}{d\epsilon}\Big|_{\epsilon=0} = \zeta_1(u_0,x,t) + (\nabla_u \zeta)_0\, u_1 .$$

Formal substitution of (1.8) in (1.6), taking also into account (1.3), to the zero order in ϵ we get:

$$\underset{1}{\zeta_0}\, \partial L(u_0) = 0 \qquad (1.9)$$

that are the invariance conditions of the hyperbolic system (1.2) with respect to a group of transformation having the generator

$$\zeta_0 = (X_0(u_0, x, t), T_0(u_0, x, t), \eta_0(u_0, x, t))$$

and

$$\underset{1}{\zeta_0} \cdot \partial = \zeta_0 \cdot \partial + \eta_0{}^{(t)} \frac{\partial}{\partial u_{0t}} + \eta_0{}^{(x)} \frac{\partial}{\partial u_{0x}} \,; \tag{1.10}$$

here $\eta_0{}^{(t)}$ and $\eta_0{}^{(x)}$ have the same expressions of $\eta^{(t)}$ and $\eta^{(x)}$ with the formal substitution of **u** with u_0.

To the first order in ϵ we obtain

$$\underset{1}{\zeta_1} \partial L(u_0) + (\underset{1}{\zeta_0} \cdot \partial K)_0 \frac{\partial^2 u_0}{\partial x^2} + K_0 \eta_0{}^{(xx)} + \Sigma(u_0, u_1) = 0 \tag{1.11}$$

where

$$\underset{1}{\zeta_1} \partial L(u_0) = \zeta_1(u_0, x,\ t) \cdot \partial + \eta_1{}^{(t)} \frac{\partial}{\partial u_{0t}} + \eta_1{}^{(x)} \frac{\partial}{\partial u_{0x}} \tag{1.12}$$

and, in components,

$$\begin{aligned}
\Sigma_A(u_0, u_1) = {} & \Delta_{0A}{}^{(t)}[\zeta] + \Theta_{0A}{}^{(t)}[\zeta] - D_{0t}\, T_0 \frac{\partial u_{1A}}{\partial t} - D_{0t}\, X_0 \frac{\partial u_{1A}}{\partial x} + \\
& + \frac{\partial^2 f_A}{\partial u_B \partial u_C} \left(\eta_C{}^0 \frac{\partial u_{1B}}{\partial x} + u_{1B}\, \eta_{0C}{}^{(x)} \right) + \frac{\partial^3 f_A}{\partial u_B \partial u_C \partial u_L} u_{1L} \eta_C{}^0 \frac{\partial u_{0B}}{\partial x} + \\
& + \frac{\partial f_A}{\partial u_B} \left[\Delta_{0B}{}^{(x)}[\zeta] + \Theta_{0B}{}^{(x)}[\zeta] - D_{0x}\, T_0 \frac{\partial u_{1B}}{\partial t} - D_{0x}\, X_0 \frac{\partial u_{1B}}{\partial x} \right] + \\
& - \frac{\partial^2 b_A}{\partial u_B \partial u_C} \eta_B{}^0\ u_{1C} - \frac{\partial}{\partial u_C} \frac{\partial b_A}{\partial x} u_{1C}\, X_0 - \frac{\partial}{\partial u_C} \frac{\partial b_A}{\partial t} u_{1C}\, T_0
\end{aligned} \tag{1.13}$$

where

$$\Delta_{0A}{}^{(t)}[\zeta] = D_{0t}\,(\nabla_u\, \eta_A\, u_1)_0 - D_{0t}\,(\nabla_u\, X\, u_1)_0 \frac{\partial u_{0A}}{\partial x} - D_{0t}\,(\nabla_u\, T\, u_1)_0 \frac{\partial u_{0A}}{\partial t}$$

$$\Delta_{0A}{}^{(x)}[\zeta] = D_{0x}\,(\nabla_u\, \eta_A\, u_1)_0 - D_{0x}\,(\nabla_u\, X\, u_1)_0 \frac{\partial u_{0A}}{\partial x} - D_{0x}\,(\nabla_u\, T\, u_1)_0 \frac{\partial u_{0A}}{\partial t}$$

$$\Theta_{0A}{}^{(t)}[\zeta] = \left(\nabla_u\, \eta_A \frac{\partial u_1}{\partial t}\right)_0 - \left(\nabla_u\, X \frac{\partial u_1}{\partial t}\right)_0 \frac{\partial u_{0A}}{\partial x} - \left(\nabla_u\, T \frac{\partial u_1}{\partial t}\right)_0 \frac{\partial u_{0A}}{\partial t}$$

$$\Theta_{0A}{}^{(x)}[\zeta] = \left(\nabla_u\, \eta_A \frac{\partial u_1}{\partial x}\right)_0 - \left(\nabla_u\, X \frac{\partial u_1}{\partial x}\right)_0 \frac{\partial u_{0A}}{\partial x} - \left(\nabla_u\, T \frac{\partial u_1}{\partial x}\right)_0 \frac{\partial u_{0A}}{\partial t}$$

$$D_{0t} = \frac{\partial}{\partial t} + \nabla_{u_0} \cdot \frac{\partial u_0}{\partial t} \quad , \quad D_{0x} = \frac{\partial}{\partial x} + \nabla_{u_0} \cdot \frac{\partial u_0}{\partial x} \ .$$

It can be easily seen that $\underset{1}{\zeta_1 \partial L(u_0)}$ in (1.12) has the same expression as (1.10) with the formal substitution of $\zeta_0(u_0, x, t)$ with $\zeta_1(u_0, x, t)$.

We say that the expression (1.8) represents an approximate symmetry to the first order in ϵ for the system (1.1) if the relations (1.9) and (1.11) are identically satisfied for u_0 and u_1, solutions of (1.2) and (1.4).

From (1.9) it follows that ζ_0 must be a generator of an invariance group for the associated hyperbolic system. The components of ζ_0 are obtained by using the classical procedure of group analysis as solutions of the determining equations.

For what concerns the equation (1.11) as it must be satisfied identically the only possibility is to have:

$$\underset{1}{\zeta_1 \partial L(u_0)} = 0 \tag{1.14}$$

$$(\underset{1}{\zeta_0 \partial K})_0 \frac{\partial^2 u_0}{\partial x^2} + K_0 \, \eta_0^{(xx)} + \Sigma(u_0, u_1) = 0 \tag{1.15}$$

because the second order space derivatives of u_0 can be expressed as a homogeneous linear combination of u_1 and its first order derivatives (see (1.4)). Consequently the relation (1.11) compose of two parts the first one being independent upon u_1 and therefore must be both, separately, zero.

We may conclude that $\zeta_1(u_0, x, t)$ in (1.8) satisfies the same determining equations like $\zeta_0(u_0, x, t)$ so that its contribution to ζ is given by the generator $\epsilon\, \zeta_0$ that can be considered inessential and omitted [1]. This amount to require at the first order in ϵ that the generator $\zeta(u, x, t, \epsilon)$ depends on ϵ only by means of $u(x, t; \epsilon)$.

Obviously the procedure outlined above can be continued to higher order terms in ϵ without any conceptual difficulty. On the other hand it is nothing else that a variant of the method developed in [1].

For what concerns the invariant surface condition [3]:

$$X \frac{\partial u}{\partial x} + T \frac{\partial u}{\partial t} = \eta$$

at the zero order terms we obtain the well know condition to determine exact solution of the hyperbolic system. For the first order terms we have essentially:

$$X_0 \frac{\partial u_1}{\partial x} + T_0 \frac{\partial u_1}{\partial t} = (\nabla_u \eta)_0 \, u_1 - (\nabla_u X)_0 \, u_1 \frac{\partial u_0}{\partial x} - (\nabla_u T)_0 \, u_1 \frac{\partial u_0}{\partial t} \qquad (1.16)$$

that allow us to characterize u_1 once that u_0 has been determined.

2. A PHYSICAL APPLICATION

We consider a non-heat axi-symmetric fluid having viscosity μ so small that it is possible to assume the entropy to be uniform and the following relation between the pressure p and the density ρ is valid

$$p = K \rho^{\Gamma} \qquad (2.1)$$

where Γ is index of the fluid.

The equations of motion in one-dimensional flow are in the form [5]:

$$u_t + A u_x + \epsilon K u_{xx} = b \qquad (2.2)$$

$$u = \begin{bmatrix} \rho \\ v \end{bmatrix} \qquad A = \begin{bmatrix} v & \rho \\ \frac{a^2}{\rho} & v \end{bmatrix} \qquad K = \begin{bmatrix} 0 & 0 \\ 0 & -1 \end{bmatrix}$$

$$b = \begin{bmatrix} -\nu \rho v \\ 0 \end{bmatrix} \qquad \epsilon = \mu \qquad (2.3)$$

where $a^2 = \frac{dp}{d\rho} = \frac{\Gamma p}{\rho}$, v the fluid velocity, $\nu = 0, 1, 2$ respectively for the plane, cylindical or spherical case.

The associated hyperbolic system is invariant with respect to the dilatation group of transformations characterized by the generators:

$$\eta_0^{(\rho)} = \alpha \rho_0 \,, \qquad \eta_0^{(v)} = \beta v_0 \,, \qquad X_0 = x \,, \qquad T_0 = \gamma t \qquad (2.5)$$

provided that $\beta = 1-\gamma$, $\alpha = \dfrac{2(1-\gamma)}{\Gamma - 1}$.

Then we try for approximate symmetries of the system (2.2) in the form:

$$
\begin{aligned}
\eta^{(\rho)} &= \alpha\,\rho_0 + \epsilon\,(\nabla_{\mathbf{u}}\,\eta^{(\rho)})_0 \cdot \mathrm{u}_1 \\
\eta^{(\mathrm{v})} &= \beta\,\mathrm{v}_0 + \epsilon\,(\nabla_{\mathbf{u}}\,\eta^{(\mathrm{v})})_0 \cdot \mathrm{u}_1 \\
\mathrm{X} &= \mathrm{x} + \epsilon\,(\nabla_{\mathbf{u}}\,\mathrm{X})_0 \cdot \mathrm{u}_1 \\
\mathrm{T} &= \gamma\,\mathrm{t} + \epsilon\,(\nabla_{\mathbf{u}}\,\mathrm{T})_0 \cdot \mathrm{u}_1
\end{aligned}
\tag{2.6}
$$

where

$$
\mathrm{u}_1{}^{\mathrm{T}} = (\rho_1, \mathrm{v}_1)\ .
$$

By using (1.11), after some computations, we are able to show that in our case we have:

$$
\begin{aligned}
&(\nabla_{\mathbf{u}}\,\eta^{(\rho)})_0 \cdot \mathrm{u}_1 = (\gamma-2)\,\rho_1\ , \qquad &&(\nabla_{\mathbf{u}}\,\eta^{(\mathrm{v})})_0 \cdot \mathrm{u}_1 = -(\alpha+1)\ \mathrm{v}_1 \\
&(\nabla_{\mathbf{u}}\,\mathrm{X})_0 \cdot \mathrm{u}_1 = 0\ , \qquad &&(\nabla_{\mathbf{u}}\,\mathrm{T})_0 \cdot \mathrm{u}_1 = 0
\end{aligned}
\tag{2.7}
$$

The conditions of invariance for v and ρ at the zero order in ϵ give:

$$
\begin{aligned}
\mathrm{r}\,\frac{\partial \rho_0}{\partial \mathrm{x}} + \gamma\,\mathrm{t}\,\frac{\partial \rho_0}{\partial \mathrm{t}} &= \alpha\ \rho_0 \\
\mathrm{r}\,\frac{\partial \mathrm{v}_0}{\partial \mathrm{x}} + \gamma\,\mathrm{t}\,\frac{\partial \mathrm{v}_0}{\partial \mathrm{t}} &= \beta\ \mathrm{v}_0
\end{aligned}
\tag{2.9}
$$

while at the first order in ϵ we have:

$$
\begin{aligned}
\mathrm{r}\,\frac{\partial \rho_1}{\partial \mathrm{x}} + \gamma\,\mathrm{t}\,\frac{\partial \rho_1}{\partial \mathrm{t}} &= (\gamma-2)\ \rho_1 \\
\mathrm{r}\,\frac{\partial \mathrm{v}_1}{\partial \mathrm{x}} + \gamma\,\mathrm{t}\,\frac{\partial \mathrm{v}_1}{\partial \mathrm{t}} &= -(\alpha+1)\ \mathrm{v}_1
\end{aligned}
\tag{2.10}
$$

Consequently, by introducing the similarity variable $\sigma = \dfrac{\mathrm{x}}{\mathrm{t}^{1/\gamma}}$ we have the following representation for the different terms in the expansion of u:

$$\rho_0 = t^{\frac{\alpha}{\gamma}} R_0(\sigma) , \qquad \rho_1 = t^{-\frac{\gamma-2}{\gamma}} R_1(\sigma)$$

$$v_0 = t^{\frac{\beta}{\gamma}} V_0(\sigma) , \qquad v_1 = t^{-\frac{\alpha+1}{\gamma}} V_1(\sigma) \tag{2.11}$$

that would give the forms of an invariant solution approximate at the first order in ϵ.

Taking into account of (2.8) and (2.11) in (2.2) we easily find that the following two systems of ordinary differential equations must be satisfied:

$$\mathcal{A}_0 \frac{d\mathbf{w}_0}{d\sigma} = \mathfrak{B}_0(\mathbf{w}_0) \tag{2.12}$$

$$\mathcal{A}_0 \frac{d\mathbf{w}_1}{d\sigma} = \mathfrak{B}_0\left(\mathbf{w}_0, \mathbf{w}_1, \frac{d\mathbf{w}_0}{d\sigma}, \frac{d^2\mathbf{w}_0}{d\sigma^2}\right) \tag{2.13}$$

where

$$\mathcal{A}_0 = \begin{bmatrix} V_0 - \frac{\sigma}{\gamma} & R_0 \\ K\Gamma R_0{}^{\Gamma-2} & V_0 - \frac{\sigma}{\gamma} \end{bmatrix}, \qquad \mathbf{w}_\alpha = \begin{bmatrix} R_\alpha \\ V_\alpha \end{bmatrix} \quad (\alpha = 0, 1) \tag{2.14}$$

$$\mathfrak{B}_0 = \begin{bmatrix} -R_0\left(\frac{\alpha}{\gamma} + \frac{\nu V_0}{\sigma}\right) \\ -\frac{\beta}{\gamma} V_0 \end{bmatrix}, \tag{2.15}$$

$$\mathfrak{B}_1 = \begin{bmatrix} -\left(\frac{dV_0}{d\sigma} + \frac{\nu}{\sigma} V_0 + \frac{\gamma-2}{\gamma}\right) R_1 - \left(\frac{dR_0}{d\sigma} + \frac{\nu}{\sigma} R_0\right) V_1 \\ \frac{\alpha+1}{\gamma} V_1 - K\Gamma(\Gamma-2) R_0{}^{\Gamma-3} \frac{dR_0}{d\sigma} R_1 - \frac{dV_0}{d\sigma} V_1 + \frac{1}{R_0} \frac{d^2 V_0}{d\sigma^2} \end{bmatrix} \tag{2.16}$$

The systems (2.12) and (2.13) both admit an associated group of invariance [4], so that can be reduced to autonomous form by the following change of variables:

$$R_0 = \sigma^a G_0 \quad , \qquad R_1 = \sigma^c G_1$$

$$V_0 = \sigma Z_0 \quad , \qquad V_1 = \sigma^b Z_1 \tag{2.17}$$

where

$$a = \frac{2}{\Gamma - 1}, \qquad b = -\frac{\Gamma+1}{\Gamma-1}, \qquad c = -2$$

obtaining, respectively,

$$\tilde{\mathcal{A}}_0 \frac{d\tilde{\mathbf{w}}_0}{d\tau} = \tilde{\mathfrak{B}}_0 \quad , \qquad \tilde{\mathcal{A}}_0 \frac{d\tilde{\mathbf{w}}_1}{d\tau} = \tilde{\mathfrak{B}}_1 \tag{2.18}$$

where

$$\tau = \ln \sigma \,, \qquad \tilde{\mathbf{w}}_\alpha = \begin{bmatrix} G_\alpha \\ Z_\alpha \end{bmatrix} \quad (\, \alpha = 0,1 \,)$$

$$\tilde{\mathcal{A}}_0 = \begin{bmatrix} Z_0 - \frac{1}{\gamma} & G_0 \\ K\,\Gamma\,{G_0}^{\Gamma\text{-}2} & Z_0 - \frac{1}{\gamma} \end{bmatrix} , \tag{2.19}$$

$$\tilde{\mathfrak{B}}_0 = \begin{bmatrix} -G_0\Big[Z_0\Big(\frac{\Gamma+1}{\Gamma-1}+\nu\Big) - \frac{2}{\Gamma-1}\Big] \\ -Z_0\,(Z_0-1) - K\,\Gamma\,{G_0}^{\Gamma\text{-}1} \end{bmatrix} , \tag{2.20}$$

$$\tilde{\mathfrak{B}}_1 = \begin{bmatrix} -\Big\{G_1\frac{dZ_0}{d\tau} + Z_1\frac{dG_0}{d\tau} + (\nu-1)(G_0 Z_1 + G_1 Z_0) + G_1\Big\} \\ K\,\Gamma\,{G_0}^{\Gamma\text{-}3}\,G_1\Big(\frac{2\,G_0}{\Gamma-1} - (\Gamma-2)\frac{dG_0}{d\tau}\Big) + \frac{2}{\Gamma-1}Z_1\,(Z_0-1) + \\ \qquad + \Big(\frac{1}{G_0} - Z_1\Big)\frac{dZ_0}{d\tau} + \frac{1}{G_0}\frac{d^2 Z_0}{d\tau^2} \end{bmatrix} \tag{2.21}$$

In conclusion by solving (2.18), we obtain a similarity solution of the hyperbolic system where the dissipation is negleded then the first order terms in the expression (1.3) can be characterized by solving a linear, in general, non homogeneous system of ordinary differential equations.

As can be seen from (2.1) the inhomogeneity is due only to the presence of the dissipation.

In the assigned variables we obtain an approximate similarity solution in the form:

$$\rho = \left(\frac{x}{t}\right)^{\frac{2}{\Gamma-1}} G_0(\tau) + \epsilon\, t\, x^{-2}\, G_1(\tau) + O(\epsilon^2)$$
$$v = \frac{x}{t} Z_0(\tau) + \epsilon\, t^{\frac{2}{\Gamma-1}} x^{-\frac{\Gamma+1}{\Gamma-1}} Z_1(\tau) + O(\epsilon^2) \tag{2.22}$$

ACKNOWLEDGMENTS

The authors wishes to thank G.N.F.M. - C.N.R. for supporting partially this research.

REFERENCES

1. V. A. BAIKOV, R. K. GAZIZOV and N. H. IBRAGIMOV, Math. USSR Sbornik 64, 427-441 (1989).
2. W. F. AMES, R. J. LOHNER and E. ADAMS, Int J. Nonlinear Mech. 16, 439-447 (1981).
3. L. V. OVSIANNIKOV, Group Analysis of Differential Equations, Academic Press (New York 1982).
4. W. F. AMES and A. DONATO, Int. J. Nonlinear Mech. 23, 167-174 (1988).
5. D. CRIGHTON, Basic Theoretical Nonlinear Acoustic, reprinted from Frontiers in Physical Acoustic SIF (Bologna 1986).

Andrea Donato
Dipartimento di Matematica
Università di Messina
Contrada Papardo, salita Sperone n° 31
98166 Sant'Agata - MESSINA
ITALY

Annunziata Palumbo
Istituto del Biennio
Facoltà di Ingegneria
Via E. Cuzzocrea 48
89128 Reggio Calabria
ITALY

J ENGELBRECHT

Solitary waves in systems with energy influ

ABSTRACT

Two basic cases for solitary waves in systems with energy influx are briefly described. Some ideas about the behaviour of solitary waves in systems with weak energy influx are presented together with the numerical results.

1. INTRODUCTION

As widely accepted, solitary waves form a paradigm in contemporary mathematical physics. In mathematical terms, solitary waves are the solutions to nonlinear but still intgrable evolution equations. The integrable equations or systems form the "building blocks" of nature /1/, related to a certain energetical background. In the processes modelled by integrable evolution equations this usually means the conservation of energy. Nature is still far from being homogeneous, conservative, etc. In reality, energy is usually unbalanced and there exists either outfux or influx of energy with respect to a certain dynamical process. The first case dealing with the outfux of energy has been a topic of discussion in dynamics for a long time and the mechanisms of attenuation are rather well analyzed. The energy influx, however, is not so well understood or, better said, modelled. It seems that these processes have become more important just during the last decades. In both cases an intriguing question arises. If solitary waves are shown to exist in conservative media, do they loose their significance immediately when dealing with more realistic situations? And if the answer is "no", then what is going to happen to them?

Below we discuss some principles in the theory of solitary waves in open systems with energy influx. Our earlier research in this field includes the derivation of certain mathematical models /2,3/ and also the analysis of some evolution equations /4,5/. In Section 2 the basic principles in the theory of solitary waves in open systems are described. Section 3 is devoted to perturbation analysis and Section 4 to the numerical results.

2. BASIC PRINCIPLES

There are two main cases in nonlinear wave dynamics of open systems: the weak energy influx and the strong energy influx. The adjectives "weak" and "strong" have been chosen to distinguish between two basically different processes, which might be characterized in the following way:
(i) the "weak energy influx" is chosen to characterize processes where solitary waves exist also in the absence of the energy influx, and in this sense the energy influx has only a perturbative character;
(ii) the "strong energy influx" is chosen to characterize processes where solitary waves cannot exist in the absence of the energy influx.

This paper is about the systems with weak energy influx.

The structure of a solitary wave in open systems is different from that in conservative systems. If we take a KdV soliton with an amplitude u as a basis for a standard solitary wave which is symmetric with respect to its central line ($\eta=0$), then we may use the following definitions:

Definition 1: Symmetric solitary pulse waves tend to the equilibrium state at $\eta\to\pm\infty$ with the equal rates either as $u\to+0$ or as $u\to-0$.

Definition 2: Asymmetric solitary pulse waves of the first kind tend to the equilibrium state at $\eta\to\pm\infty$ with different rates as $u\to+0$ or as $u\to-0$.

Definition 3: Asymmetric solitary pulse waves of the second kind tends to the equilibrium state at $\eta\to+\infty$ as $u\to\pm0$

and at $\eta\to-\infty$ as $u\to\mp 0$.

The properties of solitary waves obeying these definitions are shown in FIG.1.

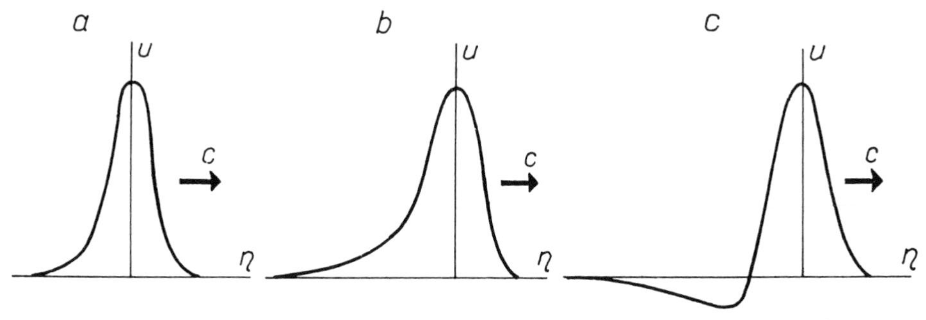

FIGURE 1 Profiles according to Defs. 1 - a, 2 - b and 3 - c.

All the numerical results known to the author demonstrate asymmetry of solitary waves in nonconservative open systems. To prove this property by rigorous analysis seems to be an interesting problem

3. PERTURBATIVE ANALYSIS

Systems (equations) with weak energy influx, i.e., with a r.h.s. of a perturbative character can often be solved by perturbation methods /6/. Here we present some results for a KdV equation with a special r.h.s. modelling the influence of additional forces /3/. Let the KdV equation be given in the canonical form

$$U_T - 6UU_X + U_{XXX} = \varepsilon F(U) \tag{1}$$

where $\varepsilon>0$ is a small parameter and $F(U)$ is a smooth function. For $\varepsilon=0$, a soliton exists, determined by

$$U_S(X,T) = -2\kappa^2 \operatorname{sech}^2 z, \quad z = \kappa(X - \xi), \tag{2}$$

where $\kappa\equiv\kappa_1$ is an eigenvalue and ξ is a phase shift. For $\varepsilon\neq 0$, the solution can be found by the perturbative inverse scattering transform (IST) /6/ like

$$U(X,T) = U_S(z,\kappa) + \delta U(z,T),$$

$$U_S(z,\kappa) = -2\kappa^2(T)\operatorname{sech}^2 z, \qquad z = \kappa(T)[X - \xi(T)], \tag{3}$$

$$\delta U(z,T) = -2\kappa^2(T)W(z,T) .$$

Leaving aside the details which can easily be found in /6/, the perturbed eigenvalue is determined by

$$\kappa_T = - \frac{\varepsilon}{4\kappa} \int_{-\infty}^{\infty} F(U_S)\operatorname{sech}^2 z \, dz \tag{4}$$

and the perturbed "tail" by

$$\frac{d}{dT} \kappa \int_{-\infty}^{\infty} W(z,T)dz = -2\varepsilon q\kappa^4 . \tag{5}$$

Here parameter q is related to F(U) by the expression

$$q = \frac{1}{4\kappa^5} \int_{-\infty}^{\infty} F(U_S)\tanh^2 z \, dz. \tag{6}$$

In FIG.2 the qualitative sketches of the deformed solitons are shown for $\varepsilon q<0$ and $\varepsilon q>0$, respectively.

For region BC (see FIG.2) the estimation of W gives almost a straight horizontal line $W = -\varepsilon q/2$. The length of this region increases with T increasing and at large negative X there are oscillations which die out /6/. Taking this fact into consideration, the Defs. 2 and 3 from Sect.2 are valid. If the straight part (plateau) for $q\neq 0$ exists, then its relative height with respect to the main part of the soliton (kern soliton) may be called eccentricity

$$e = |\delta U/\max U_S| . \tag{7}$$

Let now $F(U) = -(U+b_2U^2+b_3U^3)$, $b_2>0$, $b_3>0$. It is known /5/ that in this case a certain stable eigenvalue exists which shows the existence of a stable soliton. The eccentricity in this case is

$$e = \left|\frac{1}{6}\varepsilon \left(\frac{1}{\kappa^3} - \frac{4}{5}\frac{b_2}{\kappa} + \frac{32}{35} b_3\kappa\right)\right| . \tag{8}$$

The solution of (4) is given by an implicit expression and the question how an initial soliton is transformed to a

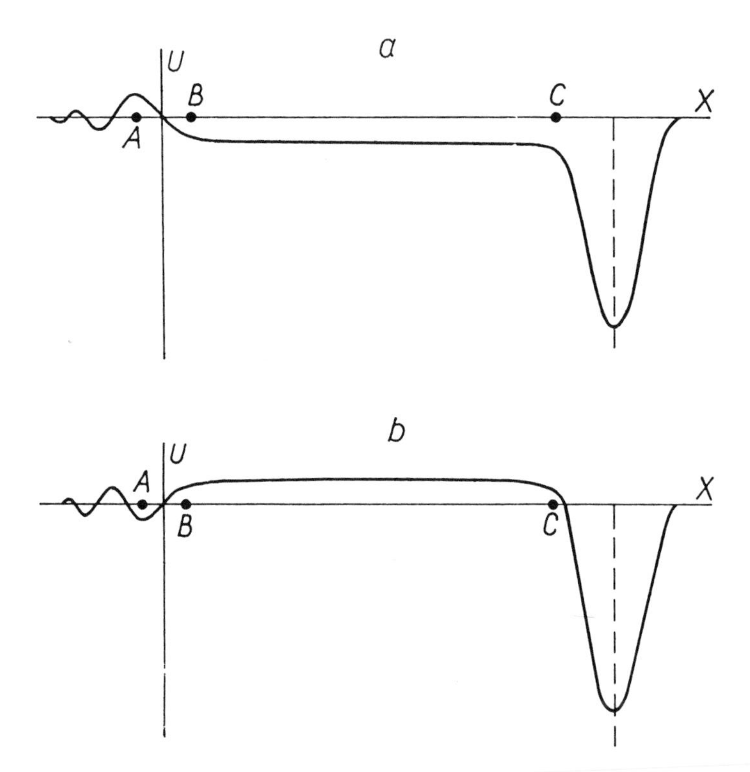

FIGURE 2 Deformed solitons for $\varepsilon q<0$ (a) and for $\varepsilon q>0$ (b).

new stable form still remains open. This is shown in the next section by numerical calculations.

4. NUMERICAL RESULTS

Let us analyze the perturbed KdV equation

$$u_t + uu_x + \mu u_{xxx} = 2f(u) \tag{9}$$

where

$$f(u) = -(u-0.5u^2+0.0556u^3), \tag{10}$$

The roots of $f(u)=0$ are $u_1=0$, $u_2=3$, $u_3=6$. Let $u=0.15$, then a soliton with the initial amplitude $u_0=6$ is stable for $\varepsilon=0$. For $\varepsilon\neq0$, another stable solution is possible but the route from a soliton with $u_0=6$ to the perturbed solution may be complicated, undergoing a certain oscillation. Nu-

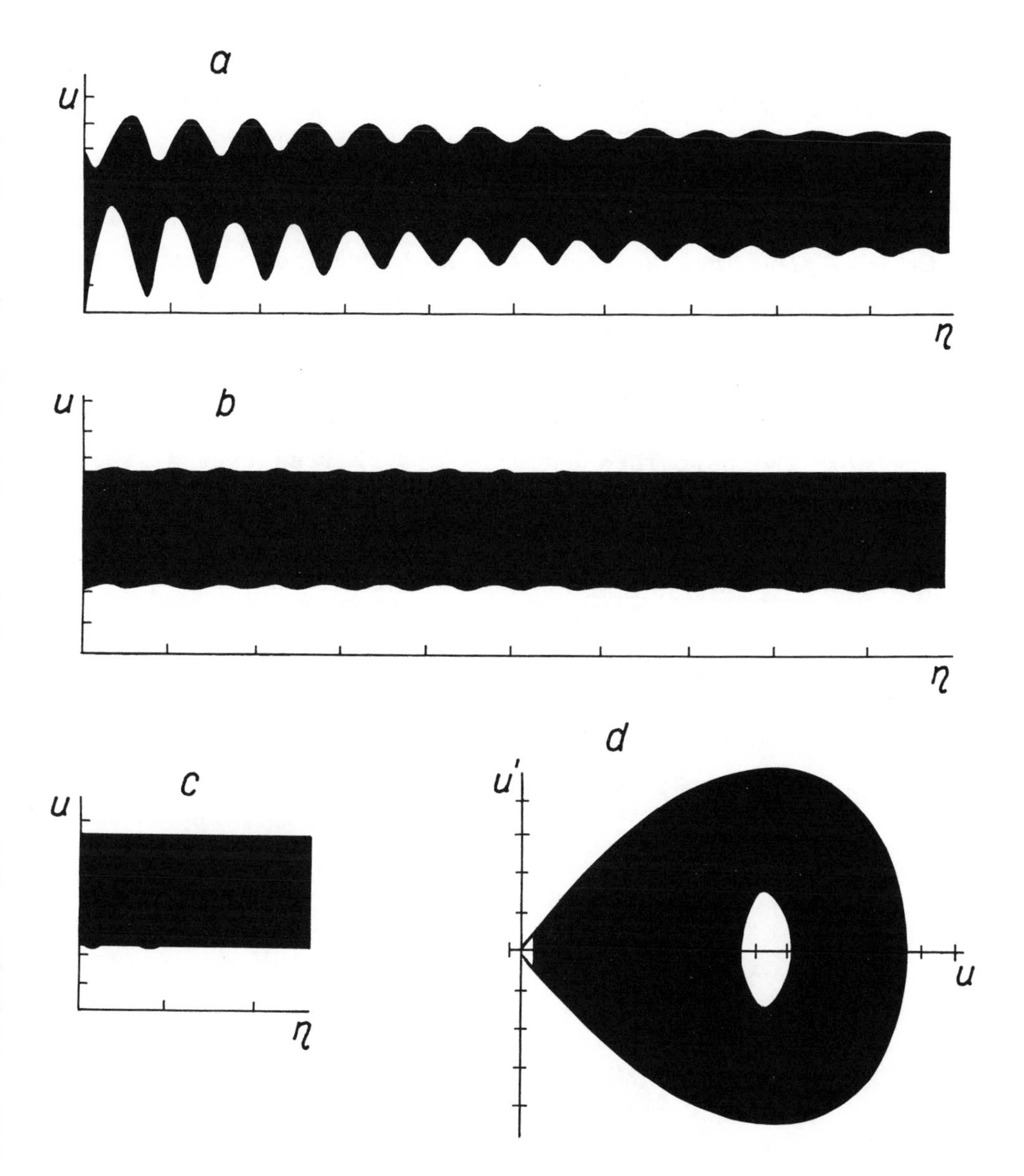

FIGURE 3 Solution to stationary equation (a) for $\eta \leqq 11000$, $\varepsilon=0.10$: a) $\eta=[0,5000]$, b) $\eta=[5000,10000]$, c) $\eta=[10000,11000]$, d) phase portrait u-u'.

merical results obtained by using a standard Runge-Kutta approach show this phenomenon explicitly. However, the time needed for establishing the stable stationary solution is large. In FIG.3 the numerical calculations are shown for $\varepsilon=0.10$ for large $\eta=x-\lambda t$. Single waves cannot be distinguished but the envelope modulation is clearly seen. Actually, the period T_e of this modulation depends upon the value of ε for given f(u) and seems to be inversely propositional in respect to ε. For very small ε's there is no oscillation, for large ε's (here $\varepsilon \geqq 0.20$), the oscillations start to grow and the stability is lost. The reason is that the trajectory goes to the repulsive region of the corresponding phase-space. The compressed phase portrait u-u' is also shown (FIG.3d), resembling a "doughnut" with a sharp corner at the original saddle point (0,0). The singular points of the perturbed equation are shifted considerably. A similar situation is noted for a quadratic r.h.s. /7/.

5. CONCLUDING REMARKS

The best example for solitary waves in systems with strong energy influx is the celebrated nerve pulse transmission. Usually it forms a different chapter in wave dynamics because the classical models of this process are based on parabolic equations /8/. Nevertheless, a unified approach /9/ seems to be inevitable (see also /10/) and it will be given in further publications of the author.

6. ACKNOWLEDGEMENTS

The support from the LOC of Euromech 270 and the excellent atmosphere during the Colloquium in Reggio di Calabria are highly appreciated.

REFERENCES

1. M.TABOR, Chaos and Integrability in Nonlinear Dynamics (Wiley New York 1989).
2. J.ENGELBRECHT,On Theory of Pulse Transmission in a Nerve Fibre. Proc. Royal Soc.London A375 (1981) 195-209.
3. J.ENGELBRECHT, Y.KHAMIDULLIN, On the Possible Amplification of Nonlinear Seismic Waves. Phys.Earth Plan.Int. 50 (1988) 39-45.
4. J.ENGELBRECHT, The Evolution of Nonlinear Waves in Active Media. Wave Motion 8 (1986) 93-100.
5. J.ENGELBRECHT, A.JEFFREY, KdV Solitons in Active Media. Wave Motion 9 (1987) 533-541.
6. V.I.KARPMAN, Soliton Evolution in the Presence of Perturbation. Phys. Scripta 20 (179) 462-478.
7. J.G.B.BYATT-SMITH, Effects of Empirical Dissipation Terms in the Solution of the Undular Bore. Quart.Appl. Math. 28 (1971) 499-515.
8. S.C.COLE, Membranes, Ions and Impulses. A Chapter of Classical Biophysics (Univ. of California Press Berkley 1972).
9. A.C.SCOTT, Birth of Paradigma. In: P.L.E.USLENGHI (ED.), Nonlinear Electromagnetics (Academic Press New York 1980).
10. J.ENGELBRECHT (ED.), Nonlinear Waves in Active Media (Springer Heidelberg 1989).

Jüri Engelbrecht
Institute of Cybernetics
Estonian Acad. Sci.
Tallinn, SU 200108
Estonia

R FAZIO

A moving boundary problem for rate-type materials

ABSTRACT

This note aims to present some preliminary results obtained in studying the mathematical model describing the longitudinal wave propagation determined by a time dependent stress impact to a bar of the so-called Maxwell material. Due to the impact a shock front propagates with a finite velocity. Here our interest is to make evident the influence of the dissipative term on the propagation of the shock front. That is shown by means of similarity and numerical approach. The influence of the dissipative term is evident: the free boundary is a decreasing function of the dissipative coefficient.

1. INTRODUCTION

We consider the longitudinal wave propagation problem, determined by a stress impact to a bar of rate-type material, governed by the following one-dimensional model, see [1-3],

$$\rho_0 \frac{\partial v}{\partial t} - \frac{\partial \sigma}{\partial x} = 0$$

$$\frac{\partial \sigma}{\partial t} - \Phi(\sigma) \frac{\partial v}{\partial x} = \Psi(\sigma) \ . \qquad (1)$$

In this model v is the normal component of velocity along the axis of the bar, σ is the tensile stress, ρ_0 and x are respectively the reference density and longitudinal coordinate at reference time $t = 0$, whereas $\Phi(\sigma) = \Phi_0 \, \sigma^{1/p}$ and $\Psi(\sigma) = - \Psi_0 \, \sigma^{1/q}$ (Φ_0, Ψ_0, p and q are constants) are material response functions, see [1]. The system (1), provided $\Phi(\sigma) > 0$, is hyperbolic.

As far as the stress impact problem is concerned we have to consider, see [1], the boundary condition

$$\sigma(0, t) = \sigma_0 \, t^{\delta} \, H(t) \tag{2}$$

where σ_0 and δ are constants and $H(\cdot)$ is the Heaviside step function. We assume that the boundary condition (2) determines the propagation of a shock front $x_s(t)$ in the bar initially at rest, i.e., $v(x > x_s(t), t) = 0$, $\sigma(x > x_s(t), t) = 0$ and $x_s(0) = 0$. At the shock front we have to apply the well known Rankine-Hugoniot conditions [4]. A similar problem for a velocity impact problem in elastic materials is considered in [5-7].

2. SIMILARITY AND NUMERICAL ANALYSIS

A similarity analysis allows us to introduce the similarity variables

$$\begin{aligned}
\eta &= \sigma_0^{-1/2p} \, \Phi_0^{-1/2} \, \rho_0^{1/2} \, x \, t^{-\gamma} \\
\eta_s &= \sigma_0^{-1/2p} \, \Phi_0^{-1/2} \, \rho_0^{1/2} \, x_s(t) \, t^{-\gamma} \\
V(\eta) &= \sigma_0^{(1-2p)/2p} \, \Phi_0^{1/2} \, \rho_0^{1/2} \, t^{-\alpha} \, v(x, t) \\
\Sigma(\eta) &= \sigma_0^{-1} \, t^{-\delta} \, \sigma(x, t)
\end{aligned} \tag{3}$$

where $\alpha = \delta - \frac{1}{2}\frac{\delta}{p}$, $\gamma = 1 + \frac{1}{2}\frac{\delta}{p}$ and if $\Psi_0 \neq 0$ it has to be $\delta = \frac{q}{q-1}$. By making use of (3) the model problem (1)-(2) and related Rankine-Hugoniot conditions can be reduced to the ordinary differential system in normal form

$$\begin{aligned}
\frac{dV}{d\eta} &= \frac{\alpha \, \gamma \, \eta \, V - \delta \, \Sigma - \Psi_0 \, \sigma_0^{(1-q)/q} \, \Sigma^{1/q}}{\gamma^2 \, \eta^2 - \Sigma^{1/p}} \\
\frac{d\Sigma}{d\eta} &= \frac{\gamma \, \eta \left(\delta \, \Sigma + \Psi_0 \sigma_0^{(1-q)/q} \, \Sigma^{1/q} \right) - \alpha \, V \, \Sigma^{1/p}}{\gamma^2 \, \eta^2 - \Sigma^{1/p}}
\end{aligned} \tag{4a}$$

along with the boundary conditions

$$\Sigma(0) = 1 \tag{4b}$$

$$\Sigma(\eta_s) = \left(\frac{p-1}{p} \frac{1}{\gamma^2 \eta_s{}^2}\right)^{-p} \tag{4c}$$

$$V(\eta_s) = -\frac{\Sigma(\eta_s)}{\gamma\, \eta_s} \ . \tag{4d}$$

Here (4c) and (4d) are obtained by the Rankine-Hugoniot relations. Since η_s is unknown the problem (4) defines a free boundary value problem.

In the present analysis our interest is to point out the role played by the dissipative term. Hence, we will consider the two possibilities $\Psi_0 \neq 0$ and $\Psi_0 = 0$. The main result is the following: if $\Psi_0 = 0$ or $q = 1$ then (4a), (4c) and (4d) result invariant under the associated group

$$V^* = \omega^{2p-1} V \quad ; \quad \Sigma^* = \omega^{2p} \Sigma \quad ; \quad \eta^* = \omega\, \eta \tag{5}$$

whereas when $\Psi_0 \neq 0$ and $q \neq 1$ such an associated group does not exist. We conclude that in order to solve numerically the free boundary value problem (4) we can use for the non-dissipative case the non-iterative transformation method [8] whereas for the dissipative case we have to use the iterative transformation method [9].

In the following we give the general outlines of the numerical solution of the free boundary value problem (4). Let us denote with ϑ the dissipative coefficient, i.e.

$$\vartheta = \Psi_0\, \sigma_0{}^{(1-q)/q} \ .$$

First we consider the non-dissipative case $\vartheta = 0$. From (5) it is easily found that

$$\begin{aligned} \eta_s &= \left(\Sigma^*(0)\right)^{-1/2p} \eta_s{}^* \\ V(0) &= \left(\Sigma^*(0)\right)^{(1-2p)/2p} V^*(0) \ . \end{aligned} \tag{6}$$

Therefore, we obtain (an approximation of) the correct value of η_s by a numerical integration backwards in $[0, \eta_s{}^*]$, where we can choose $\eta_s{}^*$ at our convenience. We found the numerical results listed in Table 1 either by setting $\eta_s{}^* = 1$ or $\eta_s{}^* = 0.5$.

Table 1. The non-dissipative case $\vartheta = 0$. We assume $p = 2$.

δ	$V(0)$	η_s
1.5	−1.357543	0.390085
2.5	−1.349733	0.302722
5.0	−1.343047	0.194976

Next let us consider the dissipative case $\vartheta \neq 0$. In order to apply the iterative transformation method we have to introduce a numerical parameter h. If we extend the stretching group (5) by

$$h^* = \omega \, h \tag{7}$$

then the differential system (4a) has to be modified, see [9], as follows

$$\frac{dV}{d\eta} = \frac{\alpha \, \gamma \, \eta \, V - \delta \, \Sigma - h^{2p\,(1-1/q)} \, \vartheta \, \Sigma^{1/q}}{\gamma^2 \, \eta^2 - \Sigma^{1/p}}$$

$$\frac{d\Sigma}{d\eta} = \frac{\gamma \, \eta \left(\delta \, \Sigma + h^{2p\,(1-1/q)} \, \vartheta \, \Sigma^{1/q}\right) - \alpha \, V \, \Sigma^{1/p}}{\gamma^2 \, \eta^2 - \Sigma^{1/p}} \; . \tag{8}$$

Moreover, along with (6) we have to use

$$h = \left(\Sigma^*(0)\right)^{-1/2p} \, h^* \; . \tag{9}$$

Here we note that (4a) is recovered from (8) by setting $h = 1$. So that we will find the correct value of η_s if we get $h = 1$ from (9). Then, fixed a value of $\eta_s{}^*$, we define h^*

iteratively by a root finder. Our purpose is to find a zero of the transformation function

$$\Gamma(h^*) = \left(\Sigma^*(0)\right)^{-1/2p} h^* - 1 \ . \tag{10}$$

As well know every iterative numerical procedure needs an appropriate criterion of convergence. In order to accept a value of $\eta_S{}^k$ we shall require

$$| \Gamma(h^*{}_k) | < \text{Tol} \qquad ; \qquad | \eta_S{}^k - \eta_S{}^{k-1} | < \text{Tol} \tag{11}$$

where Tol is a prescribed tolerance. For the proposed problem some results, related to a value of δ in Table 1, were found by a secant method. There we used Tol $=$ 1D$-$06.

Table 2. The dissipative case for $p = 2$, $\delta = 3/2$, $q = 3$ and $\eta_S{}^* = 1$.

ϑ	V(0)	η_S
$-1.$	-1.080078	0.421117
-0.5	-1.221729	0.405372
0.	-1.357543	0.390085

Table 2 shows a representative set of numerical results.

Let us remind here the functional form of the moving boundary

$$x_S(t) = \sigma_0{}^{1/2p} \, \Phi_0{}^{1/2} \, \rho_0{}^{-1/2} \, \eta_S \, t^{\gamma} \tag{12}$$

where $\gamma = 1 + \frac{1}{2}\frac{\delta}{p}$. In order to discuss the influence of the dissipative term on the shock front propagation we use the obtained values for the case $\delta = 3/2$. The influence of the dissipative term is evident: it turns out that η_S is a decreasing function of ϑ.

ACKNOWLEDGEMENT

This work was supported by the Research National Council of Italy. The author wish to thank Professor D. Fusco for several helpful discussions.

REFERENCES

1. D. B. TAULBEE, F. A. COZZARELLI and C. L. DYM, Similarity solutions to some non-linear impact problems, Int. J. Non-linear Mech., **6** (1971) 27-43.

2. W. KOSINSKI, Gradient catastrophe in the solution of nonconservative hyperbolic systems, J. Math. Analysis Appl., **61** (1977) 672.

3. D. FUSCO, Complete exceptionality, constitutive laws and symmetric form for a non-linear inelastic rod, Int. J. Non-linear Mech., **16** (1981) 459-464.

4. A. JEFFREY, Quasilinear hyperbolic systems and waves, Research Notes in Maths., 5 (Pitman London 1976).

5. L. W. DRESNER, Similarity solutions of non-linear partial differential equations, Research Notes in Maths., 88 (Pitman London 1983).

6. A. DONATO, Similarity analysis and non-linear wave propagation, Int. J. Non-linear Mech., 22 (1987) 307-314.

7. R. FAZIO, Numerical evaluation for shock and discontinuity velocity related to similarity solutions of quasilinear hyperbolic systems, J. Comput. Appl. Math., **30** (1990) 341-349.

8. R. FAZIO and D. J. EVANS, Similarity and numerical analysis for free boundary value problems, Int. J. Comput. Math., **31**, 215-220 (1990).

9. R. FAZIO, The iterative transformation method: free or moving boundary value problems, to appear.

Riccardo Fazio
Department of Mathematics, University of Messina,
Contrada Papardo, Salita Sperone, 31,
98166 Sant'Agata, Messina
Italy

J M GHIDAGLIA AND A MARZOCCHI

Finite dimensional global attractors for strongly damped wave equations

ABSTRACT

An application of some results on the asymptotic behaviour of solutions to strongly damped wave equations (such as the Sine-Gordon equation) contained in [1] is reviewed. The main theorem states the existence of a finite-dimensional global attractor for the solutions of the equations taken into consideration.

1. INTRODUCTION

We are going to show some results which are contained in a joint paper with J.M.Ghidaglia [1], referring to it concerning the most technical parts.

Let us begin by mentioning two applications that have motivated this work. Firstly, the perturbed sine-Gordon equation

$$\frac{\partial^2 u}{\partial t^2} - \Delta u + \sin u = -\beta \frac{\partial u}{\partial t} + \alpha \frac{\partial(\Delta u)}{\partial t} + f \ , \tag{1}$$

where $u(x,t)$ is the current in a Josephson junction (see e.g. [2]), x is the space variable and (1) is posed in a bounded domain Ω in $\mathbb{R}^n$ (with appropriate boundary conditions). The parameters α and β are non negative and correspond to loss effects. We are concerned in this work by the case where $\alpha > 0$ (referring to [3,4] for the case $\alpha = 0$). The function f, $f(x,t)$, is time-periodic and figures the external current that drives the device. Another example reads

$$\frac{\partial^2 u}{\partial t^2} - \Delta u + |u|^q u = -\beta \left|\frac{\partial u}{\partial t}\right|^p \frac{\partial u}{\partial t} + \alpha \frac{\partial(\Delta u)}{\partial t} + f \ , \tag{2}$$

which is also a perturbed wave equation occuring in quantum mechanics.

These partial differential equations can be seen as infinite dimensional dynamical systems, and the questions we address are related to determining whether or not these systems depend on a finite number of degree of freedom after a transient period. A mathematical approach to this type of problem has been introduced for dissipative parabolic equations, motivated by the study of turbulence in fluids. It was proved that a global attractor exists and captures all the solutions as time goes to infinity, and that this set was finite dimensional. A bound on this dimension provides an estimate on the number of degrees of freedom in the long time behaviour.

2. SETTING OF THE PROBLEM AND *A PRIORI* ESTIMATES

Consider a problem of the form

$$\frac{\partial^2 u}{\partial t^2}(x,t) - \alpha\Delta\frac{\partial u}{\partial t}(x,t) - \Delta u(x,t) + g(u(x,t)) + h\left(\frac{\partial u}{\partial t}(x,t)\right) = f(x,t) \ ,$$
$$u(x,t)_{|\partial\Omega} = 0 \ , \tag{3}$$
$$u(x,0) = u_0(x) \ , \ \frac{\partial u}{\partial t}(x,0) = u_1(x) \ .$$

where Ω is an open bounded connected domain in $\mathbb{R}^n$ with Lipschitz boundary, $u_0(x) \in H_0^1(\Omega)$, $u_1(x) \in L^2(\Omega)$, and g,h are given nonlinear real functions. We will indicate by $|\cdot|$,$\|\cdot\|$ the $L^2(\Omega)$- and $H_0^1(\Omega)$-norms, respectively.

We will furthermore suppose that these functions have suitable growth and regularity assumptions in order to ensure to (3) existence, uniqueness and regularity of the solutions. This will be the case if *e.g.*

$$g(s) = \lambda|s|^{\beta-1}s \text{ and } h(s) = \mu|s|^{\gamma-1}s$$

$$\text{with } \lambda \geq 0, \quad \mu \geq 0, \quad \beta > 0, \quad \gamma > 0 \text{ and (for } n \geq 3), \gamma, \ \beta < 1 + \frac{n^2}{n-2}. \tag{H}$$

Of course much more general functions g and h could be considered but we have restricted the exposition to homogeneous ones for the sake of simplicity (see [1] for a more complete discussion).

In this case, assuming that the external force term f satisfies $f \in L^2(0,T; H^{-1}(\Omega))$ $\forall T > 0$, it is straightforward to derive from the above hypotheses an *a priori* bound on the smooth solutions of (3). Existence of solutions can be then obtained via the usual Faedo-Galerkin method and using the fact that, thanks to the above hypotheses, the nonlinear terms are locally lipschitzian (see [1]). The result is that for every $f \in L^2(0,T;H^{-1}(\Omega))$ $\forall T > 0$ and $u_0 \in H_0^1(\Omega)$, $u_1 \in L^2(\Omega)$ there exists an unique function $u(t)$ such that

$$u \in C(\mathbb{R}^+; H_0^1(\Omega)) \qquad u_t \in C(\mathbb{R}^+, L^2(\Omega)) \cap L^2(0,T; H_0^1(\Omega)) \ \forall T > 0$$

that satisfies (3).

If f is time-independent, *i.e.* $f \in H^{-1}(\Omega)$, we can introduce the mappings

$$S(t) = \{u_0, u_1\} \rightarrow \{u(t), u_t(t)\}$$

which form a semigroup on $H_0^1(\Omega) \times L^2(\Omega)$: that is, for fixed $t > 0$, $S(t)$ is continuous on $H_0^1(\Omega) \times L^2(\Omega)$ and

$$S(0) = I \ , \ S(t_1 + t_2) = S(t_1) \circ S(t_2) \qquad t_1, t_2 > 0 \ .$$

When $f = g = h = 0$, *i.e.* in the linear homogeneous case, equation (3) defines a linear semigroup on $H_0^1(\Omega) \times L^2(\Omega)$ which we denote by $\Sigma(t)$

$$\Sigma(t) : \{u_0, u_1\} \to \{u(t), u_t(t)\} , \qquad t \geq 0 .$$

When $\alpha = 0$ we recover the usual wave propagator (which is a unitary group) while when $\alpha > 0$ (the dissipative case), the $\Sigma(t)$ are no more invertible and the trajectory tends exponentially to zero as $t \to +\infty$, as shown in the following

Proposition 1. *Let λ_1 be the first eigenvalue of $-\Delta$ and set*

$$\varepsilon_0 = \min\left(\frac{1}{\alpha}, \frac{\alpha\lambda_1}{(2(1+\lambda_1))}\right).$$

Then, for every $\varepsilon \in]0, \varepsilon_0[$

$$|\{u,v\}|_\varepsilon^2 \equiv (1-\varepsilon\alpha)\|u\|^2 + \varepsilon^2|u|^2 + |v+\varepsilon u|^2$$

induces a norm on $H_0^1(\Omega) \times L^2(\Omega)$, equivalent to the usual one, and for every $\{u_0, u_1\} \in H_0^1(\Omega) \times L^2(\Omega)$ we have

$$|\Sigma(t)\{u_0, u_1\}|_\varepsilon \leq e^{-\varepsilon t}|\{u_0, u_1\}|_\varepsilon , \quad \forall t \geq 0 . \tag{4}$$

Proof. Actually we are going to show a slightly stronger result than (4). We take $g = h = 0$ and $f \in H^{-1}(\Omega)$ and denote by $\{u(t), u_t(t)\}$ the solution of (3). Then

$$|\{u,v\}|_\varepsilon^2 \leq |\{u_0, v_0\}|^2 e^{-2\varepsilon t} + \frac{(1-e^{-2\varepsilon t})}{2\alpha\varepsilon}|f|_{-1}^2 . \tag{5}$$

In order to prove (5) we set $v = u_t + \varepsilon u$, where $\varepsilon \leq \varepsilon_0$. Thus

$$v_t = u_{tt} + \varepsilon u_t = u_{tt} + \varepsilon(v - \varepsilon u) , \tag{6}$$

and making use of eq.(3) we find

$$\begin{cases} v_t - (1-\varepsilon\ \alpha)\Delta u + \varepsilon^2 u - (\alpha\Delta + \varepsilon)v = f \\ u_t = v - \varepsilon u \ . \end{cases} \tag{7}$$

Taking now the scalar product of eq.$(7)_1$ with v one has

$$\frac{1}{2}\frac{d}{dt}|v|^2 - (1-\varepsilon\alpha)(\Delta u, v) + \varepsilon^2(u,v) - ((\alpha\Delta+\varepsilon)v, v) = (f, v) , \tag{8}$$

and using $(7)_2$, we can deduce that

$$(-\Delta u, v) = \frac{1}{2}\frac{d}{dt}\|u\|^2 + \varepsilon\|u\|^2 ,$$

$$(u,v) = \frac{1}{2}\frac{d}{dt}|u|^2 + \varepsilon|u|^2 .$$

Substituting in eq.(6) we get

$$\frac{1}{2}\frac{d}{dt}[(1-\varepsilon\alpha)\|u\|^2+|v|^2+\varepsilon^2|u|^2]+\varepsilon(1-\varepsilon\alpha)\|u\|^2-((\alpha\Delta+\varepsilon)v,v)+\varepsilon^3|u|^2=(f,v)\ . \tag{9}$$

Now, by Poincaré's inequality, we have that

$$\lambda_1|v|^2 \leq \|v\|^2\ .$$

Thus, thanks to the hypotheses made on ε, it is easy to see that

$$-((\alpha\Delta+\varepsilon)v,v) \geq \frac{\alpha}{2}\|v\|^2+\varepsilon|v|^2 \tag{10}$$

So that by (9) we get, setting $y=|\{u,v\}|_{\varepsilon}^2$,

$$\frac{1}{2}\frac{dy}{dt}+\varepsilon y+\frac{\alpha}{2}\|v\|^2 \leq |f|_{L^\infty(\mathbb{R}^+,H^{-1}(\Omega))}\|v\| \leq \frac{\alpha}{2}\|v\|^2+\frac{1}{2\alpha}|f|^2_{L^\infty(\mathbb{R}^+,L^2(\Omega))}, \tag{11}$$

by which and Gronwall's Lemma it is straightforward to derive (5) ∎

In the nonlinear case, using the growth assumptions on g,h, we obtain an analogous result: all orbits are bounded and eventually enter in a bounded set in the space $H_0^1(\Omega)\times L^2(\Omega)$, named *absorbing set* for the dynamical system:

Proposition 2. *If $f\in L^2(\Omega)$ and the nonlinearities verify* (H). *Then there exist a bounded absorbing set B_0 in the space $H_0^1(\Omega)\times L^2(\Omega)$ for the dynamical system represented by* (3).

Proof. We can prove the following result (G indicates the primitive of g): for every $\varepsilon\in\,]0,\varepsilon_1[$, where $\varepsilon_1=\min\left(\frac{1}{\alpha},\frac{\alpha}{3\lambda_1}\right)$ there exist positive constants $k=k(\varepsilon)$ and ρ (independent of ε) such that the solution $u(t)$ of (1.1) satisfies

$$\begin{aligned}(1-\varepsilon\alpha)\|u\|^2+\varepsilon^2|u|^2+|u_t+\varepsilon u|^2+2G(u) \leq [(1-\varepsilon\alpha)\|u_0\|^2+\\ \varepsilon^2|u_0|^2+|u_1+\varepsilon u_0|^2+2G(u_0)]e^{-\varepsilon\rho t}+\frac{1}{\varepsilon\rho}(1-e^{-\varepsilon\rho t})(k+\frac{1}{\alpha}|f|^2)\ .\end{aligned} \tag{12}$$

The proof follows that of Proposition 1 and makes use of the growth assumptions on the nonlinearities. The reader is referred to [1] for the technical details ∎

3.FURTHER SMOOTHNESS PROPERTIES OF THE SOLUTIONS

Now we show that if the initial data are more regular, so is the a solution of (3), and that there exists an absorbing set for the system (3) in the space $H_0^1(\Omega)\times(H^2(\Omega)\cap H_0^1(\Omega))$:

Proposition 3. *Let $f \in L^2(\Omega)$ and g,h such that there exists $C > 0$ such that*

$$|g(\xi)| \leq C(1 + |\xi|^2)^{\beta/2} \quad , \tag{13}$$

with $0 \leq \beta < \infty$ if $n = 2$, $0 \leq \beta < 5$ if $n = 3$, $0 \leq \beta < \frac{n+2}{n-2}$ if $n \geq 4$, and

$$|h(\eta)| \leq C(1 + |\eta|^2)^{\gamma/2} \quad , \tag{14}$$

with $0 \leq \gamma < \infty$ if $n = 2$, $0 \leq \gamma < 7/3$ if $n = 3$, $0 \leq \gamma \leq \frac{n+4}{n}$ if $n \geq 2$. Then there exists an absorbing set B_1 in the space $H_0^1(\Omega) \times (H^2(\Omega) \cap H_0^1(\Omega))$ for the dynamical system (3).

For the proof, see [1] in which the following result

$$\begin{aligned} (1 - \varepsilon\alpha)|\Delta u|^2 + \varepsilon\|u\|^2 + \|u_t + \varepsilon u\|^2 \leq [(1 - \varepsilon\alpha)|\Delta u_0|^2 + \|u_0\|^2 + \\ + \|u_1 + \varepsilon u_0\|^2]e^{-\varepsilon t} + \frac{1}{\varepsilon}(1 - e^{-\varepsilon t})[K_(R) + |f|] \quad . \end{aligned} \tag{15}$$

is obtained ∎

4.CONSTRUCTION AND FINITE DIMENSIONALITY OF THE GLOBAL ATTRACTOR

Here we prove and state the main results of our work.

Theorem 1. *The ω-limit set of B_0,*

$$\mathcal{A} = \omega(B_0) = \bigcap_{s>0} cl(\bigcup_{t \geq s} S(t)B_0) \quad ,$$

where cl stands for the closure with respect to the topology of $H_0^1(\Omega) \times L^2(\Omega)$, is the global attractor for $S(t)$ in that space, i.e.:
(i) $\mathcal{A}$ is a compact nonempty connected set in $H_0^1(\Omega) \times L^2(\Omega)$;
(ii) $\mathcal{A}$ is invariant under $S(t)$; $S(t)\mathcal{A} = \mathcal{A} \quad \forall t \geq 0$;
(iii) $\mathcal{A}$ is globally attracting: for every bounded set $\mathcal{B}$ in $H_0^1(\Omega) \times L^2(\Omega)$, the distance $\mathrm{dist}(S(t)B, \mathcal{A}) = \sup_{x \in S(t)B} \inf_{y \in A} d(x, y)$ *tends to zero as $t \to \infty$.*
Proof. The proof will be an easy consequence of the following abstract result ([3], [5]):

Proposition 4. *If a semigroup $S(t)$ on a metric space $\mathcal{E}$ possesses a bounded absorbing set B_a and for every bounded set B in $\mathcal{E}$, there exists a compact set K in $\mathcal{E}$ such that $\lim_{t \to +\infty}$ dist $(S(t)B, K) = 0$ then the ω-limit set $\omega(B_a)$ is the global attractor of $S(t)$.*

Let therefore $R \geq 0$ be given with $u_0 \in H_0^1(\Omega), u_1 \in L^2(\Omega)$ such that $\|u_0\| \leq R$ and $|u_1| \leq R$. We know from proposition 1 that there exist $C_0, T_0(R)$ and $K_0(R)$ such that

$$\|u(t)\|^2 + |u_t(t)|^2 \leq C_0, \quad \forall t \geq T_0(R),$$

$$\|u(t)\|^2 + |u_t(t)|^2 \leq K_0(R), \quad \forall t \geq 0.$$

Let us write now $u = v + w$ where

$$v_{tt} - \alpha \Delta v_t - \Delta v = \varphi(t),$$

$$v(0) = 0, \ v_t(0) = 0,$$

$$\varphi(t) = f - g(u(t)) - h(u_t(t)),$$

and

$$w_{tt} - \alpha \Delta w_t - \Delta w = 0,$$

$$w(0) = u_0, \ w_t(0) = u_1.$$

Proposition 1 shows that there exist $C_\varepsilon > 0$ such that

$$\|w(t)\|^2 + |w_t(t)|^2 \leq C_\varepsilon e^{-\varepsilon t} R^2 \ , \quad \forall t \geq 0. \tag{16}$$

From Proposition 3 one then has

$$|v(t)|_{H^2(\Omega) \cap H_0^1(\Omega)} + \|v_t(t)\| \leq C_{12}(R);$$

then $\bigcup_{\substack{t \geq 0 \\ \|u_0\| \leq R}} \{v(t), v_t(t)\}$ is bounded in $(H^2(\Omega) \cap H_0^1(\Omega)) \times H_0^1(\Omega)$ and being the injection $(H^2(\Omega) \cap H_0^1(\Omega)) \times H_0^1(\Omega) \to H_0^1(\Omega) \times L^2(\Omega)$ compact, we have

$$\overline{\bigcup_{\substack{t \geq 0 \\ \|u_0\| \leq R}} \{v(t), v_t(t)\}} = K \quad \text{is compact in } H_0^1(\Omega) \times L^2(\Omega) \ .$$

But from (16) it follows that $\bigcup_{\substack{t \geq 0 \\ \|u_0\| \leq R}} \{u(t) - v(t), u_t(t) - v_t(t)\}$ is contained in the ball $B_{H_0^1(\Omega) \times L^2(\Omega)}(0, Ce^{-\varepsilon t} R^2)$ which tends to zero strongly, so that the set

$$\overline{\bigcup_{t \geq 0} S(t) B}^{H_0^1(\Omega) \times L^2(\Omega)}$$

is compact in $H_0^1(\Omega) \times L^2(\Omega)$ and the theorem is proved. ∎

Finally, using an abstract result of GHIDAGLIA [6] which extends that of CONSTANTIN, FOIAS and TEMAM [5], it is possible to prove that the above attractor is finite-dimensional, *i.e.*

Theorem 2. *The global attractor constructed in Theorem 1 has finite Hausdorff and fractal dimensions.*

The proof relies on an analysis of the linearized flow, and on some abstract results (see [1] for the details)∎

REFERENCES

[1] J.M.GHIDAGLIA and A.MARZOCCHI, Long Time Behaviour of Strongly Damped Wave Equations, Global Attractors and their Dimension, to appear on SIAM *J. Math. Anal.*

[2] P.S.LANDAHL, O.H.SØRENSEN and P.L.CHRISTIANSEN, *Soliton Excitations in Josephson Tunnel Junctions*, Phys.Rev. B **25**, (1982), 5337-5348.

[3] J.M.GHIDAGLIA and R.TEMAM, *Attractors for Damped Nonlinear Hyperbolic Equations*, J.Math. pures et appl. **66**, (1987), 273-319.

[4] J.M.GHIDAGLIA and R.TEMAM, *Dimension of the Universal Attractor Describing the Periodically Driven Sine-Gordon Equations*, Transport Theory and Statistical Physics, **16**, (1987), 253-265.

[5] P.CONSTANTIN, C.FOIAS and R.TEMAM, Attractors Representing Turbulent Flows, Memories of the A.M.S., **53**, 314, (1985).

[6] J.M.GHIDAGLIA, *Weakly Damped Forced Korteweg-de Vries Equation Behave as a Finite Dimensional Dynamical System in the Long Time*, J.Diff.Eq., **74**, (1988), 369-390.

Alfredo Marzocchi
Dipartimento di Matematica
Università Cattolica del S.Cuore
Via Trieste 17, 25121 BRESCIA (Italy)

N MANGANARO

Similarity reduction of the nonlinear Schrödinger equation with variable coefficients

1. INTRODUCTION

The nonlinear Schrödinger equation (NSE) with variable coefficients

$$iu_t + h(t)\, u_{xx} + p(t)\, |u|^2 u = 0 \tag{1}$$

describes the slow modulation of a weakly nonlinear narrow band wave packet [1]. In (1) u is the field wave envelope, x and t represent, respectively, space and time coordinates while h and p are functions of t.

Within the framework of the nonlinear optics, the NSE represents a model equation describing the propagation of short pulses of quasi-monochromatic light into axially nonuniform optical fiber [2]. In a perfect, lossless fiber optical signals propagate as soliton pulses without changing their shape and amplitude. Unfortunately in a real fiber losses produce attenuation and broadening of solitons. To avoid this Tajima in [3] proposed to make axially nonuniform the fiber to study how the fiber parameters must vary with axial distance to have compression, amplification, constant amplitude or constant width in soliton propagation.

For an axially nonuniform single-mode fiber, it is possible to show, from Maxwell equations, that the slowly varying envelope of the electric field must satisfy to the generalized NSE with variable coefficients:

$$i\left(\frac{\partial q}{\partial \bar{z}} + \gamma q + k_1 \frac{\partial q}{\partial \bar{t}} + \frac{\partial F}{\partial \bar{z}}\frac{q}{2F}\right) - \frac{k_2}{2}\frac{\partial^2 q}{\partial \bar{t}^2} + \frac{\omega_0 n_2 G}{c}\, |q|^2 q = 0 \tag{2}$$

while the electric field E is given by:

$$E(\bar{x},\bar{y},\bar{z};\bar{t}) = q(\bar{z},\bar{t})\, V(\bar{x},\bar{y},\bar{z}) \exp\left\{i\left(\int k_0 d\bar{z} - \omega_0 t\right)\right\} \tag{3}$$

In (2), (3) $\bar{z}$, $\bar{x}$, $\bar{y}$ and $\bar{t}$ are, respectively, axial, transverse and time coordinates, q is the envelope of the electric field, γ the absorption coefficient which takes into account the power loss during the propagation, k_0 the local wave number carried at frequency ω_0, V the transverse field distribution, $k_1=\partial\omega/\partial k$ and $k_2=\partial^2\omega/\partial k^2$ are evaluated at the frequency ω_0, c is the light velocity in the vacuum. Furthermore n_2 is the nonlinear term in the index of refraction

$$n = n_0 + n_2 \,|E|^2 \tag{4}$$

(n_0 is the linear index), while F and G have the form

$$F(\bar{z})=n_0(\bar{z}) \int\int |V|^2 d\bar{x}d\bar{y} \;; \qquad G(\bar{z})=\int\int |V|^4 d\bar{x}d\bar{y} \Big/ \int\int |V|^2 d\bar{x}d\bar{y} \tag{5}$$

In the present case of an axially nonuniform fiber the coefficients γ, k_1, k_2, n_2, F and G are slowly varying functions of $\bar{z}$. If the fiber is axially uniform F vanishes, the other coefficients are constants and V is function of the transverse coordinates $\bar{x}$, $\bar{y}$ only.

In [2] it is proved that the equation (2), after the variable transformation

$$t=\int|k_2|dz; \qquad x=2^{1/2}\,(\bar{t}-\int k_1 d\bar{z}); \qquad u=F^{1/2}q\,\exp\!\Big(\int\gamma d\bar{z}\Big) \tag{6}$$

reduces to the NSE with variable coefficients (1), in which we assume, without loss of generality, h=1.

Unfortunately, because of the variable coefficient, the equation (1) is not integrable by means of the inverse scattering method as the NSE with constant coefficients.

Then, within the framework of the similarity analysis, it would be of a certain interest to find set of exact solutions to the equation (1). This is strictly related to the possibility of characterizing functional forms of the variable coefficient needed in order that the similarity procedure holds. For this purpose we follow a similarity approach proposed recently by Clarkson and Kruskal [4].

The paper is organized as follows. In section 2 we sketch the Clarkson method and set of similarity reductions as well as of exact solutions to (1) are obtained. In section 3 we propose a sligth generalization of the Clarkson approach to transform the NSE with variable coefficients to the NSE with constant coefficients.

Finally, in the case of propagation in optical fiber, the one-soliton solution of (2) is obtained and it is studied how the fiber parameters must
vary with axial distance to have amplification, compression or constant amplitude in soliton propagation.

2. SIMILARITY REDUCTIONS

The idea of the Clarkson approach is to look for the class of exact solutions:

$$u = U\Big(x,t;\ w(z(x,t))\Big) \tag{7}$$

of the governing partial differential equation depending not only on the independent variables x and t, but also upon an extra-variable w. Furthermore w is function of a new variable z, called "similarity variable", depending on x and t. Substituting (7) into the governing partial differential equation the request that the result be an ordinary differential equation for w imposes restrictions on the functional form of U as well as of z. Finally exact solutions to the partial differential equation are obtained integrating the ordinary differential equation for w and substituting the result into (7).

In the present case substituting (7) in (1) it results sufficient to take (7) in the form

$$u = \beta(x,t)w(z) \tag{8}$$

while the equation (1) reduces to:

$$(i\beta_t + \beta_{xx})w + p|\beta|^2\beta\ |w|^2 w + (i\beta z_t + 2\beta_x z_x + \beta z_{xx})\ w' + \beta {z_x}^2\ w'' = 0 \tag{9}$$

(here and in the following the prime means for derivative with respect to the argument).

In order that (9) be an ordinary differential equation for w the following relations must hold:

$$\begin{aligned} &p|\beta|^2 = \Gamma_1(z)\, {z_x}^2; \quad i\beta_t + \beta_{xx} = \Gamma_2(z)\ \beta {z_x}^2 \\ &i\beta z_t + 2\beta_x z_x + \beta z_{xx} = \Gamma_3(z)\ \beta {z_x}^2 \end{aligned} \tag{10}$$

where $\Gamma_{1,2,3}$ are arbitrary functions of z to be determined later.

After some algebra the consistency of the equations (10) yields the functional forms of β, z, $\Gamma_{1,2,3}$ as well as of the coefficient p(t). Then, taking into account (8) , the following similarity reductions of the equation (1) are obtained:

$$i)\quad u=u_0\,\frac{(t-t_0)^{n-\frac14}}{(t+t_0)^{n+\frac14}}\exp i\left\{\frac{x^2t}{4(t^2-t_0^{\,2})}+\frac{\lambda t_0^{\,2}}{2(t^2-t_0^{\,2})}\left(x+\frac{\lambda}{2}t\right)+\mu\lg\frac{t-t_0}{t+t_0}\right\}w(z)$$

$$z=\frac{x+\lambda t}{(t^2-t_0^{\,2})^{1/2}}\quad;\quad p=p_0\,\frac{(t+t_0)^{2(n-\frac14)}}{(t-t_0)^{2(n+\frac14)}}\tag{11}$$

$$ii)\quad u=\frac{u_0}{t^{1/2}}\exp\left\{\frac{n}{t}+i\left(\frac{x^2}{4t}+\lambda\frac{x}{t^2}+\frac{2\lambda^2}{3t^3}+\frac{\mu}{t}\right)\right\}w(z)$$

$$z=\frac{xt+\lambda}{t^2}\quad;\quad p=\frac{p_0}{t}\exp\left(-\frac{2n}{t}\right)\tag{12}$$

$$iii)\quad u=u_0\,(t^2+t_0^{\,2})^{-\frac14}\exp\left\{n\,\mathrm{tg}^{-1}\left(\frac{t}{t_0}\right)+i\left(\frac{x^2t}{4(t^2+t_0^{\,2})}-\frac{\lambda t_0^{\,2}}{2(t^2+t_0^{\,2})}\left(x+\frac{\lambda}{2}t\right)+\right.\right.\tag{13}$$

$$\left.\left.+\,\mu\,\mathrm{tg}^{-1}\left(\frac{t}{t_0}\right)\right)\right\}w(z)\;;\;z=\frac{x+\lambda t}{(t^2+t_0^{\,2})^{1/2}}\;;\;p=p_0\,(t^2+t_0^{\,2})^{-1/2}\exp\left(-2n\,\mathrm{tg}^{-1}\left(\frac{t}{t_0}\right)\right).$$

$$iv)\quad u=u_0\,t^n\exp i\left\{\frac{x^2}{8t}-\frac{\lambda}{4}\left(x+\frac{\lambda}{2}t\right)+\mu\,\mathrm{lg}t\right\}w(z)$$

$$z=\frac{x+\lambda t}{t^{1/2}}\;;\;p=p_0\,t^{-(2n+1)}.\tag{14}$$

$$v)\quad u=u_0\exp\left\{nt+i\left(\lambda xt-\frac{2}{3}\lambda^2t^3-\mu t\right)\right\}w(z)$$

$$z=x-\lambda t^2\;;\;p=p_0\exp\left(-2nt\right)\tag{15}$$

$$vi)\quad u=u_0\exp\left\{nt+i\left(\frac{\lambda}{2}x+\mu t\right)\right\}w(z)$$

$$z=x-\lambda t\;;\;p=p_0\exp\left(-2nt\right).\tag{16}$$

Furthermore the function w must satisfy the ordinary differential equation

$$w'' + h\,|w|^2 w + (\,\gamma_1 + \gamma_2 z + \gamma_3 z^2 + i\nu\,)w = 0 \tag{17}$$

where $\gamma_1=-2\mu t_0$, $\gamma_2=0$, $\gamma_3=t_0{}^2/4$, $\nu=2nt_0$ in the case *i*); $\gamma_1=\mu$, $\gamma_2=\lambda$, $\gamma_3=0$, $\nu=-n$ in the case *ii*); $\gamma_1=-\mu t_0$, $\gamma_2=0$, $\gamma_3=-t_0{}^2/4$, $\nu=nt_0$ in the case *iii*); $\gamma_1=-\mu$, $\gamma_2=0$, $\gamma_3=1/16$, $\nu=n+1/4$ in the case *iv*); $\gamma_1=\mu$, $\gamma_3=0$, $\gamma_2=-\lambda$, $\nu=n$ in the case *v*); $\gamma_1=-\mu-\lambda^2/4$, $\gamma_2=0$, $\gamma_3=0$, $\nu=n$ in the case *vi*); while in all the previous cases it is $h=p_0 u_0{}^2$ and u_0, t_0, λ, μ, p_0, n are arbitrary constants.

Some possible solutions of (17) are:

a) $w = (-2/h)^{1/2}\, z^{-1} \exp(i\,Az^2)$

which holds in the case *i*) with $A= \pm t_0/4$, $\mu=0$, $n=1/4$ and in the case *iv*) with $\mu=0$, $n=0$ or $n=-1/2$ depending if $A=\pm 1/8$.

b) $w = \rho_0 \exp(i\,Az^2)$

which holds in the case *i*) with $A=\pm t_0/4$, $\rho_0=\pm(2\mu t_0/h)^{1/2}$, $n=1/4$ and in the case *iv*) with $\rho_0=\pm(\mu/t_0)^{1/2}$, $n=-1/2$ or $n=0$ depending if $A=\pm 1/8$.

c) $w = \rho_0 z \exp(i\,Az^2)$

which holds in the case *i*) with $A=\pm(1/4)\,(t_0{}^2+4\,h\rho_0{}^2)^{1/2}$, $\mu=0$, $n=\mp(3/4t_0)$ $(t_0{}^2+4\,h\rho_0{}^2)^{1/2}$; in the case *iv*) with $A=\pm(1/8)\,(1+16\,h\rho_0{}^2)^{1/2}$, $\mu=0$, $n=\mp(3/4)$ $(1+16\,h\rho_0{}^2)^{1/2}-1/4$ and in the case *iii*) with $A=\pm(1/4)\,(4h\rho_0{}^2-t_0{}^2)^{1/2}$, $\mu=0$, $n=\mp(3/2t_0)\,(4h\rho_0{}^2-t_0{}^2)^{1/2}$.

d) $w = \rho_0 z$

which holds in the case *i*) with $\rho_0=\pm t_0/2(-h)^{1/2}$, $\mu=n=0$; in the case *iv*) with $\rho_0=\pm 1/4(-h)^{1/2}$, $\mu=0$, $n=-1/4$ and in the case *iii*) with $\rho_0=\pm t_0/2(-h)^{1/2}$, $\mu=n=0$.

Of course not all of these solutions have a physical meaning. However, as usual in the similarity approach, they can be useful to test numerical procedures to integrate (17).

3. OPTICAL PULSES PROPAGATION

First we will use a generalisation of the Clarkson approach to transform the NSE with variable coefficients to the NSE with constant coefficients. For our purpose we suppose that w depends not only on z but also upon an arbitrary function of t:

$$u = U\Big(x,t;\ w(\ z(x,t),T(t)\ )\Big) \tag{18}$$

Then, taking into account the results of the previous section, we transform the equation (1) into the partial differential equation:

$$i\,\frac{T'}{f^2}\,w_T + w_{zz} + h|w|^2w + (\gamma_1+\gamma_2 z+\gamma_3 z^2 + i\nu)w = O \tag{19}$$

with f given in terms of the function p:

$$f(t) = p^2(t)\,\Big(2\nu\textstyle\int p^{-2}dt + f_0\Big)\,;\quad f_0{=}\text{cost.} \tag{20}$$

Now, choosing $T'{=}f^2$ and $\gamma_2{=}\gamma_3{=}\nu{=}O$, then the equation (1) reduces to the NSE with constant coefficients

$$i\,w_T + w_{zz} + \tilde{h}\,|w|^2w = O \tag{21}$$

by means of the variable transformation:

$$u = u_0 t^{-1/2} w(T,z)\,\exp(i\,x^2/4t);\qquad z = xt^{-1/2};\qquad T = -t^{-1}. \tag{22}$$

iff the variable coefficient p has the form $p{=}p_0 t^{-1/2}$.

Then, owing to the transformation (22), it is possible to reduce any initial value problem for the equation (1) to an initial value problem for w. Since the equation (21) has constant coefficients, the inverse scattering method is available to solve the transformed initial value problem. In passing we note that we find again the variable transformation proposed by Grimshaw [5] within a different framework.

Finally we return to the case of fibre optics, and consider the one-soliton solution of (21). To this solution, having in mind the transformations (22) and (6), the following soliton solution for the generalized NSE (2) is obtained:

$$\begin{aligned}
q &= \eta(\bar z)\,\text{sech}\Big\{\xi(\bar z)\,\Big(\bar t - \textstyle\int k_1 d\bar z\Big)\Big\}\,\exp\Big(i\,\Phi(\bar z)\Big)\\
\eta(\bar z) &= u_0\,\Big(F\textstyle\int |k_2|d\bar z\Big)^{-1}\exp\Big(-\textstyle\int \gamma d\bar z\Big)\\
\xi(\bar z) &= m\,2^{1/2}\,\Big(\textstyle\int |k_2|d\bar z\Big)^{-1};\ m{=}\text{cost.}\\
\Phi(\bar z) &= \Big(\textstyle\int |k_2|d\bar z\Big)^{-1}\Big\{\tfrac{1}{2}\,\Big(\bar t - \textstyle\int k_1 d\bar z\Big)^2 - m^2\Big\}
\end{aligned} \tag{23}$$

with the condition

$$\frac{\omega_0 n_2 G}{cF|k_2|} \exp\left(-2\int \frac{\gamma}{|k_2|} dt\right) = \frac{p_0}{t} . \tag{24}$$

If $(F\int|k_2|d\bar{z})^{-1/2}$ increases with $\bar{z}$ more rapidly than $\exp(-\int\gamma d\bar{z})$ decreases, then the amplitude η will increase as the soliton travels down the fiber. That happens, for instance, if

$$\left(F\int|k_2|d\bar{z}\right)^{-1/2} \exp\left(-\int\gamma d\bar{z}\right) = h_1 \exp(\sigma_1\bar{z}) \tag{25}$$

with h_1 and $\sigma_1 > O$ constants. If $\sigma_1 = O$ the condition (25) assures the constant amplitude. Similarly if the condition

$$\int|k_2|d\bar{z} = h_2 \exp(-\sigma_2\bar{z}) \tag{26}$$

holds, (h_2 and $\sigma_2 > o$ are constants), the temporal width of the soliton will decrease as it moves down the fiber. Unfortunately a constant width is not compatible with the condition (26). Of course the equations (24)-(26) can be studied numerically to determine how the core radius of the fiber must vary to have compression, amplification or constant amplitude in soliton pulse.

REMARK

In a different way from other similarity methods [6-7], the Clarkson approach does not strictly require use of group theory. Actually Levi and Winternitz [8] recently obtained the same results as those of the Clarkson method by means of the non classical group analysis.

As well known the latter leads to a differential model involving an independent variable less, so that it is immediately useful for determining solutions to governing equations involving two independent variables. However the Clarkson approach, in principle, reduce a P.D.E. with more than two independent variables to an O.D.E.

REFERENCES

1. G. B. WHITHAM, Linear and Nonlinear Waves (Academic Press 1974).
2. H. H. KUEL, Solitons on an Axially Nonuniform Optical Fiber. J. Opt. Soc. Am. B, 5, 3 (1988) 709-713.

3. K. TAJIMA, Compensation of Soliton Broadening in Nonlinear Optical Fibers with Loss. Opt. Lett. 12, 54 (1987).

4. P. A. CLARKSON, M. D. KRUSKAL, New Similarity Reductions of the Boussinesq Equation. J. Math. Phys., 30, 10 (1989) 2201-2213.

5. R. GRIMSHAW, Slowly Varying Solitary Waves. II.Nonlinear Schrödinger Equation. Proc. R. Soc. Lond., A 368 (1979) 377-388.

6. G. W. BLUMAN, J. D. COLE, Similarity Methods for Differential Equations (Springer, Berlin 1974).

7. L. V. OVSIANNIKOV, Group Analysis of Differential Equations (Academic Press, New York 1982).

8. D. LEVI, P. WINTERNITZ, Non-classical Symmetry Reduction: Example of the Boussinesq Equation, J. Phys. A: Math. Gen., 22 (1988) 2915-2924.

Natale Manganaro
Dipartimento di Matematica,
Università di Messina,
Contrada Papardo, Salita Sperone 31,
98166 Sant'Agata, Messina, ITALY.

F OLIVERI

The substitution principle in magneto-gasdynamics: Lie group analysis approach

ABSTRACT

The equations of ideal magneto-gasdynamics with a separable equation of state are considered by means of Lie group approach. It is proved that the remarkable result referred to in literature as the Substitution Principle can be derived by suitably applying the Lie group machinery. Moreover, it is given a simple interpretation of the Substitution Principle in terms of a "generalized" stretching group and of a "generalized" time translation.

1. INTRODUCTION.

In this paper we shall consider the equations of ideal magneto-gasdynamics with a separable equation of state within the context of Lie group analysis in connection with a result known in literature as Substitution Principle.

Let us take the equations governing the flow of an inviscid, thermally non-conducting fluid of infinite electrical conductivity in the presence of a magnetic field and subject to no extraneous force:

$$\begin{aligned}
&\frac{\partial \rho}{\partial t}+\frac{\partial}{\partial x_k}(\rho v_k)=0\\
&\rho\left(\frac{\partial v_i}{\partial t}+v_k\frac{\partial v_i}{\partial x_k}\right)+\frac{\partial p}{\partial x_i}+\mu H_k\left(\frac{\partial H_k}{\partial x_i}-\frac{\partial H_i}{\partial x_k}\right)=0\\
&\frac{\partial H_k}{\partial x_k}=0\\
&\frac{\partial H_i}{\partial t}+v_k\frac{\partial H_i}{\partial x_k}+H_i\frac{\partial v_k}{\partial x_k}-H_k\frac{\partial v_i}{\partial x_k}=0\\
&\frac{\partial s}{\partial t}+v_k\frac{\partial s}{\partial x_k}=0
\end{aligned}\tag{1}$$

($\rho(\mathbf{x},t)$ is the mass density, $p(\mathbf{x},t)$ the pressure, $s(\mathbf{x},t)$ the entropy, μ the (constant) magnetic permeability, $v_i(\mathbf{x},t)$ and $H_i(\mathbf{x},t)$ ($i=1,2,3$) the components of velocity and magnetic fields) together with the equation of state assumed given in separable form:

$$\rho = P(p)S(s) \quad . \tag{2}$$

In 1947 Munk and Prim [1] and in 1952 Prim [2] prove that the steady equations of an inviscid thermally non-conducting fluid with a separable equation of state are invariant with respect to a family of transformations involving a smooth scalar function of the space variables wich is constant along each streamline and call this result Substitution Principle. This result is extended in 1963 by Smith [3] to the steady equations of magneto-gasdynamics. In 1964 Smith [4] enlarges the Substitution Principle to the unsteady equations of gas-dynamics having steady the pressure. In 1969 Power and Rogers [5] extend the Smith's result to the unsteady magneto-gasdynamics having steady the total magnetic pressure. Finally, in 1990 Oliveri [6] proves that the Substitution Principle for unsteady gas-dynamics can be obtained within the context of the infinitesimal Lie group analysis and finds a Substitution Principle for a class of solutions governing unsteady flows of a polytropic gas with adiabatic index equal to $\frac{5}{3}$ having unsteady also the pressure.

The main result obtained by Power and Rogers can be stated by the following

THEOREM. *If*

$$\{p(\mathbf{x},t), \mathbf{v}(\mathbf{x},t), S(s(\mathbf{x},t)), \mathbf{H}(\mathbf{x},t)\} \tag{3}$$

constitute a solution set of the basic equations of nonsteady magneto-gasdynamics when the total magnetic pressure $p + \frac{\mu}{2}\mathbf{H}\cdot\mathbf{H}$ *is steady, then the system*

$$\begin{gathered}\{p(\mathbf{x}, mt + h(m)), m(\mathbf{x})\mathbf{v}(\mathbf{x}, mt + h(m)),\\ m^{-2}(\mathbf{x})S(s(\mathbf{x}, mt + h(m))), \mathbf{H}(\mathbf{x}, mt + h(m))\}\end{gathered} \tag{4}$$

also comprises a solution set, provided $m(\mathbf{x})$ *is a steady differentiable scalar function such that*

$$v_i \frac{\partial m}{\partial x_i} = H_i \frac{\partial m}{\partial x_i} = 0 \tag{5}$$

for all t *and* $h(m)$ *is an arbitrary function of its argument.*

In the following sections we will show

i) that the transformation underlying the Substitution Principle in magneto-gasdynamics for unsteady flows having the total magnetic pressure steady can be obtained within the context of the classical infinitesimal Lie transformation group analysis provided the constraints on the velocity vector and on the magnetic field are explicitly taken into account in the basic governing equations;

ii) that, by simply interpreting the Substitution Principle as the result of the combination of a "generalized" stretching group of transformations for all variables and of a "generalized" translation of time, the needed calculations in order to derive it become easy and straightforward.

2. LIE GROUP ANALYSIS OF THE GOVERNING EQUATIONS AND DERIVATION OF THE SUBSTITUTION PRINCIPLE.

The Lie group analysis shows, after extensive analysis (for details see [7-10]), that the governing system (1) - (2) is invariant with respect to the following family of infinitesimal transformations:

$$
\begin{aligned}
t^* &= t + \epsilon[\alpha_1 t + \alpha_2] + O(\epsilon^2) \\
x_1^* &= x_1 + \epsilon[\beta_1 x_1 + \beta_2 x_2 + \beta_3 x_3 + \gamma_1 t + \delta_1] + O(\epsilon^2) \\
x_2^* &= x_2 + \epsilon[-\beta_2 x_1 + \beta_1 x_2 + \beta_4 x_3 + \gamma_2 t + \delta_2] + O(\epsilon^2) \\
x_3^* &= x_3 + \epsilon[-\beta_3 x_1 - \beta_4 x_2 + \beta_1 x_3 + \gamma_3 t + \delta_3] + O(\epsilon^2) \\
p^* &= p + \epsilon[\theta_1 \frac{P(p)}{P'(p)}] + O(\epsilon^2) \\
s^* &= s + \epsilon[(2(\alpha_1 - \beta_1 + \theta_2) - \theta_1)\frac{S(s)}{S'(s)}] + O(\epsilon^2) \\
v_1^* &= v_1 + \epsilon[(\beta_1 - \alpha_1)v_1 + \beta_2 v_2 + \beta_3 v_3 + \gamma_1] + O(\epsilon^2) \\
v_2^* &= v_2 + \epsilon[-\beta_2 v_1 + (\beta_1 - \alpha_1)v_2 + \beta_4 v_3 + \gamma_2] + O(\epsilon^2) \\
v_3^* &= v_3 + \epsilon[-\beta_3 v_1 - \beta_4 v_2 + (\beta_1 - \alpha_1)v_3 + \gamma_3] + O(\epsilon^2) \\
H_1^* &= H_1 + \epsilon[\theta_2 H_1 + \beta_2 H_2 + \beta_3 H_3] + O(\epsilon^2) \\
H_2^* &= H_2 + \epsilon[-\beta_2 H_1 + \theta_2 H_2 + \beta_4 H_3] + O(\epsilon^2) \\
H_3^* &= H_3 + \epsilon[-\beta_3 H_1 - \beta_4 H_2 + \theta_2 H_3] + O(\epsilon^2)
\end{aligned}
\tag{6}
$$

provided the following relation is fulfilled:

$$
(\theta_1 - 2\theta_2) - \theta_1 \frac{P(p)P''(p)}{P'^2(p)} = 0 \quad . \tag{7}
$$

The finite form of the transformation generated by the group constants α_1 and α_2 is given by

$$
\begin{aligned}
t^* &= t \cdot \exp(\alpha_1 \epsilon) + \frac{\alpha_2}{\alpha_1}(\exp(\alpha_1 \epsilon - 1) \\
\mathbf{x}^* &= \mathbf{x} \quad , \quad \mathbf{v}^* = \mathbf{v}\exp(-\alpha_1 \epsilon) \\
p^* &= p \quad , \quad \mathbf{H}^* = \mathbf{H} \quad , \quad s^* = S^{-1}[S(s)\exp(2\alpha_1 \epsilon)] \quad .
\end{aligned}
\tag{8}
$$

That represents a trivial version of the Substitution Principle (the function $m = \exp(\alpha_1 \epsilon) =$ constant) while the true transformation underlying the Substitution Principle is not recovered. The result remains negative even if we take into account the steadiness of the total magnetic pressure. However, this is not surprising because the Substitution Principle is valid for the governing equations provided the velocity vector $\mathbf{v}(\mathbf{x}, t)$ and the magnetic field $\mathbf{H}(\mathbf{x}, t)$ result tangent to the surface $m(\mathbf{x}) =$ constant passing through the point $\mathbf{x}$.

To recover the Substitution Principle we must use explicitly the steadiness of the total pressure and the constraints (5) into the basic equations. These constraints obviously limit the set of solutions: as a consequence a different group of transformations leaving them invariant may arise. Let us assume, without loss of generality, $\dfrac{\partial m}{\partial x_3} \neq 0$ so that:

$$v_3 = -\frac{v_1 \dfrac{\partial m}{\partial x_1} + v_2 \dfrac{\partial m}{\partial x_2}}{\dfrac{\partial m}{\partial x_3}} \quad , \quad H_3 = -\frac{H_1 \dfrac{\partial m}{\partial x_1} + H_2 \dfrac{\partial m}{\partial x_2}}{\dfrac{\partial m}{\partial x_3}} \tag{9}$$

that, substituted into the basic equations, provide a very complicated system of 9 equations, involving the function $m(x_1, x_2, x_3)$, for the unknown p, $S(s)$, v_1, v_2, H_1, H_2.

Now, let us investigate the infinitesimal transformations which leave this complicated system invariant:

$$\begin{aligned}
t^* &= t + \epsilon\tau(t, \mathbf{x}, p, s, v_1, v_2, H_1, H_2) + O(\epsilon^2) \\
x_1^* &= x_1 + \epsilon\xi_1(t, \mathbf{x}, p, s, v_1, v_2, H_1, H_2) + O(\epsilon^2) \\
x_2^* &= x_2 + \epsilon\xi_2(t, \mathbf{x}, p, s, v_1, v_2, H_1, H_2) + O(\epsilon^2) \\
x_3^* &= x_3 + \epsilon\xi_3(t, \mathbf{x}, p, s, v_1, v_2, H_1, H_2) + O(\epsilon^2) \\
p^* &= p + \epsilon\eta^1(t, \mathbf{x}, p, s, v_1, v_2, H_1, H_2) + O(\epsilon^2) \\
s^* &= s + \epsilon\eta^2(t, \mathbf{x}, p, s, v_1, v_2, H_1, H_2) + O(\epsilon^2) \\
v_1^* &= v_1 + \epsilon\eta^3(t, \mathbf{x}, p, s, v_1, v_2, H_1, H_2) + O(\epsilon^2) \\
v_2^* &= v_2 + \epsilon\eta^4(t, \mathbf{x}, p, s, v_1, v_2, H_1, H_2) + O(\epsilon^2) \\
H_1^* &= H_1 + \epsilon\eta^6(t, \mathbf{x}, p, s, v_1, v_2, H_1, H_2) + O(\epsilon^2) \\
H_2^* &= H_2 + \epsilon\eta^7(t, \mathbf{x}, p, s, v_1, v_2, H_1, H_2) + O(\epsilon^2)
\end{aligned} \tag{10}$$

Note that neither the field variables v_3 and H_3 nor their corresponding infinitesimal generators η^5 and η^8 are involved in (10).

The well established Lie's algorithm furnishes, after a large amount of cumbersome calculations, the following infinitesimal generators of the full group of invariance:

$$\begin{aligned}
&\tau = F_1[m(\mathbf{x})]t + F_2[m(\mathbf{x})] \quad , \quad \xi_1 = \xi_2 = \xi_3 = 0 \\
&\eta^1 = \alpha\frac{P(p)}{P'(p)} \quad , \quad \eta^2 = (2F_1[m(\mathbf{x})] + 2\beta - \alpha)\frac{S(s)}{S'(s)} \\
&\eta^3 = -F_1[m(\mathbf{x})]v_1 \quad , \quad \eta^4 = -F_1[m(\mathbf{x})]v_2 \\
&\eta^6 = \beta H_1 \quad , \quad \eta^7 = \beta H_2
\end{aligned} \tag{11}$$

where F_1 and F_2 are two arbitrary functions of $m(\mathbf{x})$; moreover the following condition must hold:

$$(\alpha - 2\beta) - \alpha\frac{P(p)P''(p)}{P'^2(p)} = 0 \quad . \tag{12}$$

What we need to conclude the derivation of the Substitution Principle within the context of Lie groups, is the form of the infinitesimal transformations for v_3 and H_3. To this end let us begin with

$$v_3^* = v_3 + \epsilon\eta^5 + O(\epsilon^2) \tag{13}$$

η^5 being the generator to be found. But we have

$$\begin{aligned} v_1 &= v_1^* + \epsilon F_1(m) v_1^* + O(\epsilon^2) \\ v_2 &= v_2^* + \epsilon F_1(m) v_2^* + O(\epsilon^2) \end{aligned} \tag{14}$$

that, on account of the constraint on the velocity, provide:

$$\begin{aligned} -\frac{\partial m}{\partial x_3}(v_3^* - \epsilon\eta^{5*} + O(\epsilon^2)) = & \frac{\partial m}{\partial x_1}(v_1^* + \epsilon F_1(m) v_1^* + O(\epsilon^2)) \\ & + \frac{\partial m}{\partial x_2}(v_2^* + \epsilon F_1(m) v_2^* + O(\epsilon^2)) \end{aligned} \tag{15}$$

whereupon we get

$$\begin{aligned} & v_1^* \frac{\partial m}{\partial x_1} + v_2^* \frac{\partial m}{\partial x_2} + v_3^* \frac{\partial m}{\partial x_3} = 0 \\ & \eta^5 = -F_1(m) v_3 \quad . \end{aligned} \tag{16}$$

A similar reasoning leads to

$$\eta^8 = \beta H_3 \quad . \tag{17}$$

By integrating the well known Lie's equations, we see that the finite form of the transformation characterized by the infinitesimal generators so found is, for $\alpha = \beta = 0$, the same transformation by Power and Rogers [5].

Moreover, when α and β are non-zero constants (this for example occurs in the case of a perfect gas in which $P(p) = p^{1/\Gamma}$) then the recovered finite transformation coincides with the strong version of the Substitution Principle found by Power and Rogers involving also a scaling of the pressure and of magnetic field.

3. THE SUBSTITUTION PRINCIPLE AS A GENERALIZED STRETCHING GROUP COMBINED WITH A GENERALIZED TIME TRANSLATION.

Let us consider the basic equations (1) of magneto-gasdynamics with a general equation of state:

$$\rho = R(p, s) \quad . \tag{18}$$

Let us suppose the governing equations to be invariant with respect to the following family of transformations:

$$\begin{aligned} & t^* = m(\mathbf{x})t + h(m) \quad , \quad \mathbf{x}^* = \mathbf{x} \\ & \mathbf{v}^* = m^\alpha(\mathbf{x})\mathbf{v} \quad , \quad p^* = m^\beta(\mathbf{x})p \\ & s^* = m^\gamma(\mathbf{x})s \quad , \quad \mathbf{H}^* = m^\delta(\mathbf{x})\mathbf{H} \\ & R^*(p^*, s^*) = m^\theta(\mathbf{x})R(p,s) \end{aligned} \tag{19}$$

where α, β, γ, δ and θ are suitable constants.

The family of transformations characterized by (19) can be interpreted as the combination of a "generalized" stretching group and a "generalized" time translation because the "parameters" which are responsible of the dilatations and of the time translation are not constant but depend upon an unspecified function of $\mathbf{x}$.

On account of the given transformations we have:

$$\frac{\partial}{\partial t^*} = \frac{1}{m(\mathbf{x})}\frac{\partial}{\partial t} \quad , \quad \frac{\partial}{\partial x_i^*} = \frac{\partial}{\partial x_i} - \left(\frac{t + h'(m)}{m}\right)\frac{\partial m}{\partial x_i}\frac{\partial}{\partial t} \quad . \tag{20}$$

Moreover, a necessary condition to be assumed is that:

$$R(p,s) = s^{\theta/\gamma}\hat{R}(p \cdot s^{-\beta/\gamma}) \quad . \tag{21}$$

Now, a straightforward procedure leads us to easily recognize that the restrictions:

$$\begin{aligned} & \alpha = -1 \quad , \quad \beta = \delta = 0 \quad , \quad \theta = 2 \\ & \frac{\partial p}{\partial t} + \mu\mathbf{H} \cdot \frac{\partial \mathbf{H}}{\partial t} = 0 \\ & v_i \frac{\partial m}{\partial x_i} = H_i \frac{\partial m}{\partial x_i} = 0 \end{aligned} \tag{22}$$

are necessary and sufficient to guarantee the invariance of the governing equations under the family of transformations given before. Finally, it must be noted that, since $\beta = 0$, the only admissible equation of state is the Prim's law [1].

REFERENCES.

1. M. Munk, R. Prim. On the multiplicity of steady gas flows having the same streamline pattern. Proc. Nat. Acad. Sci. U.S.A., **33**, pp. 137-143, 1947.
2. R. Prim. Steady rotational flow of ideal gases. J. Rat. Mech. Anal., **1**, pp. 425-497, 1952.
3. P. Smith. The steady magnetodynamic flow of perfectly conducting fluids. J. Math. Mech., **12**, pp. 505-520, 1963.

4. P. Smith. An extension of the substitution principle to certain unsteady gas flows. Arch. Rational Mech. Anal. Arch. Rat. Mech. Anal., **15**, pp. 147-153, 1964.
5. G. Power, C. Rogers. Substitution principles in nonsteady magneto-gasdynamics. Appl. Sci. Res., **21**, pp. 176-184, 1969.
6. F. Oliveri. Derivation of the substitution principle in gas dynamics by means of Lie group techniques. Proc. Third International Conference on Hyperbolic Problems, Uppsala (Sweden), 1990.
7. G. W. Bluman, J. D. Cole. Similarity methods for differential equations. Springer, New York, 1974.
8. W. F. Ames. Nonlinear partial differential equations in engineering, vol. 2. Academic Press, New York, 1976.
9. P. J. Olver. Applications of Lie groups to differential equations. Springer, New York, 1986.
10. G. W. Bluman, S. Kumei. Symmetries and differential equations. Springer, New York, 1989.

Francesco Oliveri
Department of Mathematics, University of Messina
Contrada Papardo, Salita Sperone 31, 98166 Sant'Agata, Messina, Italy.

T PEIPMAN

Spectral changes in formation of solitary waves

ABSTRACT

The perturbed KdV equation is solved numerically by using an algorithm based on FFT. The spectral amplitudes give essential information about the stable solutions which may exist in perturbed systems.

1. INTRODUCTION

Soliton evolution in the presence of even a small perturbation leads to the following effects: (i) deformation of its shape and (ii) formation of a soliton tail /1/. The most interesting question in this problem is whether a stable solitary wave (different from the initial soliton) can exist in the presence of perturbation. Such a possibility has been shown analytically for the KdV equation with a r.h.s. in the form of a cubic polynomial /2/ using the inverse scattering transform /1/. However, this analysis has established only the limit eigenvalues but not the evolution process itself. In this paper the attention is paid to this process by making use of numerical experiment. The task is to solve the KdV equation

$$\frac{\partial u}{\partial \tau} - u\frac{\partial u}{\partial \xi} + \mu\frac{\partial^3 u}{\partial \xi^3} = \varepsilon f(u) \tag{1}$$

where ε is a small parameter and $f(u)$ is a smooth function, specified later. Equation (1) is solved numerically subject to fixed initial conditions. The numerical calculations are based on the FFT technique /3/ which uses the leapfrog scheme in τ and the FFT in ξ for evaluating the derivatives. Beside the high accuracy in calculating, useful information in spectral language is obtained at every time-step. The discrete spectrum describes the soliton evolution in a novel way which can be more informative than traditional amplitudes and widths.

2. NUMERICAL CALCULATIONS

Equation (1) is solved for two different initial conditions

$$u(0,\xi) = u_0 \sin \xi, \tag{2}$$

$$u(0,\xi) = u_0 \mathrm{sech}^2 \xi, \tag{3}$$

i.e. we are interested in soliton formation from a monochromatic initial condition (2) and in soliton evolution from an initial soliton (3) which is stable for $\varepsilon=0$.

Function $f(u)$ is taken in the form of a cubic polynomial

$$f(u) = -(u - 0.5u^2 + 0.0556u^3). \tag{4}$$

Some numerical results are shown in FIGS.1-5 for fixed ε's indicated separately in captions. The dispersion parameter was taken $\mu=-0.3$, if not indicated otherwise. The calculations for $\varepsilon=0$ and initial condition (3) showed that for a stable soliton the spectral amplitudes were constant in τ corresponding to the energy conservation.

Soliton formation is shown in FIGS.1-3. Instead of the classical 8-soliton chain /4/ we represent here the 2-soliton chain. In FIG.1a the profiles and FIG.1b the first three spectral amplitudes are shown. As the initial state is reinstated, the changes in spectral amplitudes are periodic showing, however, a more complicated pattern. At the moment when solitons are formed, the spectral amplitudes take the following values: the first is minimal, the second is maximal and the third is locally minimal.

In the case of $\varepsilon\neq0$ this evolution is drastically different. Let us note first that function $f(u)$ has three roots: $u_1=0$, $u_2=3$, $u_3=6$. It means that for $u<3$ attenuation takes place but for $3<u<6$ a soliton (or, to be more exact,

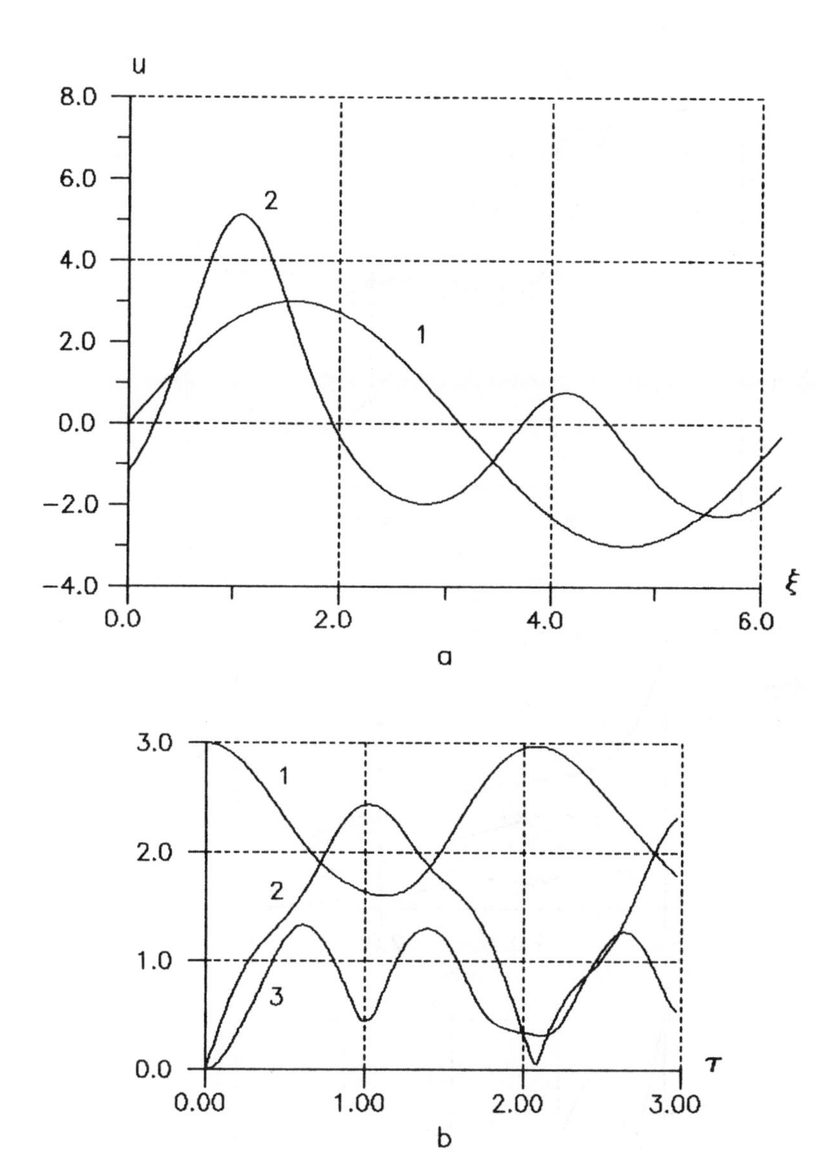

FIGURE 1 Formation of solitons from a sine-wave, $\varepsilon=0$: (a) profiles, 1 - τ = 0, 2 - τ = 1; (b) spectral amplitudes, figures show the number of the spectral amplitude.

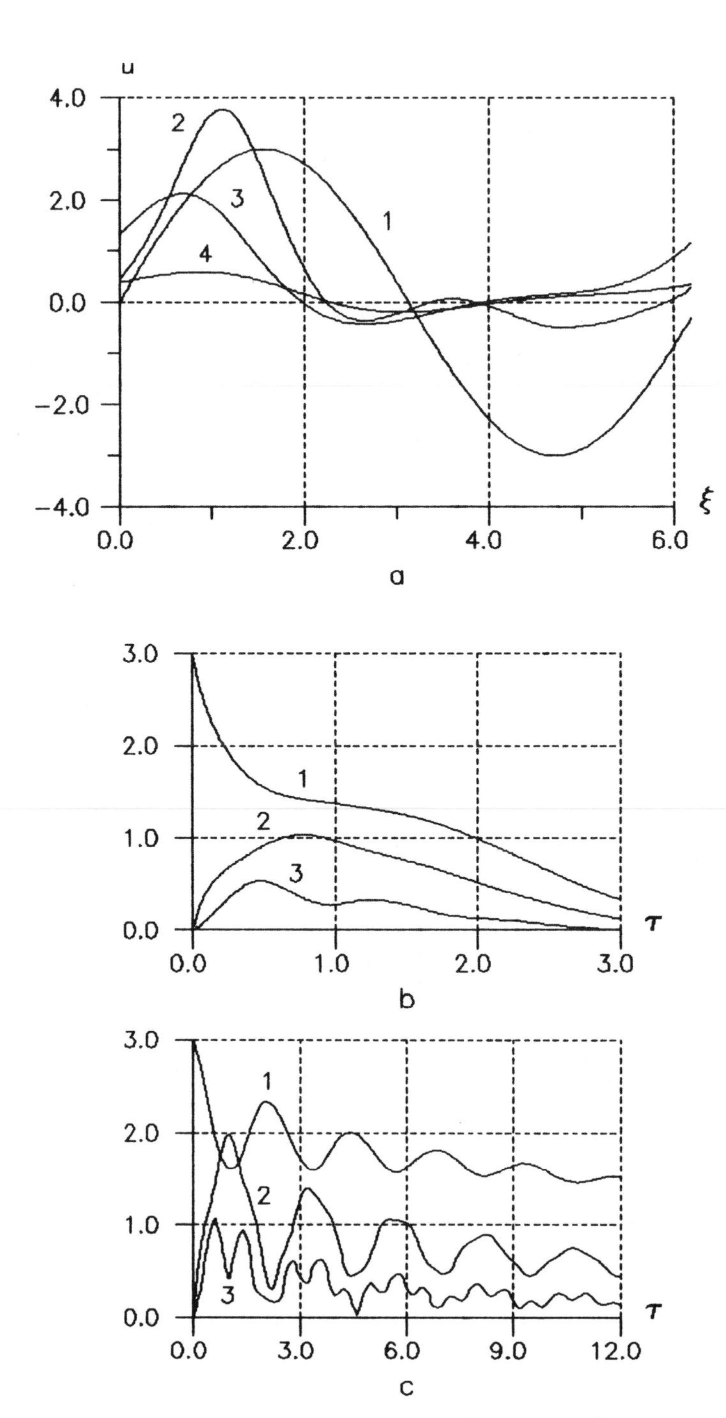

FIGURE 2 Formation of solitons from a sine wave $\varepsilon \neq 0$: (a) profiles for $\varepsilon=0.01$, $u_0=3$: 1 - $\tau=0$, 2 - $\tau=0.6$, 3 - $\tau=2$, 4 - $\tau=3$; (b) spectral amplitudes for $\varepsilon=0.01$; (c) spectral amplitudes for $\varepsilon=0.001$.

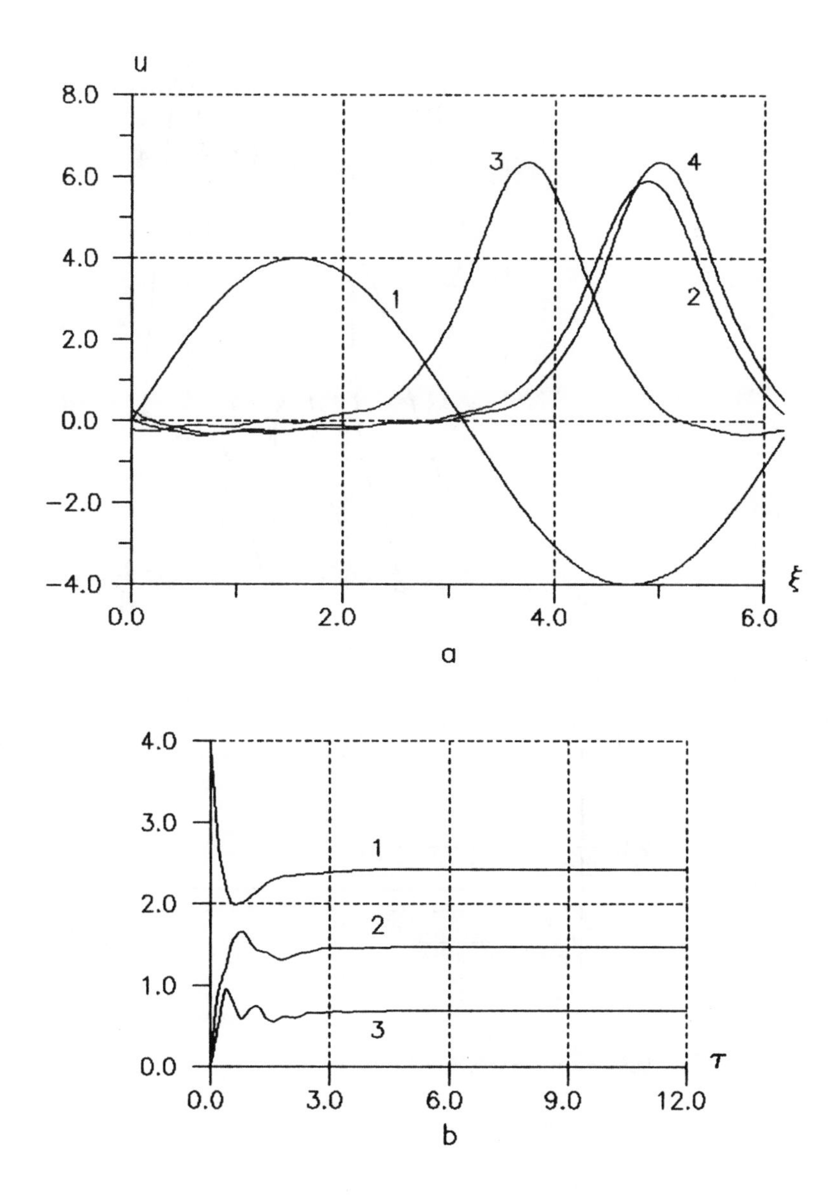

FIGURE 3 Formation of solitons from a sine wave, $\varepsilon \neq 0$: (a) profiles for $\varepsilon=0.01$, $u_0=4$: 1 - $\tau=0$, 2 - $\tau=2$, 3 - $\tau=6$, 4 - $\tau=12$; (b) spectral amplitudes.

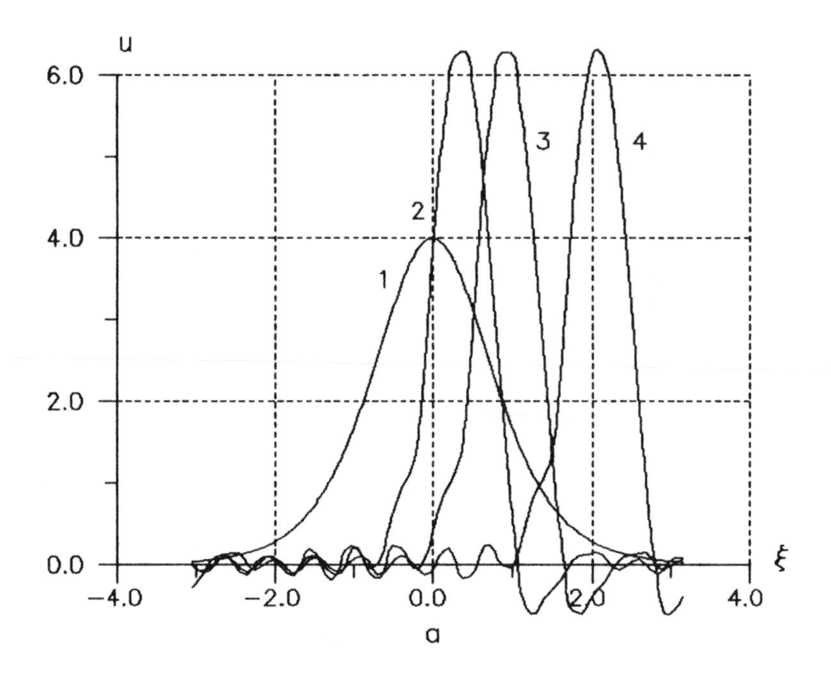

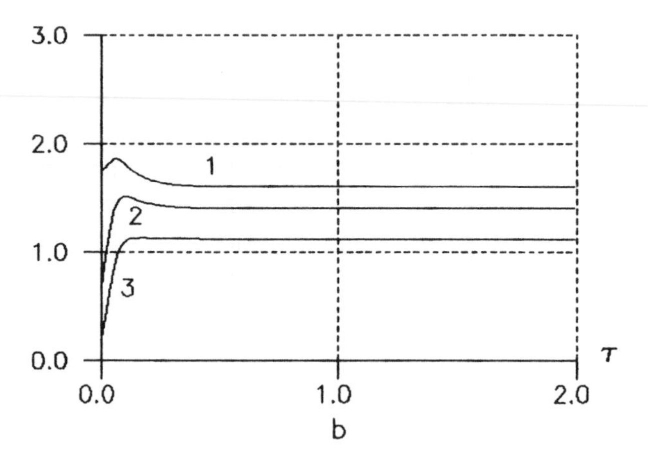

FIGURE 4 Soliton-type excitation, $\varepsilon=0.10$, $u_0=4$: (a) profiles: 1 - $\tau=0$, 2 - $\tau=0.5$, 3 - $\tau=1$, 4 - $\tau=2$; (b) spectral amplitudes.

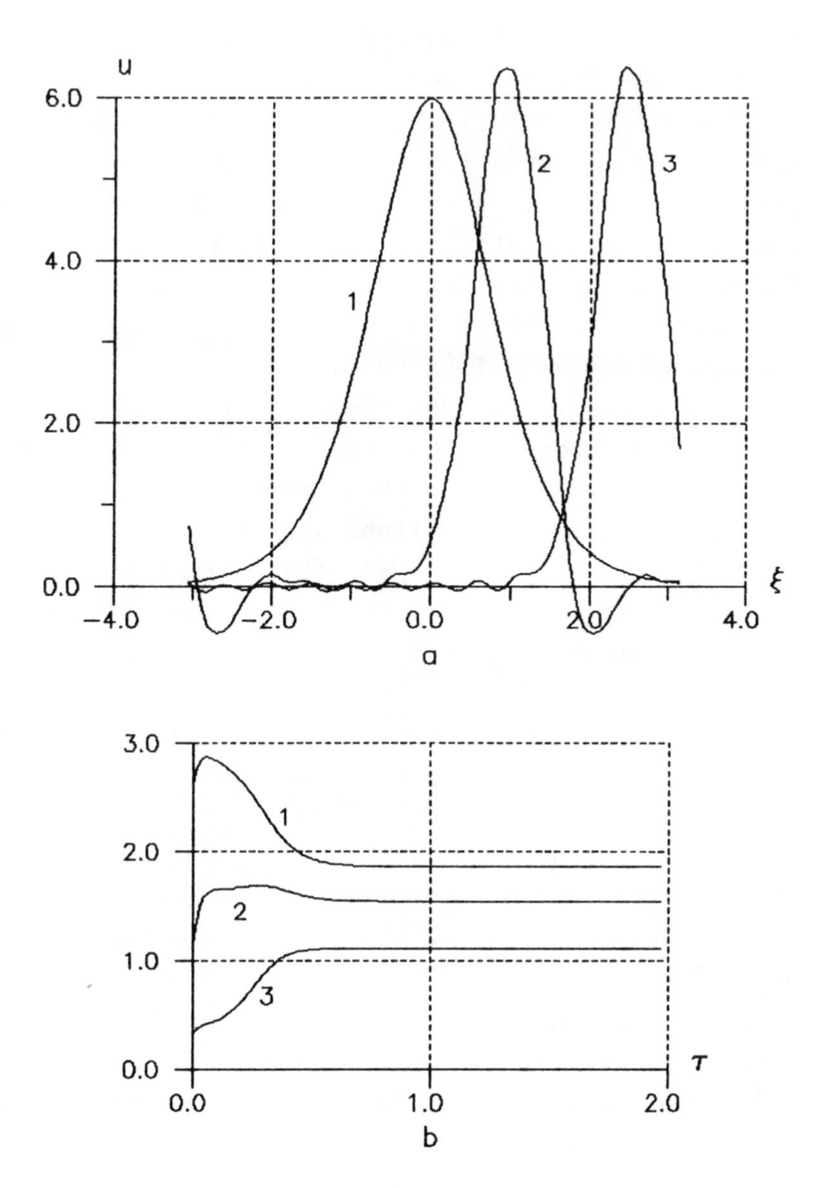

FIGURE 5 Soliton-type excitation $\varepsilon=0.10$, $u_0=6$: (a) profiles: 1 - $\tau=0$, 2 - $\tau=1$, 3 - $\tau=2$; (b) spectral amplitudes.

a certain part of it) undergoes amplification. From FIG.1 we know that for $u_0=3$ in condition (2) the larger soliton has the amplitude in the amplification region, the smaller one - in the attenuation region. However, the rate of the growth in the process of formation must be compared with the rate of attenuation which may prevent the soliton formation. The numerical experiment shows that for $u_0=3$ and $\varepsilon\neq 0$, the attenuation takes over. The structure of spectral amplitudes during a long transient period (FIG.2c) resembles the formation in the case $\varepsilon=0$ (FIG.1) but finally, in the long run, both the expected solitons are attenuated. If ε is larger (FIG.2b), then the wave is attenuated faster and the formation of solitons is interrupted from the very beginning. If, however, $u_0=4$, then out of two expected solitons in the case of $\varepsilon=0$ the larger one is formed (FIG.3). The spectral amplitudes after a certain transient time remain constant (FIG.3b). The periodicity as seen in FIG.1, is lost and one stable soliton is propagating further. In the case of a longer chain of solitons, more than one soliton can quite probably be formed.

The evolution of a soliton-type input (3) is shown in FIGS.4,5 for various u_0's. The dispersion parameter μ was chosen such that for $\varepsilon=0$ the soliton was stable. The behaviour of spectral amplitudes (FIG.4b,FIG.5b) shows that in these cases new stable solitons are formed with certain oscillating tails. The new solitons have a larger amplitude which, with the accuracy of the numerical experiment, does not depend upon the initial amplitude. This effect is known from the theory of active media /5/. The width of solitons and their velocity have also undergone changes.

3. CONCLUDING REMARKS

Seemingly, the spectral amplitudes give additional information about the evolution of solitons in the presence of perturbation. It appears to be a rather new field of research. Closely related to this are the results obtained by Osborne et al /6/. Their idea is to find the nonlinear spectrum of $u(\tau,\xi)$ related to the Floquet diagram. As a result, the solution to the unperturbed KdV equation is a linear superposition of nonlinearly interacting nonlinear waves, given by hyperelliptic functions. This approach gives a good explanation for a two-soliton collision showing explicitly the temporal evolution of the process. The approach in this paper is based on using a linear spectrum at each step of calculation. This may give more information for those who put more emphasis on wave dynamics rather than on soliton flows and on the energetical background /7/.

There are many open questions still to be answered. They are related to the velocity of perturbed solitons and also to their geometrical parameters. The author hopes to be able to explain this interesting phenomen of perturbation in his forthcoming publications more thoroughly.

4. ACKNOWLEDGEMENTS

The author expresses his thanks to the LOC of Euromech 270 for the financial support.

REFERENCES

1. V.I.KARPMAN, Soliton Evolution in the Presence of Perturbation. Phys.Scripta 20 (1979) 462-478.
2. J.ENGELBRECHT, A.JEFFREY, KdV Solitons in Active Media. Wave Motion 9 (1987) 533-541.
3. T.PEIPMAN, The Application of Fast Fourier Transform to Nonlinear Wave Propagation. Inst. of Cybernetics, Estonian Acad.Sci., Mech. Report 3/90, 1990.
4. N.J.ZABUSKY, M.D.KRUSKAL, Interaction of Solitons in a Collisionless Plasma and the Recurrence of Initial States. Phys.Rev.Lett. 15 (1965) 240-243.
5. V.S.ZYKOV, Modelling of Wave Processes in Excitable Media (Manchester University Press 1988).

6. A.R.OSBORNE, E.SERGE, Numerical Solutions of the Korteweg-de Vries Equation Using the Periodic Scattering Transform μ-Representation. Phys. D44 (1990) 575-604.
7. A.R.BISHOP, J.A.KRUMHANSL, S.E.TRULLINGER, Solitons in Condensed Matter: a Paradigm. Phys. 1D (1980) 1-44.

Tőnu Peipman
Dept. of Structural Mech.
Tallinn Technical University
Tallinn, SU 200108
Estonia

A M SAMSONOV

On some exact travelling wave solutions for nonlinear hyperbolic equations

ABSTRACT

Direct methods for solving nonlinear wave equations have extensive application in any particular problem, mainly due to the fact that many physical problems are not fully integrable by means of the IST method.

The problem of travelling wave solutions of nonlinear quasi-hyperbolic equations (NHE) with dissipative terms is considered here in the framework of a direct algebraic approach. It is shown that the main equation for the dissipative ($g \neq 0$) long wave approximation has the form $u_{tt} - u_{xx} = [P_k(u) + au_{tt} - bu_{xx}]_{xx} + (eu^2 + fu)_{xt} + gu_{xxt}$, associated with the description of problems of a collisionless cold ion plasma ($b = d = f = 0$), of strain waves in a rod ($e = f = 0$), of waves in stratified fluids ($a = e = 0$), waves in atomic lattices in the continuous limit approximation ($e = f = 0$), polarization waves in a nonlinear dielectric crystal, and so on. We reduce NHE to the nonlinear ordinary differential equation of the form: $u'' = Q_n(u) + Au'$, where $z = x \pm Vt$ is the phase variable, and $Q_n(u)$ is a polynomial. Some proper exact solutions are found for this equation in terms of the Weierstrass $\mathcal{P}$-function. For example, for problems with dissipation ($g \neq 0$) we have for $n = 3 : u = \alpha \exp(cz)\mathcal{P}'_z[\beta \exp(cz)]/\mathcal{P}[\beta \exp(cz)]$, whereas for $n = 2 : u = \alpha \exp(cz)\mathcal{P}[\beta \exp(cz/2)]$. These exact explicit travelling wave solutions seem to be new, and may be of interest in particular physical problems.

1. INTRODUCTION AND BASIC EQUATIONS

We consider long wave propagation problems in wave guides, of various physical kinds, which can be described in terms of the general nonlinear hyperbolic equation of the type

$$u_{tt} - u_{xx} = [P_k(u) + au_{tt} - bu_{xx}]_{xx} + (eu^2 + fu)_{xt} + gu_{txx} \tag{1}$$

where x is the space coordinate, t is time, $P_k(u)$ is a polynomial of order k with respect to the unknown function $u(x,t)$. Our aim is to propose an approach, suitable for generating exact explicit travelling wave solutions for this equation, which seems not to be integrable by means of the inverse scattering transform (IST) method.

There are several examples of physical problems, which can be written as the equation (1). Indeed, for the problem of the ion-acoustic wave propagation in a cold ion plasma it was shown in [1], that, for isothermal motion and for the small Debye length approximation, the governing equation has the form

$$u_{tt} - u_{xx} = (-u^3/3 + au_{tt})_{xx} - (eu^2)_{xt} + gu_{xxt} \quad . \tag{2}$$

For the classical problem of condensed matter physics of waves in stratified liquid the following equation was obtained in [2]:

$$u_{tt} - u_{xx} = (cu^3 + du^2 - bu_{xx})_{xx} + (fu)_{xt} + gu_{xxt} \quad , \tag{3}$$

whilst wave propagation in a different medium, namely, in a one-dimensional nonlinearly elastic solid wave guide, is described in [3] by the DDE with dissipation

$$u_{tt} - u_{xx} = \epsilon(cu^2 + au_{tt} - bu_{xx} + gu_t)_{xx} + O(\epsilon^2) \quad , \tag{4}$$

where u is the longitudinal strain, and $\epsilon \ll 1$. The polarization wave propagation problem for a nonlinear dielectric crystal is of a different nature, although it can be reduced to the equation, for the polarization $p(x,t)$, of the type

$$p_{tt} - p_{xx} = \epsilon[(Q_k(p) + ap_{tt})_{tt} + (Q_k(p) - bp_{tt})_{xx} + g(p_{tt} - cp_{xx})_t - cp_{xx})_t] + O(\epsilon^2) \quad , \tag{5}$$

where Q is a polynomial and $\epsilon \ll 1$; the study of this problem will be published elsewhere.

Therefore the following common features are typical for all the equations above. Indeed, from the physical viewpoint nonlinearity and dispersion are small in comparison with the terms in the main operator, which is of hyperbolic type, while the nonlinearity is of polynomial type (and, as a rule, not higher than of third order), the dispersive terms consist of the highest derivatives, and the independent variables are absent. The most famous equation of type (1) is, of course, the Boussinesq equation, which can be written in one of the following forms

$$u_{tt} - V^2 u_{xx} = \epsilon(du^2 + bu_{xx})_{xx} + O(\epsilon^2) \quad , \tag{1a}$$

$$u_{tt} - V^2 u_{xx} = \epsilon(du^2 + au_{tt})_{xx} + O(\epsilon^2) \quad , \tag{1b}$$

depending on the kind of derivative which was used for reduction of the initial hydrodynamical problem, of long wave propagation on shallow water according to the equality $u_t = -u_x + O(\epsilon)$.

As a rule, wave solutions of (1a) or (1b) differ from those of equations of type (1). Indeed, even in the corresponding linearized problem, with $k = 2, e = f = g = 0$ in (1), one can reveal the difference in dispersion properties, when either u_{ttxx} or u_{xxxx} is given.

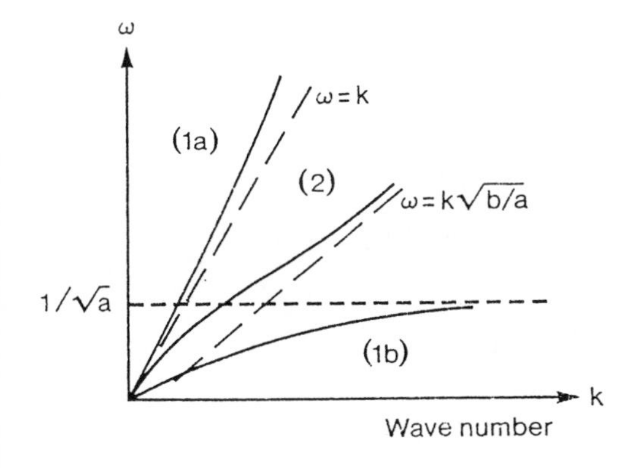

Figure 1. Frequency ω vs. wave number k of linear wave solution $u \sim \exp i(kx - \omega t)$ of linearized version of the Boussinesq equation
(1a) $a = 0$,
(1b) $b = 0$,
(2) $a \neq b \neq 0$, and $\sqrt{b/a} < 1$ in (1)

One of the remarkable aspects is the finite and non-zero value of the phase velocity for (1b), i.e., for $b = 0$ in (1), in the limit $k \to \infty$, when compared with (1a). Moreover, equation (1b) provides the separation of variables $u =$

$-V^2/2d\epsilon + T(t)X(x)$, that leads to the form $f_{tt} = \epsilon(df^2 + af_{tt})_{xx}$ and, therefore

$$T''/(\epsilon dT^2) = (X^2)''/(X - a\epsilon X)'' \equiv \lambda$$

resulting in the exact solution

$$u = -V^2/2d\epsilon + T(t)X(x) = -V^2/2d\epsilon + [(x+a)^2 + 3b\epsilon]/[2d\epsilon(t+\beta)^2] \quad ,$$

whereas the equation $(1a)$ does not provide this.

2. REDUCTION TO NEE AND FORMAL QUASI-STATIONARY SOLUTIONS OF NHE

The next problem under consideration will be the reduction of any NHE to a nonlinear evolution equation (NEE). It should be noted that the nonlinear equations (1)-(5) are quasi-hyperbolic; that is the reason why the complete set of boundary and initial conditions for the wave problem can be satisfied , in contrast with the corresponding evolution equation , which is actually a result of widespread, but not always well- grounded, reduction of (1).

Let $e = f = g = 0$ in (1) for brevity, i.e., the nondissipative problem is considered,

$$u_{tt} - u_{xx} = \epsilon[(P_k(u) + au_{tt} - bu_{xx})_{xx}] \quad . \tag{6}$$

As a rule, when one assumes $\tau = \epsilon^{n+1}t, \xi = \epsilon^n(x - Vt)$ with appropriate integer n, the problem for $k = 2$ results in the KdV equation, while for $k = 3$ it results in the modified KdV equation, following from (1) in the form

$$u_\tau + [(u^k)/k]_\xi + (a-b)u_{\xi\xi\xi} = O(\epsilon) \quad .$$

By contrast, the NHE contains ϵ, multiplying both highest derivatives. If $\partial^k u/\partial\xi^k \to 0$ for $|\xi| \to \infty$, both equations could provide the same solutions, but in the general case, for the boundary conditions $u \to u_i, i = 1,2$ at infinity ($|\xi| \to \infty$), there is no compatibility. Indeed, only one independent initial condition can be taken into account for the KdV equation, and for any problem reduced to NEE the question arises, how to satisfy $u(t=0) = u_0(x)$?

The simplest example is the refined beam deflection problem, governed by a fourth order O.D.E., the solution of which consists of the solution of the reduced problem, governed by the second order O.D.E., together with two boundary layer functions; that is a definite indication of a singular perturbation problem. Moreover, a double asymptotic expansion (with respect to ϵ and to the inhomogeneity parameter) will be necessary in any inhomogeneous problem for the NEE with slowly varying coefficients. Therefore we deal with the quasihyperbolic equation (12) in order to avoid extra difficulties resulting in the limit $\epsilon \to 0$.

Therefore we deal with the NHE:

$$u_{tt} - u_{xx} = \epsilon[(P_k(u) + au_{tt} - bu_{xx})_{xx} + (eu^2 + fu)_{xt} + gu_{txx}] \quad . \tag{1}$$

Letting $\epsilon u \to u, \epsilon a \to a, \epsilon b \to b$, etc., and introducing the phase variable $z = x \pm Vt$, after integration twice of the result, we have a nonlinear second order O.D.E with respect to the new unknown u, in the form

$$u'' = Au' + Q_k(u) \quad , \tag{7}$$

which is, indeed, the equation for a nonlinear oscillator with damping. Here primes denote differentiations with respect to $z, Q_k(u)$ is a polynomial, and two conditions are assumed to be valid: $u \to u_i, i = 1, 2$, when $z \to \pm\infty$. Two conditions are important for the approach under consideration, namely, that the differential function (DF) should be polynomial, and it should not contain z (that is, the equation (7) must be reducible to an autonomous one).

The theorem is well known (for non-dissipative problems) on the reduction of the O.D.E. of order k with polynomial DF: $d^k v/dz^k = P_m(v)$. By the substitution $v'(z) = q(v)$ to an O.D.E. of order $(k-1)$ with polynomial DF, and for any known $q(v)$, the first integral of it can be written down.

We generalize it, formulating for the problem with dissipation the following:

Theorem 1

A nonlinear O.D.E. $v'' = Av' + vP_m(v)$ of second order, with dissipative term, and with polynomial DF having at least one real root, can be reduced to an O.D.E. of the Abel type.

Proof is straightforward. We have $v''/v = Av'/v + P_m(v)$; let us introduce the logarithmic variable $y \equiv \log v$, and after that the standard substitution $y' = p(y)$ leads to the second order Abel equation

$$pp' = Ap + p^2 + Q_m(\exp(y)) \quad , \tag{8}$$

with the new polynomial Q_m, depending upon $\exp y$. Further reduction by means of the inverse function $u \equiv 1/p(y)$ results in the first order Abel equation, either in the form $u' = -u - Au^2 - u^3 Q_m(y)$, or in the normal form

$$f' = f^3 + F(z) \quad . \tag{9}$$

The study of Abel's equations was done by P. Boutroux [4] and many others at the beginning of the century.

A natural way to solve the problem seems to be the following: having the polynomial DF, we assume the polynomial form for $q(v) : q(v) = q_0 + q_1 v + q_2 v^2 +$ ..., substitute it into the equation and find the order n, annihilate coefficients at each order of v^i, and as a result we get the coupled nonlinear algebraic equations for q_i and wave velocity V.

Trying to do it even for the O.D.E. without dissipation: $u'' = Q_k(u), k = 2,3$, we substitute $u'(z) = \sqrt{q(u)}$, therefore $u'' = q'_u/2 = Q_k(u)$, and from $q'_u/2 = Q_k(u)$ we get $q(u) - q(u_0) = \int du/(2Q_k(u)) \equiv I(u)$, and on the way back we have to do the integration $z - z_0 = \int_{u(z_0)}^{u(z)} dy/q(y)$. The main problem in that process to obtain an exact explicit travelling wave solution is how to carry out all integrations and to return to the initial variables in explicit form. The theorem can be useful, not only for investigation of the problems of existence and stability of any solution, but also because of the possibility to exploit, in the dissipative case, the following fact from [5],[6]:

An autonomous equation $v'(z) = q(v)$ with polynomial $q(v)$ can be solved in terms of $\exp(\alpha z)$ and of the Weierstrass function $\mathcal{P}(z + c)$ among the special functions of mathematical physics, because any derivative of both of these can be expressed in terms of a polynomial with respect to the function itself. Indeed, from [5] we have

$$(\mathcal{P}')^2 = 4\mathcal{P}^3 - g_2\mathcal{P} - g_3; \ \mathcal{P}'' = 6\mathcal{P}^2 - g_2/2, \ \mathcal{P}''' = 12\mathcal{P}'\mathcal{P} \quad ,$$

$$\mathcal{P}''' = 120\mathcal{P}^3 - 18g_2\mathcal{P} - 12g_3 \quad ,$$

where g_2, g_3 are the invariants of the $\mathcal{P}$-function.

Therefore we propose to find some solutions of the nonlinear O.D.E. (7) in any of the following forms:

$$u = M_m(\mathcal{P}) \ , \ u = A(\mathcal{P}) + B(\mathcal{P})\mathcal{P}' \ , \quad u = \frac{M_m(\mathcal{P})}{N_n(\mathcal{P})}; \tag{10}$$

with rational A, B and polynomial M, N; these result in various periodic, doubly periodic, and localized solutions of the NHE. The procedure seems to be quite simple. Let us substitute any of the expressions from (10) into (7), annihilate independently the coefficients at each order of $\mathcal{P}$ and $\mathcal{P}'$, and obtain the set of coupled algebraic equations. When a solution of the system exists, we obtain a solution of the NHE. Moreover, any mathematical symbolic software for the PC can be used at this stage.

3. PARTICULAR TRAVELLING WAVE SOLUTIONS

For example, when $k = 2$ we have from (7)

$$u''_{zz} = au' + cu^2 + du + e \quad . \tag{11}$$

Finding the solution in the form similar to the second formula in (10), we get, with $y = \exp(\gamma z)$

$$u = \alpha y^2 \mathcal{P}(\beta y) = \alpha \exp(2\gamma z)\mathcal{P}(\beta \exp \gamma z) \quad , \alpha, \beta\text{-constants} \quad , \tag{12}$$

that is, when $d < 0, e = 0$, and in addition, $16d = -3a^2$, the solution has the form

$$u = ay^{1/4}\mathcal{P}(y^{1/2}/\alpha; 0, g_3 = \text{const}\Big) =$$

$$e_2 + hy^{1/4}\frac{1 + \text{cn}\,((2yh)^{1/2}|k)}{1 - \text{cn}\,((2yh)^{1/2}|k)} \quad , \tag{13}$$

where $h^2 = 9m^2 + n^2$, and the roots e_i of the characteristic polynomial for $\mathcal{P}(\cdot)$ are $e_1 = m + ni, e_3 = m - ni, i^2 \equiv -1$, and the modulus of the Jacobi elliptic function is defined by $k^2 = (h - 3e_2)/(2h)$.

If $g_3 = 0$ in (13), one can conclude that either $e_1 = e_2$ or $e_3 = e_2$, and this leads to the kink solution in the form

$$u = \alpha + \beta\frac{\exp 2\gamma z}{(\kappa + \exp\gamma z)^2} = a + \beta y^2/(\kappa + y)^2 \quad , \tag{14}$$

where $\alpha, \beta, \gamma, \kappa$ are constants. This solution was obtained recently [7] by means of the Painlevé equation analysis. For cubic nonlinearity it follows from (7), that $u'' = au' + bu^3 + cu^2 + du + e$, and substituting $y = \exp\gamma z$,we have instead

$$y^2\gamma^2 u'' = (a\gamma - \gamma^2)yu' + bu^3 + cu^2 + du + e \quad . \tag{15}$$

Finding a solution in the form

$$u = A(\mathcal{P}) + y^n\mathcal{P}^m(\beta y^k)\mathcal{P}' \tag{16}$$

we have : $n = 2k, \gamma = a/3, m = -1$, hence

$$u = -c/3b + (1/\sqrt{2b})y\mathcal{P}'_y/\mathcal{P}(y, g_2, 0) \quad , \tag{17}$$

and two additional conditions on the coefficients should be satisfied:

$$3c^2 - 9bd = 2a^2b \; ; \quad c(2c^2 - 9bd) + 27eb^2 = 0 \quad .$$

For the particular problem with the following restrictions on the coefficients in (15) $c = 0, e = 0, b \neq 0, 9d = -2a^2, d < 0$, we have $u'' = au' + bu^3 + du$, hence for $g_3 = 0$ it is valid that $e_1 e_2 e_3 = 0$, whilst for $g_2 = 0$ we have the expression $\mathcal{P}(y) = 1/y^2$; for $g_2 < 0 : \mathcal{P}(y) = \frac{1+\mathrm{cn}\,(2my|k)}{1-\mathrm{cn}\,(2my|k)}$; and, finally, for $g_2 > 0$, the general expression for the $\mathcal{P}$-function in terms of Jacobi's elliptic sn-function, $\mathcal{P}(y) = e_1 + 2e_1 \mathrm{sn}^{-2}(y\sqrt{2e_1}|k)$, can be used for calculations. We note here the connection between the condition $e = 0$ above and the one required on coefficients in the Theorem 1 to provide a root of the polynomial in the DF.

The same approach can be used for fourth order O.D.E. of the form

$$u'''' = Au'' + Q_k(u) \quad .$$

Following the first proposition in (10), $u = M_m(\mathcal{P}(z))$, we obtain the condition $m(k-1) = 2$ and for $k = 3$ we have $m = 1$, hence the exact explicit solution has the form $u = \alpha\mathcal{P}(\gamma z)$, while for $k = 2$ the equality $m = 2$ is valid, and therefore $u = \alpha\mathcal{P}(\gamma z) + \beta\mathcal{P}^2(\gamma z)$. The last formula can be useful for the theory of the Kuramoto-Sivashinsky equation.

4. FINAL REMARKS

We have shown that the usual routine of reduction of NHE to NEE leads to the necessity to take into account the singular perturbation problem, as well as to the difference in evaluation of the parameters in the physical problem. The theorem was proved on the reduction of any nonlinear polynomial O.D.E. with damping to the first order Abel equation. The approach was developed to obtain exact travelling wave solutions, in terms of rational functions with respect to the Weierstrass $\mathcal{P}$-function, and several new exact solutions were found explicitly. The important problem for further study seems to be the problem of stability of these solutions, and of any possible scenario for transition to chaos.

ACKNOWLEDGMENTS

The kindest hospitality of D.A.M.T.P. and St. John's College of the University of Cambridge, where the manuscript was prepared for publication, and the discussions with Prof. D.G. Crighton are gratefully acknowledged.

REFERENCES

1. P.ROSENAU, Evolution and breaking of ion-acoustic waves. Phys.Fluids, **31**, (1988), no.6, 1317-1319.
2. R.DODD et al., Solitons and Nonlinear Wave Equations, Academic Press (1984).
3. A.M. SAMSONOV, On existence of longitudinal strain solitons in an infinite nonlinearly elastic rod. Sov.Phys.-Doklady, **33** (1988), no.4, 298-300.
4. P.BOUTROUX, On multiform functions defined by differential equations of the first order, Ann. of Math., **22**, (1920-21), 1-10.
5. H. BATEMAN, A. ERDELYI, Higher transcendental functions, vol.3, Mc-Graw Hill (1953-54).
6. E.T.WHITTAKER, G.N.WATSON. A Course of Modern Analysis, vol.2, Cambridge, C.U.P. (1962).
7. A.M. SAMSONOV, Transonic and subsonic localized waves in nonlinearly elastic waveguides. In: Proceed. Intern. Conference on Plasma Physics, Kiev, Naukova dumka, 1987, vol.4, pp.88-90.

Alexander M. Samsonov
A.F. Ioffe Physical Technical Institute
of the USSR Academy of Sciences
Leningrad, 194021 USSR

A K SCHIERWAGEN

Impulse propagation in nonuniform nerve fibres: analytical treatment

Abstract

One of the mathematical models for impulse propagation in a nerve fibre is given by the hyperbolic telegraph equations. In the present study the assumption of constant coefficients in the equations (i.e. of uniform fibre geometry) has been dropped. Using a piecewise linear approximation of the FitzHugh-Nagumo equations the problem is shown to be equivalent (under certain variable transforms) to the case of an uniform fibre.
A class of fibre diameter variations could be determined by solving a special Riccati equation. For this class exact heteroclinic wave solutions describing the propagation of the "leading edge" of an impulse are shown to exist.

1. INTRODUCTION

The exact,quantitative description of excitation phenomena in nerves originated from Helmholtz' measurement of impulse conduction velocity in frog sciatic nerve in 1850. Currently, nerve impulse conduction represents one of the most fully studied nonlinear wave phenomena in excitable media. While the dominant model type in this field is that of reaction-diffusion equations, i.e. parabolic equations with a nonlinear source term (see, e.g./1-4/), the hyperbolic telegraph equations have been used in this context, too. For example, recently Dunbar and Othmer /5/ derived the equation

$$\varepsilon^2\frac{\partial^2 v}{\partial t^2} + (1+g(v))\frac{\partial v}{\partial t} = k^2\frac{\partial^2 v}{\partial x^2} + f(v) \tag{1}$$

($\varepsilon, k \in \mathcal{R}$) as the equation for voltage v in an uniform transmission line with nonlinear shunt conductance and series inductance along the length of the line.
The motivation for studying this hyperbolic equation mainly comes from models of the nerve axon which explicitly consider induction along the axon length /1,6,7/. Investigations with either model type (parabolic or hyperbolic) usually assume uniform electrical and geometric properties of the nerve axon. From these studies much insight into mechanisms of action potential (AP) propagation has been gained, suggesting constant shape and velocity of the AP. A linear or square root relationship between velocity and axonal diameter for myelinated or unmyelinated nerve fibres, respectively, was deduced /8,9/.

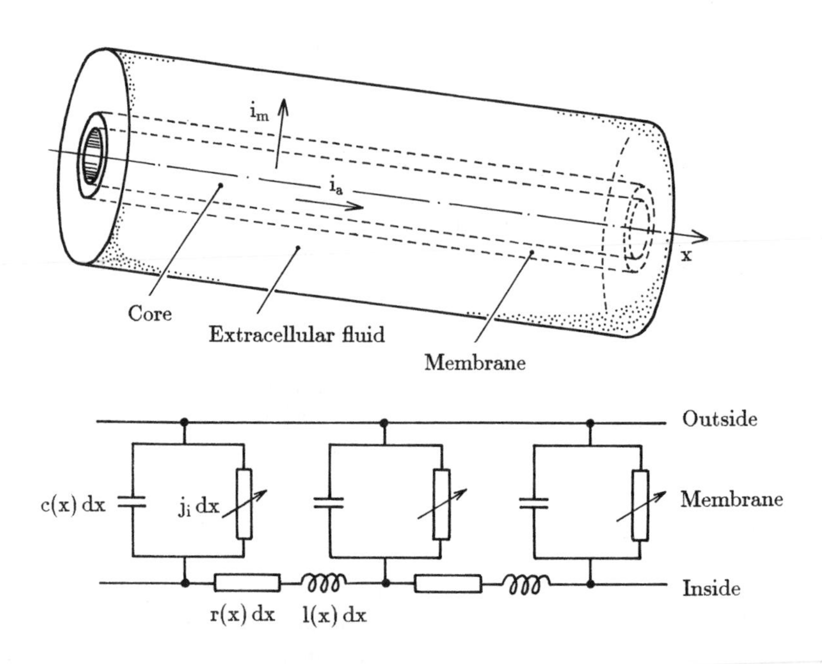

FIGURE 1 Scheme of a core conductor and its equivalent electrical circuit. Inside the membrane is the conducting core consisting of axoplasm, outside the extracellular fluid. In the equivalent circuit, the extracellular impedance is neglected. For explanation of symbols, see text.

On the other hand, experimenters have been aware of several effects which cannot be explained with this theory. Examples include blocking of impulse conduction, frequency modulation and changes of AP shape and velocity in regions of nonuniform axon geometry (for review, see e.g. /10/. Motivated by these observations, several groups of researchers have performed numerical computations in order to explore the effects of changing axonal geometry upon AP conduction /11-14/. Even if the experimental findings could be largely reproduced through the simulations, analytical studies of how the various physical parameters describing the inhomogeneous axon will affect the solutions are indispensible.
Based on the analysis in /15/, we attempt this problem in the present paper by analyzing a simple version of the FitzHugh-Nagumo equations /2,16/ as used by Scott /1/ and others /17,18/, combined with the hyperbolic telegraph equations with non-constant coefficients. Applying certain variable transformations, we demonstrate that the problem can be reduced to the case of constant coefficients if some

compatibility condition is satisfied. The condition is expressed by a *special Riccati equation* the solutions of which define a class of nonuniform axon geometries.

2. THE HYPERBOLIC NERVE CONDUCTION MODEL

A model of a nerve axon can be set up by combining the nonlinear ordinary differential equations for an excitable membrane with the hyperbolic partial differential equation for a transmission line. Here the case of a continuous, unmyelinated axon having its whole surface membrane excitable will be considered. Fig. 1 schematically displays the axon and its equivalent circuit where $r(x)$ is the longitudinal series resistance, $l(x)$ is the series inductance and $c(x)$ is

the membrane capacitance (all quantities per unit length). The membrane ionic current j_i per unit length is a rather complex nonlinear function of the voltage v across the membrane. We denote by $v(x,t)$ the voltage at location x on the axon at time t, and by $i(x,t)$ the total longitudinal current.
Applying Kirchhoff's current and voltage laws to the cicuit in Fig. 1, we find

$$\frac{\partial i}{\partial x} + c(x)\frac{\partial v}{\partial t} + j_i(v) = 0 \tag{2}$$

$$\frac{\partial v}{\partial x} + l(x)\frac{\partial i}{\partial t} + r(x)i = 0, \tag{3}$$

i.e. the telegraph equations with spatially varying coefficients.
Hadeler /19/ has shown that in the case of constant coefficients traveling wave solutions of velocity $u \leq (lc)^{-1/2}$ exist, i.e. nontrivial solutions $v(x,t) = v_u(y)$ of (2),(3) depending only on the moving coordinate $y = x - ut$ such that $v \to 0$ for $y \to \infty$.
However, here the assumption of constant coefficients is dropped, and so changes in shape and velocity of the AP must be taken into account, suspending traveling wave solutions of (2),(3).

3. DERIVATION OF CANONICAL FORM

We consider an axon of variable, circular cross-section with the specific electrical parameters (series resistivity ϱ, series inductance l_i, membrane capacitance c_m and membrane conductance g_m) assumed to be fixed along the axon. Then the corresponding coefficients per unit length of axon vary as

$$\begin{aligned} r(x) &= 4\varrho/(\pi d(x)^2), \quad c(x) = c_m \pi d(x) \\ l(x) &= 4l_i/(\pi d(x)^2), \quad g(x) = g_m \pi d(x) \end{aligned} \tag{4}$$

where $d(x)$ is the axon diameter. In the following, it will be convenient to use the quantities

$$\tau_1 = \frac{l(x)}{r(x)}, \quad \tau_2 = \frac{c(x)}{g(x)}. \tag{5}$$

From (4) it follows that both τ_1 and τ_2 are constants with the dimension of time. Adapting the ansatz for $j_i(v)$ used by Scott /1/, Rinzel and Keller /18/ and others to the present case of non-constant coefficients, we write

$$j_i(v) = g(x) \cdot f(v) \tag{6}$$

where $f(v)$ is a nonlinear function, e.g. a cubic polynomial as in the FitzHugh-Nagumo model /2,16/.
Inserting (5) into (2) and eliminating the variable $i(x,t)$ from (2), we obtain

$$\tau_1 \frac{\partial^2 v}{\partial t^2} + (1 + \frac{\tau_1}{\tau_2}\frac{\partial f}{\partial v})\frac{\partial v}{\partial t} = \frac{1}{c(x)}\frac{\partial}{\partial x}(\frac{1}{r(x)}\frac{\partial v}{\partial x} - \frac{f(v)}{\tau_2} \tag{7}$$

where obviously the only restriction imposed on the nonlinearity $f(v)$ is to be differentiable.
For negligible series inductance ($l_i = 0$) the parabolic case studied in /15/ results, i.e. the nonlinear cable equation with non-constant coefficients,

$$\frac{\partial v}{\partial t} = \frac{1}{c(x)}(\frac{\partial}{\partial x}(\frac{1}{r(x)}\frac{\partial v}{\partial x}) - g(x) \cdot f(v)). \tag{8}$$

For a piecewise linear approximation of $f(v)$, this equation could be transformed into dimensionless canonical form /15/, i.e.

$$\frac{\partial V}{\partial T} = \frac{\partial^2 V}{\partial Z^2} + F(V) \tag{9}$$

for which the problems of existence and speed of traveling fronts are known to be solved /1,3,18/.
The structure of equation (7) enables us to proceed as in /15/ in order to derive the dimensionless normal form of (7), i.e.

$$\frac{\partial^2 V}{\partial T^2} + (1 + G(V))\frac{\partial V}{\partial T} = \frac{\partial^2 V}{\partial Z^2} + F(V) \tag{10}$$

with some functions F,G depending on f in (6).
First, we define by

$$Z(x) = \int_0^x r(s) \cdot g(s)^{1/2}\, ds \tag{11}$$

a transformation of the space variable which allows to express (7) in terms of dimensionless distance Z :

$$\tau_1 \frac{\partial^2 v}{\partial t^2} + (1 + \frac{\tau_1}{\tau_2}\frac{\partial f}{\partial v})\frac{\partial v}{\partial t} = \frac{1}{\tau_2}\frac{\partial^2 v}{\partial Z^2} + q\frac{\partial v}{\partial Z} - \frac{f(V)}{\tau_2} \tag{12}$$

where

$$q(Z) = \frac{g'(Z)r(Z) - g(Z)r'(Z)}{2r(Z)c(Z)} = \frac{3}{2}\{\ln d(Z)\}', \tag{13}$$

and the prime denotes $\partial/\partial Z$.
In the following we consider the case of distortionless AP conduction along the axon for which $\tau_1 = \tau_2 = \tau$ holds. Equation (12) then reads in dimensionless form

$$\frac{\partial^2 v}{\partial T^2} + (1 + \frac{\partial f}{\partial v})\frac{\partial v}{\partial T} = \frac{\partial^2 v}{\partial Z^2} + q\frac{\partial v}{\partial Z} - f(v) \tag{14}$$

where $T = t/\tau$.
Equation (14) can be simplified by changing the variable v according to

$$v(Z,T) = \exp(-\frac{1}{2}\int q\, dZ)\, V(Z,T). \tag{15}$$

The new variable V is the solution of the equation

$$\frac{\partial V}{\partial T^2} + (1 + G(V))\frac{\partial V}{\partial T} = \frac{\partial^2 V}{\partial Z^2} + F(V) \tag{16}$$

where

$$G(V) = \frac{\partial f}{\partial V}, \quad F(V) = -pV - f(V) \tag{17}$$

and

$$p = q^2/4 + q'/2. \tag{18}$$

Obviously, equation (16) is of the same type as the equation (1) studied in /5,19/. Therefore, Hadeler's results /19/ on existence and speed of traveling fronts hold for (16). If the inverse transformations to (11), (14) and (15) are carried out, their implications for the original problem (7) can be revealed. This will be presented in a forthcoming paper.

4. PIECEWISE LINEAR APPROXIMATION OF CURRENT-VOLTAGE CHARACTERISTIC

In deriving (16) from (14) via (15), the assumption of (at least) piecewise linearity of the function f is necessary. So we choose

$$f(V) = V(1 - H(V - V_{th})) \tag{19}$$

where H is the Heaviside step function and $V_{th} = V_{th}(Z)$ is the threshold. Equation (20) has been used to approximate the cubic nonlinearity of the FitzHugh-Nagumo equations /2,3,17/. If we consider the case of no AP recovery, then the membrane will exhibit bistable behaviour,i.e. at V_{th} it switches from one functional state into the other.
Using (19), equation (16) can be written as

$$\frac{\partial^2 V}{\partial T^2} + (1 + (G(V))\frac{\partial V}{\partial T} = \frac{\partial^2 V}{\partial Z^2} - I(Z)\cdot V \tag{20}$$

where $I(Z)$ is called the *invariant* of the *normal form* (20) to (12). $I(Z)$ is related to $q(Z)$ in (13) - and so to the axon diameter $d(Z)$ - through the equation

$$I(Z) = p + H(V_{th} - V), \tag{21}$$

with p defined as in (18).

5. DIAMETER FUNCTIONS FOR AXONS

Various classes of nonuniform axon geometries are obtained by choosing the invariant $I(Z)$ in (20) and solving for the axon diameter $d(Z)$ via (13). The simplest class to consider are those axons for which p in (21) is a constant (cf. /20/ for discussion). This class can be determined by solving the *special Riccati equation*

$$2q' = 4p - q^2 \tag{22}$$

obtained from (18).
Equation (22) is integrable by separation of variables, and its solutions can be expressed by elementary functions. Depending on the constant p , six different solutions of (22) are found. Table 1 displays these solutions and the corresponding diameter variations as derived from (13).

TABLE 1: Axon geometries as defined by the solutions of the *special Riccati equation* $2\ q' = 4\ p - q^2$, $q = 3/2\ \{\ln\ d(Z)\}'$.

Geometry	p	$q(Z)$	**diameter** $d(Z)$
uniform	$p = 0$, $q = 0$	0	d_0
power	$p = 0$, $q \neq 0$	$2/(Z - C)$	$d_0\,(1 - Z/C)^{4/3}$
exponential	$p > 0$ $q_1^2 = 4p$	q_1	$d_0\ \exp(2/3\,q_1 Z)$
hyperbolic sine	$p > 0$ $\lvert q\rvert > q_1$	$q_1\ \coth[q_1(Z - C)/2]$	$d_0\,\{\sinh[q_1(Z - C)/2]\}^{4/3}$
hyperbolic cosine	$p > 0$ $\lvert q\rvert < q_1$	$q_1\ \tanh[q_1(Z - C)/2]$	$d_0\,\{\cosh[q_1(Z - C)/2\}^{4/3}$
trigonometric cosine	$p < 0$	$-\lvert q_1\rvert\ \tan[\lvert q_1\rvert(Z - C/2]$	$d_0\,\{\cos[\lvert q_1\rvert(Z - C)/2]\}^{4/3}$ $\lvert q_1\rvert(Z - C)/2 \neq \pm(2n+1)\pi$ $n \in \mathcal{N}$

REFERENCES

1. A. C. SCOTT, The electrophysics of a nerve fiber. Rev. Mod. Phys. 47(1975)487-533.

2. R. FITZHUGH, Mathematical models of excitation and propagation in nerve. In: Biological Engineering (H. P. Schwan, Ed.) New York, MacGraw-Hill 1969, pp.1-85.

3. V. S. ZYKOV, Simulation of Wave Processes in Excitable Media. Manchester, Manchester University Press 1987.

4. J. RINZEL, Simple model equations for active nerve conduction and passive neuronal integration. Lect. Math. Life Sci. 8(1976)125-164.

5. S. DUNBAR and H. OTHMER, On a nonlinear hyperbolic equation describing transmission lines, cell movements, and branching random walks. In: Nonlinear Oscillations in Biology and Chemistry (H. G. Othmer, Ed.). Berlin, Springer-Verlag 1986, pp. 274-289.

6. H. M. LIEBERSTEIN, On the Hodgkin-Huxley partial differential equation. Math. Biosci. 1(1967)45-69.

7. J. ENGELBRECHT, On the theory of pulse transmission in a nerve fibre. Proc. Roy. Soc. Lond. A 375(1981)195-209.

8. L. GOLDMAN and J. S. ALBUS, Computation of impulse conduction in myelinated fibres : theoretical basis of the velocity-diameter relation. Biophys. J. 8(1968) 596-607.

9. P. J. HUNTER, P. A. MCNAUGHTON and D. Noble, Analytic models of propagation in excitable cells. Progr. Biophys. Molec. Biol. 30(1975)99-144.

10. S. G. WAXMAN (Ed.): Physiology and Pathobiology of Axons. New York, Raven Press 1978.

11. F. A. DODGE and J. W. COOLEY, Action potential of the motoneuron. IBM J. Res. Devel. 17(1973)219-229.

12. S. S. GOLDSTEIN and W. RALL, Changes of action potential shape and velocity for changing core conductor geometry. Biophys. J. 14(1974)731-757.

13. B. I. KHODOROV and E. N. TIMIN, Nerve impulse propagation along nonuniform fibres (investigations using mathematical models). Prog. Biophys. Molec. Biol. 30(1975)145-184.

14. I. PARNAS, S. HOCHSTEIN and H. PARNAS, Theoretical analysis of parameters leading to frequency modulation along an inhomogeneous axon. J. Neurophysiol. 39(1976) 909-922.

15. A. K. SCHIERWAGEN, Travelling wave solutions of a simple nerve conduction equation for inhomogeneous axons. In: Nonlinear Wave Processes in Excitable Media (A. V. Holden, M. Markus, H. Othmer, Eds.). Manchester, Manchester University Press 1991, pp. 107-114.

16. J. NAGUMO, S. ARIMOTO and S. YOSHIZAWA, An active pulse transmission line simulating nerve axon. Proc. IRE 50(1962)2061-2070.

17. J. P. PAWELUSSEN, Heteroclinic waves of the FitzHugh- Nagumo equations. Preprint, Amsterdam 1980.

18. J. RINZEL, and J. B. KELLER, Traveling wave solutions of a nerve conduction equation. Biophys. J. 13(1973) 1313-1337.

19. K. P. HADELER, Hyperbolic travelling fronts. Proc. Edinburgh Math. Soc. 31(1988)89-97.

20. A. K. SCHIERWAGEN, A non-uniform equivalent cable model of membrane voltage changes in a passive dendritic tree. J. theor. Biol. 141(1989)159-179.

Andreas K. Schierwagen
Fachbereich Informatik der Universität Leipzig
Augustusplatz 10/11

D- O - 7010 Leipzig
Germany

J SUVOROVA

Numerical analysis for two-dimensional problem of shock wave dynamics for hereditary-type media

Abstract

The objective of this work is the analysis of two-dimensional wave propagation processes in the isotropic media under impact loading, the media is assumed to possess viscous properties and assumed dispersely damaged. The constitutive integral equation is a non-linear one.

1. INTRODUCTION

In the modern structures' analysis the evaluation of non-linear effects in the loading of materials is of great importance. It is especially important for the problems of impact loading since the loads are usually rather high.

To evaluate the structures' resourse and their behaviour under the various conditions of loading a thorough analysis of the sourses and the degree of non-linearity is necessary. For instance, substantial viscous effects, bulk damage accumulation, local failure which may lead to macrocrac initiation and other effects and phenomena should be taken into consideration.

This work consists of three parts: the first one deals with the non-linear hereditary-type model formulation; the second one concisely describes the method of calculation, whereas the third part presents some numerical results for a certain problem of impact loading of a plate by a striker.

2. NON-LINEAR CONSTITUTIVE EQUATION FOR A MEDIUM

For one-dimensional stress-state instead of the conventional linear hereditary-type equation:

$$E\,\epsilon = \sigma + \int_0^t K\,(t-\tau)\,\sigma\,(\tau)\,d\tau$$

where $K(t-\tau)$ is a Kernel of integral equation, the following equation should be taken

$$\phi(\epsilon) = \sigma + \int_0^t K\,(t-\tau)\,\sigma\,(\tau)\,d\tau$$

where $\phi(\epsilon)$ is the instantaneous stress-strain curve which actually is an upperbound on the stress-strain diagrams. At present, there is a sufficient amount of papers showing that such an approach is quite valid.

For three-dimensional stress state the hereditary-type model formulation is extremely difficult. An approach of Volterra double integral series presentation could be used but it could hardly be utilized in practice. Here, a model of the type of the one-dimensional Rabotnov equation is proposed, the hereditary effects are presumed to be not very pronounced.

The formulation of linear model is based on the generalization of the Hooke's law by virtue of involving operators instead of elasticity module. This approach provides the following relationship between stress and strain intensities

$$E\,\epsilon_i = (1 + K^*)\,\sigma_i \tag{1}$$

which is obviously analogous to the usual one-dimensional case.

Here, the relationship between components of stress and strain tensors is expressed as

$$E\,(\epsilon_{ij} - \epsilon\,\delta_{ij}) = (1 + K^*)\,(\sigma_{ij} - \sigma\,\delta_{ij}) \tag{2}$$

where σ and ϵ are the mean stress and strain respectively whereas an operator in eq. (2) is the same as in eq. (1).

Such an approach allows generalization of the eq. (1), (2) (model) for the non-linear case, similar to the Rabotnov approach for the one-dimensional equation.

The relationship within stress and strain intensities may be written as

$$\phi(\epsilon_i) = (1 + K^*)\,\sigma_i \tag{3}$$

whereas the relationship between the components of stress and strain are the following:

$$(\epsilon_{ij} - \epsilon\,\delta_{ij})\,\frac{2\,\phi(\epsilon_i)}{3\,\epsilon_i} = (1 + K^*)\,(\sigma_{ij} - \sigma\,\delta_{ij}) \tag{4}$$

The eqs. (3) and (4) are apparently analogous to the generalization of non-linear theory of infinitesimal elastoplastic deformations by means of operators presentation. On the other hand, the whole procedure is fully correspondent to that utilized by Rabotnov for non-linear hereditary media for the case of one-dimensional problem.

This approach allows to proceede to the analysis of practically one-dimensional model instead of multidimensional, that is the model in terms of intensities, namely the eq. (3), as well as to obtain the relantionship between the components of stress and strain tensors (eq. (4)).

Similar to the one-dimensional case [1] (see also references there quoted), the kernel K^* in eq. (3) can be replaced by two operators, that is reversible creep $\mathcal{L}^*$ and damage accumulation M^*:

$$\phi(\epsilon_i) = \sigma_i + \mathcal{L}^* \sigma_i + M^* \sigma_i \tag{5}$$

with further failer criterion formulation as follows:

$$\sigma_i + M^* \sigma_i = \sigma_o{}^* \tag{6}$$

The validity of such a presentation has been proved experimentally for some types of polymer materials.

3. THE DESCRIPTION OF NUMERICAL ANALYSIS

For convenience, the constitutive system of equations may be written as follows (the relaxation in the bulk volume is assumed absent):

$$\begin{aligned} \dot{\sigma}_{ij} &= \lambda\, \dot{\epsilon}_{kk}\, \delta_{ij} + 2\,\mu\, \dot{\epsilon}_{ij} - (\mathcal{L}^* + M^*)\, \dot{S}_{ij} \\ \mathcal{L}^* S_{ij} &= \int_0^t \mathcal{L}(t - \tau)\, S_{ij}\, d\tau \\ M^* S_{ij} &= \int_0^t M(t - \tau)\, S_{ij}\, d\tau \end{aligned} \tag{7}$$

$$\rho \dot{v}^i = \nabla_j \sigma^{ij}$$

$$\lambda = \lambda(\epsilon_i) \quad , \qquad \mu = \mu(\epsilon_i) \tag{8}$$

where S_{ij} are the components of stress deviator.

Obviously, the $\lambda(\epsilon_i)$ and $\mu(\epsilon_i)$ values allows to pass to the relationship $E(\epsilon_i) = \frac{\phi(\epsilon_i)}{\epsilon_i}$, if one of the two assumptions is adopted:

$$K(\epsilon_i) = \text{const} \qquad , \qquad \nu(\epsilon_i) = \text{const} \tag{9}$$

For the numerical analysis of the wave propagation processes a routine of splitting by physical parameters was used. The first step was the presentation of the apparent net-characteristic scheme, which had been developed for numerical analysis of the deformable solid mechanichs dynamics.

The second step of the scheme:

$$\frac{\sigma_{ij}^{n+1} - \sigma_{ij}^{n}}{\tau} = F_{ij}^{n+1} (S_{ij}^{n+1}, S_{ij}^{n}, \dots, S_{ij}^{n-N}) \tag{10}$$

where N is the number of layers in time taken into account in the hereditary integral. The integral to be approximated in the following way, for example, the Abel-type kernel is taken:

$$\mathcal{L}(t - \tau) = \frac{1}{(t - \tau)^{\alpha}} \quad ; \qquad M(t - \tau) = \frac{m}{(t - \tau)^{\alpha}}$$

$$F_{ij}^{n+1} = \frac{1 + m}{\tau} \left(\frac{S_{ij}^{n+1}}{1 - \alpha} + \sum_{k=n-N+1}^{n} \frac{S_{ij}^{k} - S_{ij}^{k-1}}{(n - k + 1)^{\alpha}} \right)$$

The first number in brackets corresponds to the approximatiion of the integral on the interval $[t-\tau, t]$, the second one - on the residual interval of the integration $[t-N\tau, t-\tau]$. Both of the steps must be used since there are two processes with substantially different characteristic times, that is acoustic and relaxation processes.

4. SOME NUMERICAL RESULTS

Here the problem of disc-shaped striker impact on an infinite plate of a visco-elastic material is under consideration (Fig.1).

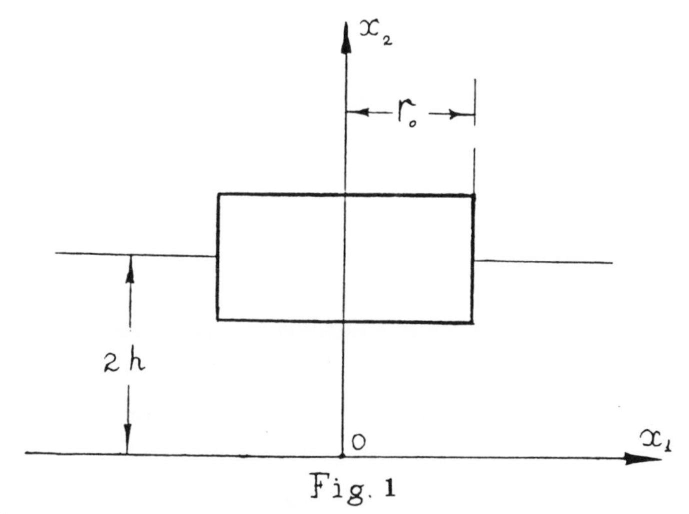

Fig. 1

The computation has been carried out for three types of kernels: a constant, an exponential and weakly singular Abel-type kernels.

The figure shows the results obtained for the Abel-type kernel, which is the most difficult case from the computational point of view.

One part of analysis has been carried out for the case of undamaged material, M(t)=0, the other part - when damage does have place under loading, M(t)≠0, but is neglected under unloading. Vanishing of damage (that is the "healing" of the material) could also be accounted for if the M* operator where taken decreasing in time under unloading but this problem was beyond the scope of our research.

The non-linearity was account for by virtue of approximation of the istantaneous stress-strain curve. Three ways of approximation have been considered:

1) $$\phi(\epsilon_i) = E\,\epsilon_i$$

2) $$\phi(\epsilon_i) = \begin{cases} E_1\,\epsilon_i & \epsilon_i \le \epsilon_{i0} \\ E_1\,\epsilon_{i0} + E_2\,(\epsilon_i - \epsilon_{i0}) & \epsilon_i > \epsilon_{i0} \end{cases}$$

3) $$\phi(\epsilon_i) = \begin{cases} E\,\epsilon_i & \epsilon_i \le \epsilon_{i0} \\ \left[\gamma\,\epsilon_i + (1-\gamma)\left(2\,\epsilon_{i0} - \frac{\epsilon_{i0}^2}{\epsilon_i}\right)\right] E & \epsilon_i > \epsilon_{i0} \end{cases}$$

where γ is the non-linearity parameter.

The non-disturbance state was taken for the initial conditions, there is

$$v_1 = v_2 = 0 \quad ; \qquad \sigma_{11} = \sigma_{22} = \sigma_{33} = \sigma_{12} = 0 \ .$$

The boundary conditions were taken as follows:

$$\sigma_{22} = \begin{cases} P & t \leq t_0 \\ 0 & t > t_0 \end{cases}$$

when $r > r_0$, $\sigma_n = \sigma_\tau = 0$, σ_n - being the normal, σ_τ - the tangential stresses. The back surface of the plate was assumed free of the stress or rigidly fixed. With the increasing of viscous properties, as it was seen elsewhere, the stresses σ_{11} also increase, while the stresses σ_{22} are tend to decrease according to the equation. Thus we might say that with th increasing of viscous properties of the yelding of the media becomes more "hydrodynamics".

The non-linearity appears to decrease the velocity and stress values, naturally enough; the fact of the damage accumulation (the operator M^*), gives after the unloading some level of the residual deformation.

The presentation of damage in the constitutive equation of a medium appears important not only to provide evaluation of residual strains, but it also facilitates the failure criterion formulation, which, in turn, allows to evaluate the rupture strength and to analyse the problem of dynamic failure of hereditary elastic materials in general.

Fig. 2 presents the lines of equal values of stress σ_{22}. The figures correspond to the moment when the impulse of compression, being reflected by the free surface and changed for the impulse of tension started propagation in the opposite direction. One can identify the localization of maximum stresses which may lead to the rupture near the back surface. If the corresponding failure criterion is orded to the above analysis the site and the moment of failure can be obtained.

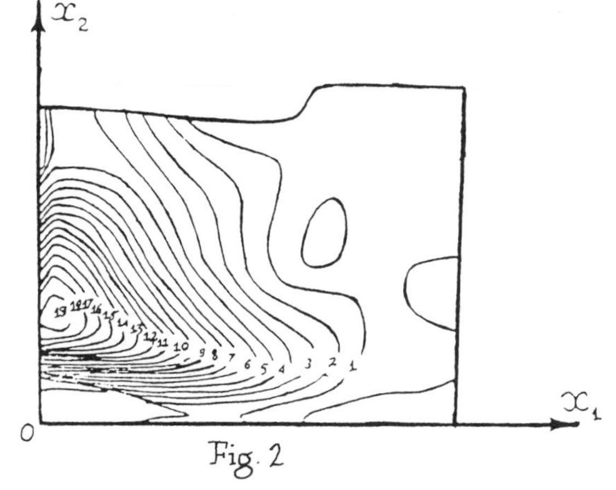

Fig. 2

REFERENCES

1. J. SUVOROVA, *The influences of time and temperature on the reinforced plastic strenght.* In book: *Failure mechanics of composites*, North-Holland 1985, V 3 177-214.

Julia Suvorova
Mechanical Engineering
Research Institute
101830, MOSCOW
USSR

Part III

Hyperbolic Dissipative Models from Kinetic Theory and Extended Thermodynamics

A M ANILE AND M SAMMARTINO

Moment method and entropy principle in radiative transfer

Abstract

We discuss second order closures for the moment equations of the radiative transfer equation. We bring to light the correlation between the existence of an entropy principle and the existence of a reference frame in which the radiation field has special symmetry properties. Imposing that the multipole series expansion of the radiation intensity stops at second order we derive a new fourth order closure.

1 Introduction

During last decades there has been an increasing interest in the field of linear transport phenomena. The wide class of physical areas in which these phenomena occurr includes neutron transport theory, plasma physics, astrophysics and cosmology.

The physical model described by linear transport theory is a two component fluid in which the first one (called material fluid), is composed of massive particles, the second one of non interacting fast moving particles, which can be absorbed, emitted and scattered by the particles of the first medium.

In the following we shall be mainly interested in radiative transfer theory, the material fluid being then composed mainly of atoms or ions of hydrogen and helium. The particles propagating through this fluid will be photons or neutrinos.

The radiative field is described by the radiation intensity $I(\mathbf{r}, t, \boldsymbol{\Omega}, \nu)$, which gives, at time t and in the spatial position $\mathbf{r}$,the energy of photons, of frequency ν, traveling along the direction determined by the unit vector $\boldsymbol{\Omega}$.The evolution equation for I_ν is (see [1][2])

$$\frac{\partial I_\nu}{\partial t} + \boldsymbol{\Omega} \cdot \nabla I_\nu + \sigma I_\nu = (\sigma_a B_\nu + \sigma_s E_\nu)/(4\pi) \tag{1}$$

where σ_a and σ_s are, respectively, the absorption and scattering coefficient, giving the rate at which photons are removed or scattered because of the presence of matter, and $\sigma = \sigma_a + \sigma_s$ is the total interaction coefficient; B_ν is the blackbody energy density relative to frequency ν , while E_ν is the radiative energy density given by

$$E_\nu(\mathbf{r},t) = \int_{4\pi} I_\nu(\mathbf{r},t,\mathbf{\Omega})d\Omega \tag{2}$$

where the integral is performed over the unit sphere. In eq.(1) we have supposed that scattering is coherent and isotropic, so that no integral term appears. Eqs.(1) and (2) can be assumed to hold for frequency integrated quantities (with the subscript ν omitted); in this case, σ_a and σ_s must be considered as average (over frequencies) absorption and scattering coefficients.

Although (1) is the basis of the radiative transfer theory, its applicability to concrete situations is very restricted. Because of the great number of variables (time, space and angles), it is almost prohibitive, in terms of numerical cost, to solve numerically the transfer equation,except in cases in which the geometry of the problem is highly symmetric; on the other hand, theoretical enquiries (e.g. on shock wave propagation or turbulence) are better performed using fluid-like variables.

The moment method (see [3]) has revealed itself to be a powerful tool in dealing with transport phenomena. Taking the moments of the transfer equations one obtains, for the moments of the distribution function, a set of coupled equations in which the angular dependence has been suppressed.

The main difficulty affecting this method is the presence, in the mth order moment equation of the $(m+1)th$ moment of the distribution function. One needs a way to close the system, at some order, by postulating a relationship linking the $(m+1)th$ moment to the lower ones.

Usually, in radiative transfer theory, one postulates a closure at the second order, i.e. one writes a relation linking the radiative pressure to the energy flux and energy density. The most usual of such closure is the Eddington approximation

$$P^{ij} = \frac{1}{3}E\delta^{ij} \tag{3}$$

but, as shown in [4], this closure does not describe radiative stresses (even to first order in the photon mean free path). Therefore (3) is inadequate in all situations in which there is even a slight deviation from thermodynamical equilibrium.

The scope of this paper is to discuss briefly some second order closures which have been recently investigated (see [5],[6] and [7]), and to suggest a complitely new fourth order closure. The leading idea of this paper will be that of the existence of a preferred reference frame in which some symmetry property holds for radiation, thus generalizing the hypothesis leading to Levermore's closure [8].

The plan of the paper is as follows. In the second paragraph we shall state some transformation rules which will be useful later. In the third we shall rediscover Levermore's closure, using a phenomenological approach [8], and discuss it in the light of the entropy principle [5]. In the fourth we shall find a fourth order closure that we believe will be very useful in all situations in which the radiation field has a strong degree of anisotropy.

2 Transformation Formulas

Let (k^μ) be the four momentum of a photon; if we have two different observers, whose four-velocities are n^μ and n_0^μ, we can write

$$\begin{aligned} n_0^\mu &= \gamma(1, \mathbf{v}) \\ k^\mu &= \nu(1, \boldsymbol{\Omega}) = \nu_0(1, \boldsymbol{\Omega}_0), \end{aligned} \tag{4}$$

where $\mathbf{v}$ is the 3-velocity of the observer n_0^μ with respect to n^μ. The frequencies ν and ν_0 are given by

$$\nu = -n^\mu k_\mu \ , \quad \nu_0 = -n_0^\mu k_\mu,$$

the "boost" factor is

$$\gamma = (1 - v^2)^{-\frac{1}{2}},$$

and the vectors $\boldsymbol{\Omega}$ and $\boldsymbol{\Omega}_0$, giving the directions of flight of photons in the two systems are linked by the aberration relation

$$\boldsymbol{\Omega}_0 = \frac{\nu}{\nu_0}\left[\boldsymbol{\Omega} + \left(\frac{\gamma - 1}{v^2}\boldsymbol{\Omega}\cdot\mathbf{v} - \gamma\right)\mathbf{v}\right]. \tag{5}$$

The two frequencies ν and ν_0 are related by the Doppler formula

$$\nu = \nu_0\gamma(1 + \mathbf{v}\cdot\boldsymbol{\Omega}_0) \tag{6}$$

which also gives

$$\frac{d\nu}{d\nu_0} = \gamma(1 + \mathbf{v}\cdot\boldsymbol{\Omega}_0). \tag{7}$$

Using the fact that the invariant volume element in phase space dP can be written in the form (see [9])

$$dP = \nu d\nu d\Omega = \nu_0 d\nu_0 d\Omega_0 \ , \tag{8}$$

one can write, also using (6) and (7),

$$d\Omega = [\gamma(1 + \mathbf{v} \cdot \mathbf{\Omega_0})]^{-2} d\Omega_0. \tag{9}$$

Let us now introduce the first three moments of the integrated radiation intensity I_ν, evaluated in the reference frame of the observer n^μ

$$\begin{aligned} E &= \int\int I_\nu d\nu d\Omega = \int_0^\infty E_\nu d\nu \\ F^i &= \int\int I_\nu \Omega^i d\nu d\Omega = \int_0^\infty F^i_\nu d\nu \\ P^{ij} &= \int\int I_\nu \Omega^i \Omega^j d\nu d\Omega = \int_0^\infty P^{ij}_\nu d\nu. \end{aligned} \tag{10}$$

In the reference frame of n_0^μ analogous formulas hold for the moment of I_{ν_o}, namely E_0, $\mathbf{F}_0$, $\mathbf{P}_0$. Using the fact that E, $\mathbf{F}$ and $\mathbf{P}$ are the components of the stress-energy tensor $T^{\mu\nu}$, one can find the transformation laws [10]

$$\begin{aligned} E_0 &= \gamma^2(E - 2\mathbf{v}\cdot\mathbf{F} + \mathbf{v}\mathbf{v}\cdot\mathbf{P}) \\ F_0^i &= \gamma\{F^i + [(\gamma + \tfrac{\gamma-1}{v^2})\mathbf{v}\cdot\mathbf{F} - \gamma E - \tfrac{\gamma-1}{v^2}\mathbf{v}\mathbf{v}\cdot\mathbf{P}]v^i - P^{ij}v_j\} \\ P_0^{ij} &= P^{ij} + \tfrac{\gamma-1}{v^2}(P^{kj}v^i + P^{ik}v^j)v_k + (\tfrac{\gamma-1}{v^2})^2\mathbf{v}\mathbf{v}\cdot\mathbf{P}v^iv^j - \\ &-\gamma(v^iF^j + v^jF^i) - 2\gamma(\tfrac{\gamma-1}{v^2})v^jv^i\mathbf{v}\cdot\mathbf{F} + \gamma^2v^jv^iE. \end{aligned} \tag{11}$$

Finally we write the transformation rule for the integrated radiation intensities

$$I = \int_0^\infty I_\nu d\nu \ , \ I_0 = \int_0^\infty I_{\nu_o} d\nu_0.$$

Using the relation

$$f = \frac{I_{\nu_o}}{\nu_o^3} = \frac{I_\nu}{\nu^3},$$

where f is the invariant (for an exaustive discussion about approximation involved in this assertion, see [11]) photon distribution function, one can find

$$I = \left(\frac{\nu}{\nu_0}\right)^4 I_0 = \gamma^{-4}(1 - \mathbf{v}\cdot\mathbf{\Omega})^{-4} I_0. \tag{12}$$

3 Levermore's Closure and Entropy Principle

Let us now suppose that the observer, whose four-velocity is n_μ, detects an isotropic radiation intensity. We stress that this hypothesis does not necessarily imply thermodynamical equilibrium, for which it is also necessary that particle velocity distribution is a Maxwellian whose temperature is equal to the radiation temperature.

If I is isotropic then we have $F^i = 0$ and $P^{ij} = (1/3)E\delta^{ij}$. Using these relations in eqs.(11a,b) we have

$$E_0 = \frac{1}{3}E\frac{3+v^2}{1-v^2} \ , \ f_0^i = \frac{4}{3+v^2}v^i, \tag{13}$$

where we have defined the normalized flux $\mathbf{f}_0 = \mathbf{F}_0/E_0$.

Relations (13) are simple enough to be inverted, in order to have the relative velocity $\mathbf{v}$ as function of the first two moments of the radiation intensity in the "anisotropic" reference frame. We have

$$v^2 = \left(8 - 3f_0^2 - 4\sqrt{4 - 3f_0^2}\right) / f_0^2 \tag{14}$$

which, inserted in (13*b*) gives the relation between $\mathbf{v}$ and $\mathbf{f}_0$. Using eqs.(13) and (14), we obtain, from (11 *c*) the relation

$$P_0^{ij} = \left(\frac{1 - f_0^2}{1 + \sqrt{4 - 3f_0^2}}\delta^{ij} + \frac{3}{2 + \sqrt{4 - 3f_0^2}} f_0^i f_0^j\right) E_0. \tag{15}$$

Equation (15) is a second order closure for the moment equations. It was found by Levermore [8] using our same hypothesis but a slightly different procedure.

The hypothesis which the closure (15) is based upon, seems to be ad hoc and artificial. But a strong improvement of (15) has been found in [5], where the authors suggested a fundamental approach to second order closures in radiative hydrodynamics. Writing the system of moments, truncated at first level

$$\begin{cases} \partial_t E_0 + \nabla \cdot \mathbf{F}_0 = \sigma_a (B - E_0) \\ \partial_t \mathbf{F}_0 + \nabla \cdot \mathbf{P}_0 - \sigma \mathbf{F}_0 = 0 \end{cases} \tag{16}$$

and imposing, on mathematical and physical ground, the existence of an entropy density S and an entropy flux Q^i (see [12][13] and [14]), so that along the solutions of the system (16) the balance law holds

$$\partial_t S + \nabla \cdot \mathbf{Q} = g, \tag{17}$$

where g is a function of E_0 and $\mathbf{F}_0$ only, giving the rate of entropy production, they found (15) as the only closure compatible with the entropy principle.

The fact that Levermore's closure can be derived with hypothesis which to us seem to be quite reasonable and consistent, not only gives new dignity to (15), but also suggests that there is a deep connection between imposing that radiation can be consistently described thermodynamically, and the existence of a preferred reference frame. The fact that moment equations are truncated at some level, which is equivalent to ignoring kinetic details of radiative fields, has been usually interpreted as closeness to Thermodynamical Equilibrium; we believe that it should mean also that in some reference frame for radiation special symmetry properties hold.

In the next paragraph we shall exploit this conjecture to obtain a fourth order closure.

4 A Fourth Order Closure

The quasilinear system (16), together with the closure (15), has been shown in [5] to be symmetric hyperbolic (then the relative Cauchy problem is well posed). Although mathematically satisfactory, it is, in many respects, inadequate in describing phenomena which are common enough in radiative transfer theory.

If one tries to get from (16) a diffusion equation (i.e. a relation linking radiative flux to energy gradient), one obtains, as shown in [8], the flux-limited diffusion equation as found in [15]. The fundamental limit of this theory is that it predicts an energy flux parallel to energy gradient, while it has been experimentally observed the occurence of an "inertia" of flux to follow the direction of gradient.

The fact that (15) gives rise to a one-dimensional (thus unsatisfactory) diffusion theory, seems to be connected with another limit inherent to second order closures. To suppose that pressure tensor can be expressed in terms of energy flux and density is equivalent to suppose that the radiation intensity is symmetric with respect to some direction. Although this hypothesis is verified in a great number of phenomena of interest for radiative transfer theory, there are many other situations in which it is too restrictive, such as stellar binary systems or anisotropic cosmologies or some laboratory plasmas.

In our opinion the only way to obtain an essentially non one-dimensional theory is to extract from the radiation intensity more informations than those contained in the first two moments, seeeking for higher order closures. Let us generalize Levermore's hypothesis and suppose that in the reference frame n^μ the moments of integrated radiation intensity whose order is greater than two, contain no trace-freee part, i.e. that the expansion of radiation intensity in terms of its symmetric trace-free (STF) spherical harmonics [3], stops at second order

$$I = \frac{1}{4\pi}(E + 3F^i\Omega_i + \frac{15}{2}P^{<ij>}\Omega_i\Omega_j), \tag{18}$$

where brackets enclosing indices of a tensor mean its STF part. Now we exploit (18) to find a closure in the n_0^μ reference frame.

Let us write $\mathbf{A}_0$, the third order moment of I_0, and use (12) and (5)-(9)

$$\mathbf{A}_0 = \int I_0\boldsymbol{\Omega}_0\boldsymbol{\Omega}_0\boldsymbol{\Omega}_0 d\boldsymbol{\Omega}_0 =$$
$$= \int I\left(\tfrac{\nu_0}{\nu}\right)^4\left(\tfrac{\nu}{\nu_0}\right)^3\left[\boldsymbol{\Omega} + \left(\tfrac{\gamma-1}{v^2}\mathbf{v}\cdot\boldsymbol{\Omega} - \gamma\right)\mathbf{v}\right]^3 \tfrac{\nu}{\nu_0}\tfrac{d\nu}{d\nu_0}d\Omega.$$

Now, to make the algebra more tractable, we suppose that $v \ll 1$ and neglect terms which are non-linear in v. Then

$$A_0^{ijk} = \int I(\Omega^i\Omega^j\Omega^k - \Omega^i\Omega^j v^k - \Omega^i\Omega^k v^j - \Omega^j\Omega^k v^i + \Omega^i\Omega^j\Omega^k\Omega^m v_m)d\Omega.$$

Inserting in the above relation the expression (18) for I, we obtain

$$A_0^{ijk} = -\frac{12}{15}Ev^{(i}\delta^{jk)} + \frac{3}{5}F^{(i}\delta^{jk)} - \frac{18}{7}P^{(<ij>}v^{k)} + \frac{3}{7}P^{<m(k>}v_m\delta^{ij)}. \qquad (19)$$

Let us consider relations (11), taking only linear terms in v; with some manipulations they give

$$\begin{aligned} E &= E_0 + 2F_0^i v_i \\ F^i &= F_0^i + E_0 v^i + P_0^{ij} v_j = F_0^i + \tfrac{4}{3}E_0 v^i + P_0^{<ij>} v_j \\ P^{ij} &= P_0^{ij} + (v^i F^j + v^j F^i) = P_0^{ij} + (v^i F_0^j + v^j F_0^i). \end{aligned} \qquad (20)$$

Inserting (20) in (19), we finally obtain for A_0^{ijk}, the expression

$$A_0^{ijk} = \frac{3}{5}F_0^{(i}\delta^{jk)} - \frac{18}{7}P_0^{(<ij>}v^{k)} - \frac{36}{35}P_0^{<m(k>}v_m\delta^{ij)}. \qquad (21)$$

Now, as we did in Sec.3, where we found $\mathbf{v}$ as a function of E_0 and $\mathbf{F}_0$, we want to invert (21) to obtain $\mathbf{v}$ as a function of E_0, $\mathbf{F}_0$, $\mathbf{P}_0$ and $\mathbf{A}_0$. First of all we find the values of the four scalar quantities a,b,c,d, namely

$$a = \mathbf{v}\cdot\mathbf{F}_0\ ,\ b = P_0^{<ij>} v_i F_{0j}\ ,\ c = P_0^{<ij>} v_i H_j\ ,\ d = \mathbf{v}\cdot\mathbf{H} \qquad (22)$$

where $H^k = P_0^{<kj>} F_{0j}$.

Successively contracting (21) with $\mathbf{F}_0\mathbf{F}_0\mathbf{H}$, $\mathbf{F}_0\mathbf{H}\mathbf{H}$, $\mathbf{P}_0^{TF}\mathbf{F}_0$, $\mathbf{F}_0\mathbf{F}_0\mathbf{F}_0$, we obtain a system of four equations for the unknowns a,b,c,d

$$\left\{\begin{array}{l} -\frac{12}{7}\mathbf{P}_0^{TF}\cdot\mathbf{F}_0\mathbf{H}a + \frac{24}{35}\mathbf{F}_0\cdot\mathbf{H}b + \frac{12}{35}F_0^2 c - \frac{6}{7}\mathbf{P}_0^{TF}\cdot\mathbf{F}_0\mathbf{F}_0 d = \\ = \mathbf{A}_0\cdot\mathbf{F}_0\mathbf{F}_0\mathbf{H} - \frac{3}{5}\mathbf{F}_0\cdot\mathbf{H}F_0^2 \\ -\frac{6}{7}\mathbf{P}_0^{TF}\cdot\mathbf{H}\mathbf{H}a + \frac{12}{35}H^2 b + \frac{24}{35}\mathbf{F}_0\cdot\mathbf{H}c - \frac{12}{7}\mathbf{P}_0^{TF}\cdot\mathbf{F}_0\mathbf{H}d = \\ = \mathbf{A}_0\cdot\mathbf{F}_0\mathbf{H}\mathbf{H} - \frac{1}{5}\left[2\left(\mathbf{F}_0\cdot\mathbf{H}\right)^2 + F_0^2 H^2\right] \\ -\frac{6}{7}\mathbf{P}_0^{TF}\cdot\mathbf{P}_0^{TF}a - \frac{36}{35}c = \mathbf{A}_0\cdot\mathbf{P}_0^{TF}\mathbf{F}_0 - \frac{2}{5}\mathbf{P}_0^{TF}\cdot\mathbf{F}_0\mathbf{F}_0 \\ -\frac{18}{7}\mathbf{P}_0^{TF}\cdot\mathbf{F}_0\mathbf{F}_0 a - \frac{36}{35}F_0^2 b = \mathbf{A}_0\cdot\mathbf{F}_0\mathbf{F}_0\mathbf{F}_0 - \frac{3}{5}F_0^4\ . \end{array}\right. \qquad (23)$$

Once easily solved the system (23) in terms of a,b,c and d, we are ready to invert (21). Successively contracting it with $\mathbf{F}_0\mathbf{H}$ and $\mathbf{F}_0\mathbf{F}_0$, we obtain two vector equations for the two vector unknowns $\mathbf{v}$ and $\mathbf{P}_0^{TF}\cdot\mathbf{v}$, i.e.

$$\left\{\begin{array}{l} -\frac{6}{7}(\mathbf{P}_0^{TF}\cdot\mathbf{F}_0\mathbf{H})\mathbf{v} + \frac{12}{35}(\mathbf{F}_0\cdot\mathbf{H})\mathbf{P}_0^{TF}\cdot\mathbf{v} = \mathbf{A}_0\cdot\mathbf{F}_0\mathbf{H} - \\ -\frac{1}{5}(F_0^2\mathbf{H} + 2\mathbf{F}_0(\mathbf{F}_0\cdot\mathbf{H})) + \frac{6}{7}(a\mathbf{P}_0^{TF}\cdot\mathbf{H} + d\mathbf{P}_0^{TF}\cdot\mathbf{F}_0) - \\ -\frac{12}{35}[c\mathbf{F}_0 + b\mathbf{H}] \\ -\frac{6}{7}(\mathbf{P}_0^{TF}\cdot\mathbf{F}_0\mathbf{F}_0)\mathbf{v} + \frac{12}{35}(\mathbf{F}_0\cdot\mathbf{F}_0)\mathbf{P}_0^{TF}\cdot\mathbf{v} = \mathbf{A}_0\cdot\mathbf{F}_0\mathbf{F}_0 - \\ -\frac{3}{5}F_0^2\mathbf{F}_0 + \frac{12}{7}a\mathbf{P}_0^{TF}\cdot\mathbf{F}_0 - \frac{24}{35}b\mathbf{F}_0\ . \end{array}\right. \qquad (24)$$

Equations (24) readily give $\mathbf{v}$

$$
\begin{aligned}
\mathbf{v} = & -\tfrac{6}{7}\left[\mathbf{P}_0^{TF}\cdot\mathbf{F}_0\mathbf{H} - \left(\mathbf{P}_0^{TF}\cdot\mathbf{F}_0\mathbf{F}_0\right)(\mathbf{H}\cdot\mathbf{F}_0)/F_0^2\right]^{-1}\cdot \\
& \cdot\{\mathbf{A}_0\cdot\mathbf{F}_0\mathbf{H} - (\mathbf{A}_0\cdot\mathbf{F}_0\mathbf{F}_0)(\mathbf{F}_0\cdot\mathbf{H}/F_0^2) + \\
& +\tfrac{6}{7}\left[a\left(2\left(\mathbf{P}_0^{TF}\cdot\mathbf{F}_0\right)(\mathbf{F}_0\cdot\mathbf{H}/F_0^2) + \mathbf{P}_0^{TF}\cdot\mathbf{H}\right) + d\mathbf{P}_0^{TF}\cdot\mathbf{F}_0\right] - \\
& -\mathbf{F}_0\left[\mathbf{F}_0\cdot\mathbf{H} - \tfrac{24}{35}b\mathbf{F}_0\cdot\mathbf{H}/F_0^2 + \tfrac{12}{35}c\right] - \tfrac{1}{5}\left[F_0^2 + \tfrac{12}{7}b\right]\mathbf{H}\} .
\end{aligned}
\tag{25}
$$

Now we get an expression for the fourth order moment $\mathbf{Q}_0$

$$
\mathbf{Q}_0 = \int I_0\boldsymbol{\Omega}_0\boldsymbol{\Omega}_0\boldsymbol{\Omega}_0\boldsymbol{\Omega}_0 d\Omega_0 = \int I\frac{\nu}{\nu_0}\frac{d\nu}{d\nu_0}\left[\boldsymbol{\Omega} + \left(\frac{\gamma-1}{v^2}\mathbf{v}\cdot\boldsymbol{\Omega} - \gamma\right)\mathbf{v}\right]^4 d\Omega.
$$

Writing in components, and to first order in v, we have

$$
\begin{aligned}
Q_0^{ijkm} = & \frac{1}{15}E\left(\delta^{ij}\delta^{km} + \delta^{ik}\delta^{jm} + \delta^{jk}\delta^{im}\right) - \frac{12}{7}\delta^{(ik}F^m v^{j)} \\
& + \frac{2}{35}\mathbf{v}\cdot\mathbf{F}\left(\delta^{ij}\delta^{km} + \delta^{ik}\delta^{jm} + \delta^{jk}\delta^{im}\right) + \frac{6}{7}P^{(ij}\delta^{km)}
\end{aligned}
$$

Using (20) in the above formula we finally obtain the expression

$$
\begin{aligned}
Q_0^{ijkm} = & \tfrac{1}{15}E_0\left(\delta^{ij}\delta^{km} + \delta^{ik}\delta^{jm} + \delta^{jk}\delta^{im}\right) + \\
& +\tfrac{4}{21}\mathbf{v}\cdot\mathbf{F}_0\left(\delta^{ij}\delta^{km} + \delta^{ik}\delta^{jm} + \delta^{jk}\delta^{im}\right) + \tfrac{6}{7}P_0^{(<ij>}\delta^{km)} - \tfrac{12}{7}\delta^{(ij}F_0^k v^{m)},
\end{aligned}
\tag{26}
$$

which is a fourth order closure giving the fourth order moment $\mathbf{Q}_0$ as function of the moments E_0, $\mathbf{F}_0$, $\mathbf{P}_0$ and $\mathbf{A}_0$. Then we can write the moment equation

$$
\begin{cases}
\partial_t E_0 + \nabla\cdot\mathbf{F}_0 = \sigma_a(B - E_0) \\
\partial_t \mathbf{F}_0 + \nabla\cdot\mathbf{P}_0 = -\sigma_a\mathbf{F}_0 \\
\partial_t \mathbf{P}_0 + \nabla\cdot\mathbf{A}_0 = \sigma_a\left(\tfrac{1}{3}\mathbf{I}B - \mathbf{P}_0\right) \\
\partial_t \mathbf{A}_0 + \nabla\cdot\mathbf{Q}_0 = -\sigma_a\mathbf{A}_0
\end{cases}
\tag{27}
$$

as a quasilinear first order system in conservation form. Here the right hand sides are simple source terms involving only moments of the energy intensity. This is a consequence of having supposed that the scattering, as observed in the reference frame n_0^μ, is frequency-coherent and isotropic.

5 Conclusions

In this paper we have proposed a new fourth order closure for radiative hydrodynamics moment equations. The resulting system (27), although involving a great number of variables, is however simpler to solve than the original transfer equation. Moreover we believe that if one has to handle situations in which radiation has a high degree of anisotropy, the only way to avoid to solve the transfer equation is to resort to moments of radiation intensity whose order is greater than two. As far as we know, all high level closures that have appeared in the literature are based on truncation at some order of the expansion of multipole series of radiation intensity in the reference frame of matter (see e.g.[3] or [4]). Imposing that the preferred reference frame for radiation is the comoving frame can be a good choice in many cases. However there are many situations in which such a choice is unnatural.

If one considers a flow driven by radiation, in the sense that radiation dominates energy and momentum, as happens in stellar atmospheres or in plasma laboratories exposed to high-energy X-rays beams, the matter rest frame is a preferred reference frame only in the sense that scattering of photons through matter appears to be isotropic (if no effect such as Thomson or Raman scattering is present), as well photon redistribution through emission processes.

But in optically thin media since scattering and emission have negligible effects in determining the local properties of radiation, it is preposterous to impose that in the matter comoving frame one finds the minimum of the radiation anisotropy.

In our approach we have only supposed the existence of a frame in which the expansion (18) holds, which generalizes the hypothesis of Levermore. We believe that the resulting closure (26), holding in another frame, which can be the comoving frame, can correctly describe dissipative effects in radiation, taking into account also its non one-directional character.

The system (27) together with the closure (26) needs more theoretical investigation. In particular hyperbolicity and existence of a supplementary conservation law, but also its behaviour in the limit of slowly varying (in time) radiation intensity must be thouroughly investigated and will be the subject of future work.

References

[1] G.C.Pomraning,*The Equation of Radiation Hydrodynamics* (Pergamon,New York,1973).

[2] D.Mihalas and B.Weibel Mihalas,*Foundations of Radiation Hydrodynamics* (Oxford U.P.,New York,1984).

[3] K.S.Thorne,Mon.Not.R.astr.Soc.,**194**,439,(1981).

[4] J.L.Anderson and E.A.Spiegel,Astrophys.J.,**171**,127,(1972)

[5] A.M.Anile,S.Pennisi and M.Sammartino, J.Math.Phys., **32**,544,(1991).

[6] A.M.Anile,S.Pennisi and M.Sammartino,*Covariant Radiation Hydrodynamics*, to appear in Ann.Inst.H.Poincaré.

[7] S.Pennisi and M.Sammartino,*A Mathematical Model for Radiation Hydrodynamics*, submitted to Le Matematiche.

[8] C.D.Levermore,J.Quant.Spectrosc.Radiat.Transfer,**31**,149,(1984).

[9] R.W.Lindquist,Ann.Phys.,**37**,487,(1986).

[10] S.H.Hsieh and E.A.Spiegel,Astrophys.J.,**207**,244,(1976).

[11] J.Oxenius,*Kinetic Theory of Particles and Photons* (Springer Verlag,Berlin,1986).

[12] G.Boillat,C.R.Acad.Sc.Paris,**278-A**,909,(1974).

[13] T.Ruggeri and A.Strumia,Ann.Inst.H.Poincaré,**34**,165, (1981).

[14] A.M.Anile,*Relativistic Fluids and Magnetofluids* (Cambridge U.P.,Cambridge,1989).

[15] C.D.Levermore and G.C.Pomraning,Astrophys.J.,**248**,321,(1981).

Authors' address: *Dipartimento di Matematica, Universitá di Catania, Viale A.Doria 6,Catania,Italy.*

V C BOFFI

Nonlinear evolution problems in extended kinetic theory

ABSTRACT

Classical and shape-preserving solutions to a semilinear hyperbolic system, arising in the extended kinetic theory of gas mixtures, are presented for various physical situations. The effects of both the interaction nature and the drift velocity are, in particular, commented.

1. INTRODUCTION

We shall be dealing with the semilinear hyperbolic system $(i=1,2,\ldots,N)$

$$\begin{cases} L_i\rho_i = \Phi_i(\rho_1,\rho_2,\ldots,\rho_N) \; ; \\ L_i \equiv \partial_t + (\underline{v}_i+\underline{w}_i t)\cdot\partial_{\underline{x}} \; , \end{cases} \tag{1}$$

$$(\underline{x},\underline{v}_i,\underline{w}_i \in R_3;\; t \in R_1^+)$$

subjected to the initial conditions

$$\rho_i(\underline{x},0) = Q_i(\underline{x}) \; . \tag{1a}$$

A system like (1) is encountered in extended kinetic theory of gas mixtures, as will be next shown. In this context $\rho_i(\underline{x},t)$ is the mass number of the ith species, $\Phi_i(\rho_1,\rho_2,\ldots,\rho_N)$ is a nonlinear term including interactions of general type and nature (e.g., removal, creation, chemical, polynomial,...), $\underline{v}_i+\underline{w}_i t$ is a drift velocity, sum of the velocity $\underline{v}_i$, at which the ith species is injected in the mixture at t=0, plus the acceleration term $\underline{w}_i t$, $\underline{w}_i$ being the ratio between the external constant force $\underline{F}_i$ acting on the species i and the mass m_i of the same species. We bear in mind that physically both ρ_i and Q_i must be nonnegative in their own domain of definition.

The paper is structured as follows.

We first show how the hyperbolic system (1) can be rigorously derived in the frame of the extended kinetic theory of gas mixtures. Then we present either classical and shape-preserving solutions to it for the case N=2, by distinguishing the two main situations when the two drift velocities are or not equal each other. The cases, also, when $\Phi_i(\rho_i,\rho_j)$ is a general quadratic nonlinear term or a nonlinear term of Lotka-Volterra type are considered. In the conclusions we finally discuss future extensions of the present work with particular regard to the problem concerning the fully nonlinear hyperbolicity of the system (1), which is actually only semilinear, for now.

2. A SIMPLE MODEL LEADING TO SYSTEM (1)

Let us refer to a mixture of N monoatomic gases, whose particles interact only through binary removal collisions; the relevant Cauchy initial value problem describing the evolution of the distribution function $f_i(\underline{x},\underline{v},t)$ of the general species i reads as (i=1,2,...,N)

$$\partial_t f_j + \underline{v}\cdot\partial_{\underline{x}} f_i + (\frac{\underline{F}_i}{m_i})\cdot\partial_{\underline{v}} f_i = - \sum_{j=1}^{N} f_i(\underline{x},\underline{v},t) \int_{R_3} g_{ij}^R(|\underline{v}-\underline{w}|) f_j(\underline{x},\underline{w},t) d\underline{w} \quad (2)$$

with

$$f_i(\underline{x},\underline{v},0) = S_i(\underline{x},\underline{v}) > 0 \ . \quad (2a)$$

This is a system of nonlinear integro-partial differential extended Boltzmann equations as derivable in the frame of the so-called scattering kernel formulation [1][2]. A possible approach to the solution of the sysytem (2)+(2a) consists of the following steps:

-i) the particles of the gases of the mixture considered are assumed to be maxwellian, that is, the microscopic collision frequency $g_{ij}^R(|\underline{v}-\underline{w}|)$ of the removal collisions is independent of the modulus of the relative velocity $\underline{v}-\underline{w}$, and it is a constant for all i and j, namely

$$g_{ij}^R(|\underline{v}-\underline{w}|) = \text{const} = g_{ij}^R \ ; \quad (3a)$$

-ii) integrating then both sides of (2) over the domain of the velocities, and setting for the zero- and first-order moment of f_i with respect to $\underline{v}$

$$\rho_i(\underline{x},t) = \int_{R_3} f_i(\underline{x},\underline{v},t) d\underline{v} \ , \quad (3b)$$

which is the already-mentioned mass density, and

$$\underline{J}_i(\underline{x},t) = \int_{R_3} \underline{v} \ f_i(\underline{x},\underline{v},t) \ d\underline{v} \ , \quad (3c)$$

which is, instead, the current density, respectively, we are left with

$$\partial_t \rho_i + \partial_{\underline{x}} \cdot \underline{J}_i(\underline{x},t) = - \sum_{j=1}^{N} g_{ij}^R \rho_i \rho_j \ . \tag{4}$$

As suggested by continuum mechanics, we formulate now the constitutive equation

$$\underline{J}(\underline{x},t) = (\underline{v}_i + \frac{\underline{F}_i}{m_i} t) \rho_i = (\underline{v}_i + \underline{w}_i t) \rho_i \ , \tag{5}$$

so that we end up with

$$\partial_t \rho_i + (\underline{v}_i + \underline{w}_i t) \cdot \partial_{\underline{x}} \rho_i = - \sum_{j=1}^{N} g_{ij}^R \rho_i \rho_j \ , \tag{6}$$

which is just of the form as (1) with a nonlinear quadratic term

$$\Phi_i(\rho_1, \rho_2, \ldots, \rho_N) = - \sum_{j=1}^{N} g_{ij}^R \rho_i \rho_j \quad \text{and} \quad \rho_i(\underline{x},0) = \int_{R_3} S_i(\underline{x},\underline{v}) d\underline{v} = S_i^*(\underline{x}) \ .$$

We can even justify rigorously (5) by just adding the further hypothesis that the initial datum (2a) be separable and monochromatic. Indeed, upon hypothesis (3a), the system (2)+(2a) can be rewritten as

$$\begin{cases} \partial_t f_i + \underline{v}_i \cdot \partial_{\underline{x}} f_i + (\dfrac{\underline{F}_i}{m_i}) \cdot \partial_{\underline{v}} f_i = - f_i(\underline{x},\underline{v},t) \displaystyle\sum_{j=1}^{N} g_{ij}^R \rho_j(\underline{x},t) \ ; \\ f_i(\underline{x},\underline{v},0) = S_i(\underline{x},\underline{v}) \ . \end{cases} \tag{7}$$

Integrating along the characteristics, this system is formally solved by

$$f_i(\underline{x},\underline{v},t) = S_i(\underline{x} - \underline{v}t + \frac{\underline{F}_i}{2m_i} t^2, \underline{v} - \frac{\underline{F}_i}{m_i} t) \times$$

$$\exp \left\{ - \sum_{j=1}^{N} g_{ij}^R \int_0^t \rho_i \left[\underline{x} - \underline{v}(t-u) + \frac{\underline{F}_i}{2m_i} (t-u)^2, u \right] du \right\} . \tag{8}$$

If now the initial datum $S_i(\underline{x},\underline{v})$, (2a), is taken to be separable and monochromatic, that is, we set

$$S_i(\underline{x},\underline{v}) = Q_i(\underline{x}) \ \delta(\underline{v}-\underline{v}_i) \ , \tag{9}$$

then, integrating both sides of (8) over $\underline{v} \in \mathbf{R}_3$ would yield

$$\rho_i(\underline{x},t) = Q_i(\underline{x} - \underline{v}_i t - \frac{\underline{F}_i}{2m_i} t^2) \times$$

$$\exp \left\{ - \sum_{j=1}^{N} g_{ij}^R \int_0^t \rho_i \left[\underline{x} - \underline{v}_i(t-u) - \frac{\underline{F}_i}{2m_i} (t^2-u^2), u \right] du \right\} . \tag{10}$$

On the other hand, combining (8) and (9), we verify that

$$\underline{J}_i(\underline{x},t) = \int_{R_3} \underline{v} f_i(\underline{x},\underline{v},t)\, d\underline{v} = \left[\underline{v}_i + \frac{\underline{F}_i}{m_i} t\right] \rho_i(\underline{x},t) \ . \tag{11}$$

which is just (5). Finally, differentiating both sides of (10) with respect to t and accounting for (11), we are again led back to (6) (we assume, without loss of generality, that $S_i^*(\underline{x}) \equiv Q_i(\underline{x})$). We comment that, once the system (10) is solved for the ρ_i's, then, with $S_i(\underline{x},\underline{v})$ given by (9), the sought distribution functions f_i (i=1,2,.. .,N) would follow from (8). Analogously, one could even evaluate the higher moments $\int_{R_3} \underline{v}^n f_i(\underline{x},\underline{v},t)\, d\underline{v}$ ($n \geq 2$).

3. THE SYSTEM (6) FOR N=2

For N=2 we have (i,j=1,2; i≠j)

$$\begin{cases} L_i \rho_i = \Phi_i(\rho_i,\rho_j) \ ; \\ L_i \equiv \partial_t + (\underline{v}_i + \underline{w}_i t) \cdot \partial_{\underline{x}} \ ; \\ \rho_i(\underline{x},0) = Q_i(\underline{x}) \ . \end{cases} \tag{12}$$

As for the nonlinear term Φ_i, we take

$$\Phi_i(\rho_i,\rho_j) = a_i\rho_i^2 + b_i\rho_i\rho_j + c_i\rho_j^2 + d_i\rho_i + e_i\rho_j + f_i \ , \tag{13}$$

which describes a general quadratic interaction. In the limit of $c_i, e_i, f_i \to 0$ it reduces to

$$\Phi_i(\rho_i,\rho_j) = \rho_i(a_i\rho_i + b_i\rho_j + d_i) \ , \tag{14}$$

which is of Lotka-Volterra type. The presence of the linear term $d_i\rho_i$ indicates that there is a coupling with an external bath (of fixed mass density), acting as a support for the diffusion of the considered particles.

In the case when $\underline{v}_i = \underline{v}$, $\underline{w}_i = \underline{w}$, that is, when the two drift velocities are equal, the characteristics method applies. Parametrically, the characteristic curves are defined in the space $(t,\underline{x},\rho_1,\rho_2)$ by

$$\frac{dt}{1} = \frac{dx_k}{(\underline{v}+\underline{w}t)_k} = \frac{d\rho_1}{\Phi_1(\rho_1,\rho_2)} = \frac{d\rho_2}{\Phi_2(\rho_1,\rho_2)} \ . \qquad (k=1,2,3) \tag{15}$$

We get thus

$$\begin{cases} t(\tau) = \tau \\ \underline{x}(\tau) = \underline{x}(0) + \underline{v}\tau + \frac{\underline{w}}{2}\tau^2 ; \\ \dot{\rho}_i = \Phi_i(\rho_i, \rho_j) , \end{cases} \tag{16}$$

where

$$\begin{aligned} &\rho_i(\tau) = \rho_i[\underline{x}(\tau), \tau] \\ &\dot{\rho}_i \equiv d_\tau \rho_i = \partial_\tau \rho_i + (\partial_\tau \underline{x}) \cdot (\partial_{\underline{x}} \rho_i) . \end{aligned} \tag{17}$$

Once these ODE's are solved, the sought solution to the given semilinear hyperbolic system (12) is

$$\rho_i[\underline{x}(\tau), \tau] \Rightarrow \rho_i(\underline{x} - \underline{v}t - \frac{\underline{w}t^2}{2}, t) . \tag{18}$$

4. FRONT-WAVE AND SOLITONIC SOLUTIONS

As an application of the case with equal drift velocities, let us consider the similarity variable

$$\underline{\zeta} = \underline{x} - \underline{\lambda}(t)t . \tag{19}$$

Then, with

$$\underline{\lambda}(t) = (\underline{v} - \underline{c}) + \frac{\underline{w}\ t}{2} + \underline{d} , \tag{20}$$

where $\underline{c}$ and $\underline{d}$ are two arbitrary vectors of $\mathbf{R}_3$, let us consider the scaling (k=1,2,3)

$$\xi_k \equiv \frac{\zeta_k}{c_k} = \frac{x_k}{c_k} + (1 - \frac{v_k}{c_k})t - \frac{w_k t^2}{2c_k} - \frac{d_k}{c_k} . \tag{21}$$

The given semilinear hyperbolic system (12) can be then recast as

$$\underline{\Omega} \cdot \partial_{\underline{\xi}} \rho_i = \Phi_i(\rho_i, \rho_j) , \tag{22}$$

where $\underline{\Omega}(\underline{i}_1, \underline{i}_2, \underline{i}_3)$ is a unit vector of unit components $\underline{i}_1, \underline{i}_2, \underline{i}_3$ along the axes.

The form (22) is very convenient for studying solutions as

$$\rho_i = \frac{A_i + B_i \exp(\underline{\alpha}_i \cdot \underline{\xi}_i)}{1 + \exp(\underline{\beta}_i \cdot \underline{\xi}_i)} , \tag{23}$$

that is, shape-preserving solutions.

Explicit examples of such kind of solutions will be now obtained in the monodimensional case. With c and d constant and

$$\xi = \frac{x - d}{c} + \frac{(c - v)t - \frac{1}{2} wt^2}{c} , \tag{24}$$

the monodimensional system (i=1,2)

$$\partial_t \rho_i + (v+wt)\partial_x \rho_i = \Phi_i(\rho_1,\rho_2) \tag{25}$$

becomes

$$\partial_\xi \rho_i = \Phi_i(\rho_1,\rho_2) \ . \tag{26}$$

As a front-wave solution, let us try [3]

$$\rho_i = \frac{A_i + B_i \exp(\gamma\xi)}{1 + \exp(\gamma\xi)} \ , \tag{27}$$

having an amplitude A_i at $\xi\to-\infty$ and B_i at $\xi\to\infty$, and propagating with speed (v-c)+wt along the x-axis, provided the initial condition be compatible with the form $\rho_i(x,0) = Q_i(x)$.

Inserting (27) (with $\gamma=1$) in (26) and equating the coefficients of similar exponential term we get

$$\begin{aligned}
&a_1A_1^2+b_1A_1A_2+c_1A_2^2+d_1A_1+e_1A_2+f_1 = 0 \ ;\\
&2a_1B_1A_1+b_1(A_1B_2+A_2B_1)+2c_1A_2B_2+d_1(A_1+B_1)+e_1(A_2+B_2)+2f_1 = B_1-A_1 \ ;\\
&a_1B_1^2+b_1B_1B_2+c_1B_2^2+d_1B_1+e_1B_2+f_1 = 0 \ ;\\
&a_2A_2^2+b_2A_1A_2+c_2A_1^2+d_2A_2+e_2A_1+f_2 = 0 \ ;\\
&2a_2B_2A_2+b_2(A_2B_1+A_1B_2)+2c_2A_1B_1+d_2(A_2+B_2)+e_2(A_1+B_1)+2f_2 = B_2-A_2 \ ;\\
&a_2B_2^2+b_2B_1B_2+c_2B_1^2+d_2B_2+e_2B_1+f_2 = 0 \ .
\end{aligned} \tag{28}$$

There are six algebraic equations in the four unknowns A_1,A_2,B_1,B_2 to be expressed in terms of the twelve coefficients (a_i,b_i,c_i,d_i,e_i,f_i; i=1,2), up to now completely arbitrary. There are thus two compatibility conditions between these twelve coefficients. Then, in general, we can obtain front-wave solutions parametrized by ten free parameters. Supplementary restrictions imposed by the positivity of ρ_i may reduce this freedom, or even forbid the existence of relevant front-wave solutions. As an explicit example, let us take $a_i=b_i=c_i=1/3$; $d_i+e_i=1$; $f_i=0$ and $\rho_i(\underline{x},0)=\exp x/(1+\exp x)$. Then the system

$$\begin{aligned}
\partial_\xi\rho_1 &= -\frac{1}{3}\rho_1^2 - \frac{1}{3}\rho_1\rho_2 - \frac{1}{3}\rho_2^2 + d_1\rho_1 + e_1\rho_2 \ ;\\
\partial_\xi\rho_2 &= -\frac{1}{3}\rho_2^2 - \frac{1}{3}\rho_2\rho_1 - \frac{1}{3}\rho_1^2 + d_2\rho_2 + e_2\rho_1
\end{aligned} \tag{29a}$$

is solved by

$$\rho_i(\xi) = \frac{\exp(\xi)}{1 + \exp(\xi)} \ . \tag{29b}$$

This solution is of the type (27), and the initial condition is reproduced once we set c=1, d=0 in (24).

As for a monodimensional solitonic (pulse) solution, we seek then for

$$\rho_i = \frac{A_i + B_i \exp(\alpha_i \xi)}{1 + \exp(\beta_i \xi)} \tag{30}$$

with $\alpha_i \neq \beta_i$ [3].

Proceeding like in the case of the front-wave solutions, due to the fact that $\alpha_i \neq \beta_i$, the compatibility conditions to be satisfied are more numerous. Together with the positivity condition, they impose more severe restrictions on the freedom of choosing the twelve coefficients $(a_i, b_i, \ldots, f_i;\ i=1,2)$. As an explicit example, let us take $a_i = b_i = e_i = f_i = 0$ $(i=1,2)$; $c_1 = -3$; $d_1 = 1$; $c_2 = 3$; $d_2 = -1$ and $\rho_1(\underline{x},0) = \exp x/(1+\exp 3x)$; $\rho_2(\underline{x},0) = \exp 2x/(1+\exp 3x)$. Then the system

$$\partial_\xi \rho_1 = -3\rho_2^2 + \rho_1 ; \qquad \partial_\xi \rho_2 = 3\rho_1^2 - \rho_2 \tag{31a}$$

is solved by

$$\rho_1(\xi) = \frac{\exp(\xi)}{1 + \exp(3\xi)} ; \qquad \rho_2(\xi) = \frac{\exp(2\xi)}{1 + \exp(3\xi)} . \tag{31b}$$

These are just of the type as (30), and become compatible with the initial conditions if again we set c=1, d=0 in (24).

5. SOLUTIONS FOR THE LOTKA-VOLTERRA CASE

We apply now the characteristics method to the system

$$\dot{\rho}_i = \rho_i (a_i \rho_i + b_i \rho_j + d_i) , \tag{32}$$

just holding in the case of (14).

Proceeding by substitution, the system (32) is seen to be amenable to a single nonlinear second-order ODE, namely $(i,j=1,2;\ i \neq j)$

$$\rho_i \ddot{\rho}_i = (1 + a_j/b_i)\, \dot{\rho}_i^2 + (a_i + b_j - 2a_i a_j/b_i)\, \dot{\rho}_i \rho_i^2 + (d_j - 2a_j d_i/b_i)\, \dot{\rho}_i \rho_i + (a_i^2 a_j/b_i - a_i b_j)\, \rho_i^4 + (2a_i a_j d_i/b_i - b_j d_i - a_i d_j)\, \rho_i^3 + (a_i d_i^2/b_i - d_i d_j)\, \rho_i^2 . \tag{33}$$

An extensive variety of methods is available for solving an equation like (33) [4] [5]. Some of them are reviewed in [3]. Here we recalculate exactly the new solution for the case $a_1=a_2=0$; $d_1=d_2=d$. The system (32) now becomes

$$\dot{\rho}_i = b_i\rho_i\rho_j + d\rho_i \ . \tag{34}$$

From $\rho_i\rho_j = (\dot{\rho}_i - d\rho_i)/b_i$ there follows that

$$b_j\rho_i - b_i\rho_j = K_1 \exp(d\tau) \ , \tag{35}$$

K_1 being an integration constant. Eliminating ρ_2 between (34) and (35) leads to the self-consistent equation for ρ_1

$$\dot{\rho}_1 = b_2\rho_1^2 + [d - K_1\exp(d\tau)]\rho_1 \ , \tag{36}$$

which is a solvable Bernouilli equation. We find thus

$$\rho_1(\tau) = \frac{K_1\exp(d\tau)}{b_2 - (K_1K_2+b_2)\exp\{K_1d^{-1}[\exp(d\tau)-1]\}} \tag{37a}$$

and

$$\rho_2(\tau) = \frac{1}{b_1}\,\frac{K_1(K_1K_2+b_2)\exp(d\tau)}{b_2\exp\{-K_1d^{-1}[\exp(d\tau)-1]\} - (K_1K_2+b_2)} \ , \tag{37b}$$

respectively. The two constants of integration K_1 and K_2 are then determined from the initial conditions, and they result as $K_1=b_2Q_1-b_1Q_2$ and $K_2=-1/Q_1$, respectively. Reassembling everything and returning to the original independent variables $\underline{x}$ and t, defining $\underline{y}=(\underline{x}-\underline{v}t-\underline{w}t^2/2)$, we get eventually the explicit solutions to our system as

$$\rho_1(\underline{x},t) = \frac{Q_1(\underline{y})\ [b_2Q_1(\underline{y})-b_1Q_2(\underline{y})]\ \exp(dt)}{b_2Q_1(\underline{y})-b_1Q_2(\underline{y})\exp\{[b_2Q_1(\underline{y})-b_1Q_2(\underline{y})][\exp(dt)-1]/d\}} \tag{38a}$$

and

$$\rho_2(\underline{x},t) = \frac{Q_2(\underline{y})\ [b_1Q_2(\underline{y})-b_2Q_1(\underline{y})]\ \exp(dt)}{b_1Q_2(\underline{y})-b_2Q_1(\underline{y})\exp\{[-b_2Q_1(\underline{y})+b_1Q_2(\underline{y})][\exp(dt)-1]/d\}} \tag{38b}$$

respectively.

6. THE CASE OF DIFFERENT DRIFT VELOCITIES

In particular, we take $\underline{v}_i\neq\underline{v}_j$; $\underline{w}_i=\underline{w}$. The solution to our semilinear hyperbolic system becomes then a very arduous task, in general, and also the characteristics method does not apply any longer. However, an operational approach allows to solve explicitly the following special case of (12), namely

$$\begin{cases} L_i\rho_i \equiv [\partial_t + (\underline{v}_i+\underline{w}t)\cdot\partial_{\underline{x}}]\rho_i = b_i\rho_i\rho_j \; ; \\ \rho_i(\underline{x},0) = Q_i(\underline{x}) \end{cases} \tag{39}$$

The same system like (40), but for the monodimensional case and with $\underline{w}$=0, has been solved by Lie group method in [6] and [7]. Following [3], let us suppose that

$$\Psi(\underline{x},t) = \sum_{i=1}^{2} G_i(\underline{x} - \underline{v}_i t - \underline{w}t^2/2) \; , \tag{40}$$

where G_i are arbitrary functions, and define

$$\rho_i(\underline{x},t) = \frac{\underline{v}_i-\underline{v}_j}{b_j\Psi} \cdot \partial_{\underline{x}} G_i(\underline{x} - \underline{v}_i t - \underline{w}t^2/2) \; . \tag{41}$$

It is now verified, since the following algebra holds, namely

$$\begin{aligned} & L_iG_i = L_i\partial_{\underline{x}}G_i = 0 \; ; \\ & L_j\Psi = L_jG_i = (\underline{v}_j-\underline{v}_i) \cdot \partial_{\underline{x}}G_i \; , \end{aligned} \tag{42}$$

that (41) just satisfies the system (39). In order to determine G_1 and G_2 we impose the initial conditions $\rho_i(\underline{x},0) = Q_i(\underline{x})$. From (41) we have thus

$$b_jQ_i(\underline{x}) = \frac{(\underline{v}_i-\underline{v}_j)\cdot\partial_{\underline{x}}G_i(\underline{x})}{G_1(\underline{x})+G_2(\underline{x})} \tag{43a}$$

or

$$b_iQ_j(\underline{x}) - b_jQ_i(\underline{x}) \equiv q(\underline{x}) = (\underline{v}_i-\underline{v}_j)\cdot\partial_{\underline{x}}\ln[G_1(\underline{x})+G_2(\underline{x})] \; . \tag{43b}$$

Integrating along the characteristics, with the normalization $G_1(0)+G_2(0) = 1$, we obtain

$$G_1(\underline{x})+G_2(\underline{x}) = \exp\left[\frac{\underline{v}_j-\underline{v}_i}{(\underline{v}_j-\underline{v}_i)^2} \cdot \int_0^{\underline{x}} q(\underline{x}')d\underline{x}' \right] \; . \tag{44}$$

From (43a)

$$\partial_{\underline{x}}G_i(\underline{x}) = \frac{\underline{v}_i-\underline{v}_j}{(\underline{v}_i-\underline{v}_j)} b_jQ_i(\underline{x})[G_1(\underline{x})+G_2(\underline{x})] \; , \tag{45}$$

which is combined with (44) to yield

$$G_i(\underline{x}) = \frac{\underline{v}_i-\underline{v}_j}{(\underline{v}_i-\underline{v}_j)^2}\cdot\int_0^{\underline{x}} b_jQ_i(\underline{x}')\exp\left[\frac{\underline{v}_j-\underline{v}_i}{(\underline{v}_j-\underline{v}_i)^2}\cdot\int_0^{\underline{x}'} q(\underline{x}'')d\underline{x}'' \right] d\underline{x}'+G_i(0) \; . \tag{46}$$

Finally, entering in the definition (41) with both (45) and (46), we find the sought explicit solution to (39), namely (i,j=1,2;

$i \neq j$)

$$\rho_i(\underline{x},t) = \frac{Q_i(\underline{x}-\underline{v}_i t-\underline{w}t^2/2)\exp\left[\frac{\underline{v}_j-\underline{v}_i}{(\underline{v}_j-\underline{v}_i)^2}\cdot\int_0^{\underline{x}-\underline{v}_i t-\underline{w}t^2/2} q(\underline{x}')d\underline{x}'\right]}{1+\sum_{\substack{i=1\\ i\neq j}}^{2}\frac{\underline{v}_i-\underline{v}_j}{(\underline{v}_i-\underline{v}_j)^2}\cdot\int_0^{\underline{x}-\underline{v}_i t-\underline{w}t^2/2} b_j Q_i(\underline{x}')\exp\left[\frac{\underline{v}_j-\underline{v}_i}{(\underline{v}_j-\underline{v}_i)^2}\cdot\int_0^{\underline{x}'} q(\underline{x}'')d\underline{x}''\right]d\underline{x}'} . \quad (47)$$

In the limit of the monodimensional case with $\underline{w} = 0$, (47) reproduces the result quoted in [6] and in [7].

7. CONCLUSIONS

All the present discussion has been based on the semilinear hyperbolic system

$$\partial_t \rho_i + (\underline{v}_i + \underline{w}_i t)\cdot\partial_{\underline{x}}\rho_i(\underline{x},t) = \Phi_i(\rho_1,\rho_2,\dots,\rho_N) , \quad (48)$$

and not on a fully nonlinear hyperbolic system, like, for instance,

$$\partial_t \rho_i + \underline{\Psi}_i(\rho_i;\underline{x},t)\cdot\partial_{\underline{x}}\rho_i(\underline{x},t) = \Phi_i(\rho_1,\rho_2,\dots,\rho_N) , \quad (49)$$

that would have better suited the title and the purposes of this Euromech Colloqium # 270. In this respect an open problem is the one of finding physical situations, if any, for which the nonlinear integro-partial differential extended Boltzmann system for the distribution functions $f_1(\underline{x},\underline{v},t), f_2(\underline{x},\underline{v},t),\dots,f_N(\underline{x},\underline{v},t)$ of a mixture of N monoatomic gases can be converted into a fully nonlinear hyperbolic system for the corresponding mass densities $\rho_i(\underline{x},t) = \int_{R3} f_i(\underline{x},\underline{v},t)d\underline{v}$. Actually, a more realistic model including not only removal, like done here (cft.(2)), but also scattering and creation effects, has been investigated in [8]. Upon appropriate specialization of both scattering and creation kernels, and the assumptions of maxwellian particles and of a separable and monochromatic initial datum, like here in (3a) and (9), it has been shown that the constitutive equation for $\underline{J}_i(\underline{x},t)$, as given by (5), must be actually corrected as

$$\underline{J}_i(\underline{x},t) = (\underline{v}_i + \underline{w}_i t)\rho_i(\underline{x},t) + \underline{J}_i^{SC}(\underline{x},t) , \quad (50)$$

where $\underline{J}_i^{SC}(\underline{x},t)$ is a nonclassical, rather complicated term, accounting for, just, both scattering and creation effects. However, in the limit of only removal, it turns out that $\underline{J}_i^{SC}(\underline{x},t)\to 0$, so that (50) reduces just to (5). When $\underline{J}_i^{SC}(\underline{x},t)\neq 0$, the system (6) would be actually

$$\partial_t\rho_i + (\underline{v}_i+\underline{w}_i t)\cdot\partial_{\underline{x}}\rho_i(\underline{x},t) + \partial_{\underline{x}}\cdot\underline{J}_i^{SC}(\underline{x},t) = \Phi(\rho_1,\rho_2,\ldots,\rho_N), \quad (51)$$

which is not yet of the fully nonlinear type as (49). The point is that $\underline{J}_i^{SC}(\underline{x},t)$ is an integral of a highly nonlinear function containing all the ρ_i's so that (51) still reflects the nonlinear integro-partial differential character as the original extended Boltzmann system for the distribution functions $f_i(\underline{x},\underline{v},t)$. Are there different models for scattering and creation, or even other kinds of interactions, by which (51) be amenable to a fully nonlinear hyperbolic form? This will be object of future work.

ACKNOWLEDGEMENTS

The author wishes to thank GNFM-CNR and the related Project IPPMI for supporting this research.

REFERENCES

1. V.C. BOFFI and A. ROSSANI, On the Boltzmann System for a Mixture of Reacting Gases. ZAMP 41 (1990) 254-269.
2. V.C. BOFFI, V. PROTOPOPESCU and G. SPIGA, On the Equivalence between the Probabilistic, Kinetic and Scattering Kernel Formulation of the Boltzmann Equation. Physica 164A (1990) 400-409.
3. V.C. BOFFI, V. PROTOPOPESCU and Y.Y. AZMY, Exact Solutions for a Semilinear Hyperbolic System with General Quadratic Interactions. Il Nuovo Cimento 12D (1990) 1153-1163.
4. E.L. INCE, Ordinary Differential Equations (Dover New York 1926).
5. G.M. MURPHY, Ordinary Differential Equations and Their Solutions (Van Nostrand Princeton 1960).
6. G. DUKEK and G. SPIGA, Similarity Solutions in Space Dependent Extended Kinetic Theory. ZAMP 39 (1988) 924-929.
7. V.C. BOFFI and M. TORRISI, Methods of Similarity Analysis in the Study of Nonlinear Dynamics of a Gas Mixture. TTSP 19 (1990) 139-150.
8. V.C. BOFFI, Systems of Conservation Equations in Nonlinear Particle Transport Theory, in Transport Theory, Invariant Imbedding and Integral Equations, Edited by P. Nelson et al., Lecture Notes in Pure and Applied Mathematics, Vol.115 (M. Dekker New York 1988) 123-145.

Vinicio Boffi
Dipartimento di Metodi e Modelli
Matematici per le Scienze Applicate
Università di Roma "La Sapienza"
via A. Scarpa, 10
00161 Roma (Italy)

I BONZANI AND N BELLOMO

Mathematical aspects of nonlinear shock waves propagation by the discrete Boltzmann equation

ABSTRACT

This paper provides a unified presentation and a methodological approach of the mathematical aspects concerning the analysis of nonlinear shock wave phenomena by the discrete Boltzmann equation.

1. INTRODUCTION

The analysis of the structure of shock waves and of the existence and stability of travelling shock waves is one of the classical problems of the kinetic theory of gases. This kind of flows is similar to the ones which arise near blunt bodies at small Knudsen numbers. Therefore the study of shock waves is certainly of great interest for the applications.

It is well known that shock waves can be also studied in the framework of the Navier-Stokes description of fluid-dynamics. This mathematical model certainly provides accurate description of the flow conditions in the case of weak shock waves at small Knudsen numbers. On the other hand for strong shock waves and large Knudsen numbers the description provided by the Boltzmann equation appears to be more accurate if compared with experimental observations [1].

This holds true also in the case of the discrete Boltzmann equation which provides an accurate description, both at a qualitative and a quantitative level, of several interesting features of the gas dynamic model.

This paper deals with a unified presentation and organization of the methodological aspects of the problem, with particular attention on the problem of existence of shock waves solutions and of the onset of shock waves in the initial value problem. Moreover this paper will also indicate the mathematical problems which are still open and the main difficulties to be tackled to face them.

The paper is in four sections. The first one is this introduction. The second one deals with a brief description of the discrete Boltzmann equation and of a specific model. The third section deals with the methodology to be followed in order to deal the analysis of the existence problem for travelling shock waves in the case of general discrete velocity models and with the problem of onset of shock waves. Finally a discussion follows in the last section.

2. THE DISCRETE BOLTZMANN EQUATION

The discrete Boltzmann equation is a mathematical model in nonlinear kinetic theory which describes the time and space evolution of the number densities $N_i = N_i(t, \underline{x})$ joined to the velocities $\vec{v}_i, \quad i = 1, \ldots, n$ of a gas particle which can attain only a finite number of velocities.

Specific models are available in literature either in the case of one-component gases [2] or in the case of gas mixtures [3]. The model may include binary collisions [2,3] or multiple collisions [4,5]. Reviews of the mathematical aspects of this class of equations are due to Platkowski and Illner [6] and to Bellomo and Gustafsson [7]. Modelling and fluid-dynamics applications are dealt with in the book by Monaco and Preziosi [8].

The general structure of the discrete Boltzmann equation is the following

$$\left(\frac{\partial}{\partial t} + \vec{v}_i \cdot \nabla_{\underline{x}}\right) N_i = J_i^{(2)}[N] + J_i^{(3)}[N], \tag{2.1}$$

for $i = 1, \ldots, n$ and where $\vec{v}_i$ denote the admissible velocities, $J_i^{(2)}$ and $J_i^{(3)}$ are the collision operators for the binary and triple collisions respectively. The number densities are positive defined functions of time and space

$$N_i = N_i(t, \underline{x}) \quad : \quad [0, T] \times \mathbf{R}^d \mapsto \mathbf{R}^+, \quad d = 1, 2, 3. \tag{2.2}$$

The operators J_i can be written, in their general form, as follows

$$J_i^{(2)} = \frac{1}{2} \sum_{jhk} A_{ij}^{hk} (N_h N_k - N_i N_j) \tag{2.3}$$

and

$$J_i^{(3)} = \frac{1}{3!} \sum_{jlhkp} A_{ijl}^{hkp} (N_h N_k N_p - N_i N_j N_l), \tag{2.4}$$

where the transition rates A_{ij}^{hk} and A_{ijl}^{hkp} , which can be computed in details once the velocity discretization has been stated, are positive defined constants characterized by the classical *indistinguishability* and *reversibility* properties [2,8]. Such properties essentially consist in the invariance of the terms A under permutations of the indexes and permutations of the subscripts with the superscripts. Considering that the analysis of the abstract problem will be applied to the six velocity model with binary and triple collisions [4], such a model is here reported for completness,

$$
\begin{aligned}
\left(\frac{\partial}{\partial t'} + 2\frac{\partial}{\partial x}\right) f_1 &= 2K'B[f] + K''T[f] \\
\left(\frac{\partial}{\partial t'} + \frac{\partial}{\partial x}\right) f_2 &= -K'B[f] - K''T[f] \\
\left(\frac{\partial}{\partial t'} - \frac{\partial}{\partial x}\right) f_3 &= -K'B[f] + K''T[f] \\
\left(\frac{\partial}{\partial t'} - \frac{\partial}{\partial x}\right) f_4 &= +2K'B[f] - K''T[f],
\end{aligned}
\tag{2.5}
$$

where $f_i = N_i/n_0$, (n_0 is a reference numerical density), $t' = ct$ ($2c$ is the velocity modulus), $B[f] = f_2 f_3 - f_1 f_4$, $T[f] = f_2^2 f_4 - f_3^2 f_1$, $K' = 2Sn_0/3$ and $K'' = 6\sqrt{6}n_0^2 S^{5/2}/\pi$, (S is the cross sectional area of the gas particles).

3. STATEMENT AND ANALYSIS OF THE PROBLEM

Consider eq.(2.1) which is rewritten in one space dimension as

$$
\left(\frac{\partial}{\partial t} + w_i \frac{\partial}{\partial x}\right) N_i = J_i[N], \qquad i = 1, \ldots, m, \tag{3.1}
$$

where w_i is the x-component of the velocity $\vec{v_i}$ and J_i is the collision operator which may include binary and triple collisions. We are concerned with existence and stability of travelling wave solutions $N_i = N_i(z)$, invariant with respect to the change of variables

$$
z = x - \beta t, \qquad \beta = cost, \tag{3.2}
$$

and linking two Maxwellian states M_i^+ and M_i^-.

It will be assumed, without loss of generality, that the Maxwellian state can be written as

$$
M_i = ae^{bw_i}; \qquad a, b \in \mathbf{R} \tag{3.3}
$$

and it is uniquely defined by the macroscopic observables, which is the case of several well structured models. This analysis is presented for a simple gas with a discretization characterized by one velocity modulus. As a matter of fact the analysis can certainly be developed for gas mixtures. However such a problem is still open.

The general methodology which will be here developed is the one conceivably referred to the pertinent literature with particular attention to the paper by Kawashima and Bellomo [9] which deals specifically with methodological aspects.

Keeping this in mind the various steps for the analysis of the problem can be listed as follows:

Step 1: The original system of m differential equations is written in terms of two conservation equations and $(m-2)$ equations in not conservative form, respectively

$$\begin{aligned} \frac{\partial \rho}{\partial t} + \frac{\partial(\rho u)}{\partial x} &= 0 \\ \frac{\partial(\rho u)}{\partial t} + \frac{\partial(\rho\sigma)}{\partial x} &= 0 \\ \frac{\partial f_i}{\partial t} + \frac{\partial g_i}{\partial x} &= h_i, \end{aligned} \tag{3.4}$$

where $\rho = \rho[N]$ is the mass density, $\rho u[N] = \sum_i w_i N_i$ is the momentum, $\rho\sigma[N] = \sum_i w_i^2 N_i$ and $f_i = f_i[N]$, $g_i = g_i[N]$, $h_i = h_i[N]$ are suitable (to be determined) functions of the densities.

Remark 3.1: The first two equations in (3.4) do not represent a closed system; in fact σ is not in general a function of ρ and u. Therefore one cannot solve the first two equations independently from the remaining ones.

Step 2: The second step consists in the analysis of the properties of the Euler equations. These equations are deduced by the conservative ones as the first terms of the Chapman-Enskog expansion (i.e. the quantities ρ , ρu and $\rho\sigma$ are calculated in the case of a Maxwellian distribution (3.3)); the following system is obtained

$$\begin{pmatrix} \rho \\ u \end{pmatrix}_t + \begin{pmatrix} u & \rho \\ (\sigma(u) - u^2)/\rho & (\sigma'(u) - u) \end{pmatrix} \begin{pmatrix} \rho \\ u \end{pmatrix}_x = 0, \tag{3.5}$$

where σ can be regarded as a function of u only

$$\sigma = \sigma(u), \quad u = u(b) \quad \text{and} \quad \sigma' = d\sigma/du. \tag{3.6}$$

Then one has to study the properties of the coefficient matrix

$$A(\rho, u) = \begin{pmatrix} u & \rho \\ (\sigma(u) - u^2)\rho & \sigma'(u) - u \end{pmatrix} \tag{3.7}$$

in order to prove genuine nonlinearity in the Lax sense

$$\forall(\rho, u), \qquad \nabla\lambda_j(u) \cdot r_j(\rho, u) \neq 0, \qquad j = 1, 2 \tag{3.8}$$

and strict hyperbolicity

$$\forall u, \quad \lambda_1(u) < u < \lambda_2(u), \quad 2\lambda_1(u) < \sigma'(u) < 2\lambda_2(u); \tag{3.9}$$

in eq.(3.8) λ_1 and λ_2 are the eigenvalues of matrix A and r_j the eigenvectors. Properties (3.8) and (3.9) will be used in the analysis, which follows, of the problem.

Step 3: The third step consists in a further (rather crucial) analysis of the Euler equation, in order to prove the existence of elementary shock waves. In fact [9] the existence of elementary shock waves for the Euler equation (3.5) is a necessary compatibility condition to solve the shock wave problem for the discrete Boltzmann equation (3.1). First one applies the change of variables (3.2) in order to obtain a system of ordinary differential equations

$$\frac{d}{dz}(-\rho\beta + \rho u) = 0 \qquad \frac{d}{dz}(-\beta\rho u + \rho\sigma) = 0, \tag{3.10}$$

then integrating between the two Maxwellian states yields the Rankine-Hugoniot relations

$$\begin{aligned} -\beta(\rho_+ - \rho_-) + (\rho_+ u_+ - \rho_- u_-) = 0 \\ -\beta(\rho_+ u_+ - \rho_- u_-) + (\rho_+ \sigma(u_+) - \rho_- \sigma(u_-)) = 0. \end{aligned} \tag{3.11}$$

It is well known [9] that the weak solutions of the Euler equation (3.5) are expressed by

$$(\rho, u) = \begin{cases} (\rho_-, u_-), & x/t < \beta \\ (\rho_+, u_+), & x/t > \beta, \end{cases} \tag{3.12}$$

where (ρ_-, u_-) are given and (ρ_+, u_+) are derived by (3.11) at fixed values of β. Moreover these conditions can be regarded physically reasonable if the following Lax shock conditions hold

$$\lambda_1(u_+) < \beta < \lambda_1(u_-) \quad \text{and} \quad \beta < \lambda_2(u_+) \tag{3.13}$$

or

$$\lambda_2(u_+) < \beta < \lambda_2(u_-) \quad \text{and} \quad \lambda_1(u_-) < \beta. \tag{3.14}$$

The discontinuity at $x/t = \beta$ is called *shock wave* when (3.11, 3.13 or 3.14) hold true.

The conditions defined above assure the existence of elementary shock waves for the Euler equation and have to be verified for each particular model , making use of properties (3.8) and (3.9) previously stated in step 1 and following the line of paper [9].

Step 4: This step consists in the analysis of the existence of shock solutions of eq.(3.4) once the analysis developed in the third step is completed. In fact the Rankine-Hugoniot equations can be used to obtain by eqs. (3.4) a system of $(m-$

2) equations in $(m-2)$ unknowns. This system can be put in terms of ordinary differential equations by means of the change of variable (3.2), then existence of solutions can be obtained by the qualitative analysis of these equations .

Step 5: The fifth step consists in a qualitative analysis of the solution of the initial value problem for eq.(3.1) with initial conditions given by

$$N_i(x; t=0) = N_i^s(x + \phi_i(x); x = z), \tag{3.15}$$

where N_i^s is the solution of the shock wave problem, obtained after Step 4 and ϕ_i is a *small* perturbation . The shock solution is stable if one can prove that for sufficiently small ϕ_i the solution of the problem decay to the travelling wave solution.

After having listed all the sequential steps for studying the shock wave problem we first consider the analysis described in Steps 1-3 (which is certainly a relatively easier problem with respect to the analysis of the full existence problem defined in Steps 1-4); after paper [9] it is known that such analysis can be developed for a large class of models in a general form and in particular it includes the one required by the Longo Monaco model (2.5), developed in [10] for the case $u_+ = 0$ and in [11] for the more general case $u_- > u_+ > 0$.

The conservation equations for model (2.5) are the following

$$\begin{gathered}\left(\frac{\partial}{\partial t'} + 2\frac{\partial}{\partial x}\right)f_1 + 2\left(\frac{\partial}{\partial t'} + \frac{\partial}{\partial x}\right)f_2 + 2\left(\frac{\partial}{\partial t'} - \frac{\partial}{\partial x}\right)f_3 + \left(\frac{\partial}{\partial t'} - 2\frac{\partial}{\partial x}\right)f_4 = 0 \\ \left(\frac{\partial}{\partial t'} + 2\frac{\partial}{\partial x}\right)f_1 + \left(\frac{\partial}{\partial t'} + \frac{\partial}{\partial x}\right)f_2 - \left(\frac{\partial}{\partial t'} - \frac{\partial}{\partial x}\right)f_3 - \left(\frac{\partial}{\partial t'} - 2\frac{\partial}{\partial x}\right)f_4 = 0.\end{gathered} \tag{3.16}$$

As it can be easily verified the above eqs.(3.16) satisfy the requirements indicated in Steps 1-3 .

Remark 3.2 : Not all models satisfy the conditions required by the solution of the shock wave problem. For instance, if one deals with the model (2.5) in the case in which the triple collision term is put equal to zero $(T[f] = 0)$, then one has an additional conservation law,

$$\left(\frac{\partial}{\partial t'} + 2\frac{\partial}{\partial x}\right)f_1 - \left(\frac{\partial}{\partial t'} + \frac{\partial}{\partial x}\right)f_2 + \left(\frac{\partial}{\partial t'} - \frac{\partial}{\partial x}\right)f_3 - \left(\frac{\partial}{\partial t'} - 2\frac{\partial}{\partial x}\right)f_4 = 0, \tag{3.17}$$

which provides an additional constraint preventing the solution of the problem.

Remark 3.3 : It is important calculating in the shock wave analysis the Mach number before the shock, i.e. the ratio

$$M = \frac{u_-}{c_s} \tag{3.18}$$

between the mass velocity u_- and the speed of the sound c_s. The mass velocity is a given quantity of the problem and c_s is computed by the eigenvalues of the Euler equations,

$$c_s = 2c\lambda_2(u) = c\{\sigma'(u) + \sqrt{\sigma'^2(u) + 4(\sigma(u) - u\sigma'(u))}\}, \tag{3.19}$$

where $2c$ is the velocity modulus.

If the compression rate $\eta = \rho_+/\rho_-$ is now considered, it is known that it reaches its largest value when the downstream mass velocity is equal to zero, $u_+ = 0$. In this case it is generally possible to obtain simple expressions of the propagation velocity β and of the compression rate η. Simple calculations, of algebric type, involving the conservation equations (3.16) lead to the following results for the six velocity model, $(b' = bc)$,

$$\begin{aligned}\beta =& \frac{\cosh b' - \cosh 2b'}{4(\sinh 2b' + \sinh b')} + \\ &+ \frac{\sqrt{(\cosh 2b' - \cosh b')^2 + 8(\sinh 2b' + \sinh b')^2}}{4(\sinh 2b' + \sinh b')} \\ \eta =& \frac{3(\cosh 2b' + \cosh b')}{2(\cosh 2b' + 2\cosh b')} + \\ &+ \frac{\sqrt{(\cosh 2b' - \cosh b')^2 + 8\sinh^2 b'}}{2(\cosh 2b' + 2\cosh b')}.\end{aligned} \tag{3.20}$$

As already mentioned, this result cannot be obtained in the case $B[f] = 0$.

The formation of shock waves can be observed in the analysis of an initial value problem with specialized initial conditions . In particular one deals with eq. (2.1) joined to the following initial conditions

$$\begin{aligned} x \leq 0: &\qquad N_i(0, x) = M_i^- \\ x > 0: &\qquad N_i(0, x) = M_i^+ \end{aligned}, \tag{3.21}$$

where M_i^- and M_i^+ are constant Maxwellians with mass velocities u_- and u_+ , respectively, and such that $u_- > u_+ \geq 0$. If two Maxwellians are given then, as shown in the third step, the Rankine-Hugoniot relations yield that only one speed of propagation of shock wave corresponding to such Maxwellians. The analysis of the initial value problem should hopefully show that for t tending to infinity a shock profile is reached travelling with the prescribed propagation velocity β. As a matter of fact it is important computing the velocity of propagation of the shock wave, in order to verify that for large times the steady propagation velocity which can be computed by solution of the Rankine-Hugoniot system is effectively reached. This is a relevant control of the accuracy of the numerical calculations.

Considering that the solution of the initial value problem is reached by numerical methods, one should have the support, besides suitable convergence criteria, of existence theorems. In addition the existence theorems should provide suitable informations on the regularity of the solution, related to the regularity required by the application of the algorithms. Unfortunately, as documented in the review paper [6], no global existence theorem is available for such a problem. In fact global existence in one space dimension has been proven (even without smallness assumptions) only for the gas with finite mass , i.e. for $L_1(\mathbf{R})$ initial data. Consequently one has to use local existence theorems which provide a detailled estimate of the existence interval.

This is the case of the theorem by Lachowicz and Monaco [12]:

Theorem :*Let* $N_0 \in B_+^m$ $(m \geq 2)$ *and* $T = (8\alpha\|N_0\|_0)^{-1}$, *where* B_+^m *is the set of nonnegative functions whose m-derivatives are continuous and bounded in* $\mathbf{R^d}$,

$$\alpha = \max_{i \leq n} \sum_{jkl} A_{ij}^{kl}, \qquad \|N_0\|_0 = \max_{i \leq n} ess \sup |N_{0i}|.$$

Then there exists a unique positive differentiable in time solution N to eq. (2.1) in the space B^{m-2} *on the time interval* $[0, T]$.

This result applies to models with binary collisions only. Bellomo and Toscani have proven a similar result for models with binary and triple collisions [13]. The estimate of the existence interval slightly differs; one has

$$T = \frac{1}{15\lambda}, \qquad \lambda = \frac{2}{\epsilon}\|N_0\| + \frac{4\eta}{\epsilon}\|N_0\|^2, \tag{3.22}$$

where , in the case of model (2.5),

$$\frac{2}{\epsilon} = \frac{5cS}{\pi}, \qquad \frac{4\eta}{\epsilon} = \frac{45\sqrt{6}cS^{5/2}}{2\pi}.$$

It is important to point out that the existence time interval can be enlarged reducing the size, in norm, of the initial data. This is always possible if one deals with models with binary collisions only; in fact the solution essentially depends upon the density ratio. On the other hand for models with multiple collisions high density effects are included and the solution is also affected by the size.

4. DISCUSSION

This paper provides a detailed presentation of the various sequential steps to be followed in a general analysis of the problem. As a matter of fact no paper provides all aspects of the methodological analysis, but only particular aspects related (as usual) to the content of the paper itself. Without adding further comments to the ones already given in the preceding sections, we only need mentioning that rigorous mathematical results are available only for very simple mathematical models of the discrete Boltzmann equation. This despite the fact that the general methodology is now well understood starting from paper [9] as well as in this present paper.

Moreover, as documented in [8], the mathematical modelling of the discrete Boltzmann equation has made a lot of improvements in several directions: modelling with a large number of velocity moduli, modelling of gas mixtures, modelling of gas kinetics with chemical reactions. Therefore one should expect that the analysis of shock wave phenomena is performed on the basis of advanced models of the discrete Boltzmann equation. Such a problem is , on the other hand , still open and serious difficulties have to be overcomed before new original results can be provided. Therefore we hopefully believe that this paper may contribute to those who will attempt the analysis of the various mathematical problems posed in this paper.

AKNOWLEDGMENTS This paper has been partially supported by the Minister for Scientific Research and by the National Council for Research, GNFM Project MMAIT.

REFERENCES

1. M. KOGAN, Rarefied Gas Dynamics, Plenum (New York 1969).
2. R. GATIGNOL, Theorie Cinetique des Gas a Repartition Discrete des Vitesses, Lecture Notes in Physics, Springer 36 (Berlin 1975).
3. E. LONGO and R. MONACO, On the Thermodynamics of Discrete Models of the Boltzmann Equation for Gas Mixtures, Transp. Theory Stat. Phys.,17 (1988) 423-429.
4. E. LONGO and R. MONACO, Discrete Kinetic Theory with Multiple Collisions : Plane Six Velocity Model and Unsteady Couette Flow, Rarefied Gas Dynamics, 118 of Progress in Aeronautics and Astronautics AIAA Publ.,(Washington 1989) 118-130.

5. N.BELLOMO and S. KAWASHIMA, The Discrete Boltzmann Equation with Multiple Collisions : Global Existence and Stability for the Initial Value Problem, J. Math. Phys., 31 (1980) 245-253.
6. T. PLATKOWSKI and R. ILLNER, Discrete Velocity Models of the Boltzmann Equation : a Survey of the Mathematical Aspects of the Theory SIAM Review, 30 (1988) 213-255.
7. N. BELLOMO and T. GUSTAFSSON, The Discrete Boltzmann Equation: a Review of the Mathematical Aspects of the Initial and Initial-Boundary Value Problem, Internal Report Dip. Mat. Politecnico Torino (1990).
8. R. MONACO and L. PREZIOSI, Application of the Discrete Boltzmann Equation to Fluid Dynamics, World Scientific, London, New Yersey, Singapore (1990).
9. S. KAWASHIMA and N. BELLOMO, On the Euler Equations arising in Discrete Kinetic Theory, Progress en Mecanique, R. Gatignol and Soubaramayer Eds., Springer Lect. Notes in Phys. New York (to appear).
10. N. BELLOMO and E. LONGO, Shock Profile in One Dimension by the Discrete Boltzmann Equation with Multiple Collisions, Waves and Stability in Continuous Media, S. Rionero ed., World Scientific, London, New Yersey, Singapore (1990).
11. N. BELLOMO and I. BONZANI, Nonlinear Shock Waves by the Discrete Boltzmann Equation, Internal Report n.14 Dip. Mat. Politecnico Torino (1990).
12. M. LACHOWICZ and R. MONACO, Existence and Quantitative Analysis of the Solutions to the Initial Value Problem for the Discrete Boltzmann Equation in All Space, SIAM J. Appl. Math.,49 n.4 (1989), 1231-1241.
13. N. BELLOMO and G. TOSCANI, On the Cauchy Problem for the Discrete Boltzmann Equation with Multiple Collisions: Existence, Uniqueness and Stability, Stability and Appl. Anal. 1 (1990).

Ida Bonzani and Nicola Bellomo
Dipartimento di Matematica Politecnico
10129 TORINO (Italy).

G CAMERA-RODA, F DOGHIERI, T A RUGGERI AND G C SART

Thermodynamically consistent rate-type constitutive equations for the diffusive mas flux

INTRODUCTION

The most common constitutive equation for the diffusive flux is represented by Fick's law, which is widely and successfully used in a broad variety of fields. It is well known, on the other hand, that a theoretical criticism has been developed at Fick's law, based essentially on the following points:

i) the mass balance equation obtained by using the Fick's law is parabolic and thus the instantaneous onset of a disturbance at the boundaries leads immediately to a change in the concentration profile in the entire field; this implies an unrealistic "infinite propagation rate".

ii) derivations of an expression for the diffusive flux based on molecular theories lead rather to a rate type constitutive equation including also a relaxation term [1,2].

The problem of a physically consistent constitutive equation has been approached within the framework of heat transfer, in which heat conduction was modelled consistently with a high but finite propagation speed [1,3,4].

In the case of heat transfer, experimental evidence in favour of this conclusion refers to conditions which are hardly ever encountered in common operating conditions and represent the effect of relaxation times which are very small when compared with usual experimental time scales.

When mass transfer is considered, on the contrary, one is rather commonly faced with non Fickian behaviours due to the relaxation phenomena associated with relaxation times which are quite large with respect to the usual experimental time scale and may well be even in the order of hours and sometimes of days.

Indeed, in the case of mass transport of low molecular weight liquids or vapours through solid polymers a wide variety of behaviours is observed, by changing the operating conditions or the polymer-penetrant pair [e.g. 5-12].

It is well accepted that several of the non Fickian transport behaviours are associated with the simultaneous stress and deformation build up which takes place as the mass transport is occurring. Therefore in its generality the description of the phenomenon is rather complex both in its formulation and in its solution, since one needs to consider mass and momentum balances for materials which are usually endowed with a viscoelastic response. Several models are now becoming available in this respect [13-18] all of them ultimately lead to non Fickian constitutive equations for the mass flux. Parallel to this, however, it has been pointed out that the use of viscoelastic models for the diffusive mass flux may be per se sufficient to encompass in a single theory different behaviours, such as the Fickian limit at high concentration, anomalous diffusion, Case II transport, super Case II, and the overshoots in the weight uptake [19-23]. It is worth mentioning, in addition, that the models presented in [19] and in [22] can also reproduce the Fickian limiting behaviour observed at low penetrant concentration values, while the models by Camera-Roda and Sarti [22], and by Kalospiros et al. [23] can also represent a two stage sorption behaviour. Therefore it seems rather attractive to lump all the relaxation effects into a viscoelastic constitutive equation for the mass flux thus reducing the transport problem to the mass species balance equation. In this way the simultaneous evaluation of the displacement, deformation, and stress fields is explicitly bypassed. In this framework it is rather frequent to neglect also the convective bulk transport term, as a first approximation, insofar as it does not introduce qualitatively different features into the problem, although there are cases for which it may be significant [24].

In the present work we analyse the thermodynamic constraints which are imposed by the second law of thermodynamics on the rate type constitutive equations for the mass flux which have been considered so far in the field. For the sake of simplicity our attention will be focused on the case of uniform and constant temperature, in the absence of stress fields.

RATE TYPE CONSTITUTIVE EQUATIONS FOR THE DIFFUSIVE FLUX

Viscoelastic constitutive equations for the diffusive mass flux, consistent with material responses changing with time according to a relaxation time have recently been introduced and analysed following different approaches. Neogi [19] introduced first an integral formulation with material properties (diffusivities and relaxation time) constant with changing composition. On the other side Camera Roda and Sarti [21] analysed a rate constitutive equation of Cattaneo type, which includes material coefficients changing with concentration. Camera-Roda and Sarti developed an other rate model [22] including a retardation term, which reduces also to Neogi's formulation when simplified to the case of concentration independent material properties. Different integral type viscoelastic constitutive equations based also on the relaxation of the stress field [13,14,18,25] will not be considered in the following since we explicitly disregard here any stress field.

For the sake of clarity, it is convenient to briefly recall the basic features of the two basic rate type constitutive models.

The Cattaneo-Maxwell type model for the diffusive mass flux of the penetrant, $\underline{J}$, developed in ref.[21] is embodied by the following equation:

$$\tau \frac{\partial \underline{J}}{\partial t} + \underline{J} = - \mathcal{D} \ \underline{\nabla\phi} \qquad (1)$$

In ref.[21] the diffusivity $\mathcal{D}$ and the relaxation time τ were taken to vary with the concentration of the low molecular weight penetrant according to common relationships based on the well established free volume theory [26-29] which were simplified according to usual procedures as:

$$\mathcal{D} = \mathcal{D}_1 \ e^{K_d \ \phi} = \mathcal{D}_{eq} \ e^{K_d(\phi - \phi_{eq})} \qquad (2)$$

$$\tau = \tau_1 \ e^{-K_\tau \ \phi} = \tau_{eq} \ e^{-K_\tau(\phi - \phi_{eq})} \qquad (3)$$

Typical values for the material property $\mathcal{D}$ are obtained from data for steady state diffusion while, for the relaxation time τ, values were obtained just through analogy with the relaxation response of mechanical properties, so that one assumes

$$\frac{K_d}{K_\tau} \in (0.5 \div 1.0)$$

The quantities $\mathcal{D}_{eq}$ and τ_{eq} represent the diffusivity and relaxation time at the volume fraction ϕ_{eq} reached at equilibrium with the external environment. When the constitutive equation Eq.(1) is coupled with the species balance equation:

$$\frac{\partial \phi}{\partial t} + \underline{\nabla} \cdot \underline{J} = 0 \qquad (4)$$

a hyperbolic mass transport problem is obtained which, in spite of its simplicity, is indeed consistent with several features which are observed in the mass transport of low molecular species through polymers, as it was discussed in ref.[21].

In particular the above Cattaneo-Maxwell model gives rise to shock concentration waves under Riemann jump boundary conditions, with a shock speed whose value is of the typical order of magnitude of the experimentally observed shocks and is decreasing when the concentration jump diminishes, as it is observed in the so called anomalous diffusion. By changing the temperature, or the penetrant activity or even the sample semithickness, δ, the model is able to span a variety of behaviours ranging from the limiting Fickian behaviour at high values of thickness, temperatures or concentration, to Case II transport at low values of the same quantities. The model response is determined by the value of the diffusive Deborah number D_d

$$D_d = \frac{\tau_{eq} \mathfrak{D}_{eq}}{\delta^2} \qquad (5)$$

Overshoots in the weight uptake are also accommodated during transient sorption in finite samples in qualitative agreement with experimental observation.

On the other side the previous model does not predict a Fickian diffusion at very low penetrant concentrations, when the relaxation time is extremely high for solid polymers, as, on the contrary, it is actually observed. However a stringent drawback is associated to the extremely high relaxation time at low concentrations: indeed this may lead to mass fluxes with no significant relaxation during desorption and thus to the unrealistic situation in which an outward solvent flux may be predicted even when no solvent is left within the polymer matrix.

In order to overcome the above difficulties, the viscoelastic Cattaneo-Maxwell term given by Eq.(1) has been considered as a contribution to the diffusive mass flux, to be added to a basic term of Fickian type. The resulting model is formulated as [22]

$$\underline{J} = \underline{J}_R + \underline{J}_F \tag{6}$$

with

$$\underline{J}_F = - \mathfrak{D}_F \, \underline{\nabla} \phi \tag{7}$$

$$\underline{J}_R = - \mathfrak{D}_R \, \underline{\nabla} \phi - \tau \frac{\partial \underline{J}_R}{\partial t} \tag{8}$$

The relaxation time τ is concentration dependent and is represented by Eq.(3). Two diffusivities $\mathfrak{D}_F$ and $\mathfrak{D}_R$ are thus included into the model. $\mathfrak{D}_F$ is the initial diffusivity, $\mathfrak{D}_{in}$, and for steady state initial conditions it is concentration dependent according to Eq.(2), so that:

$$\mathfrak{D}_{in} = \mathfrak{D}_1 \; e^{K_d \phi_{in}} \tag{9}$$

where $\mathfrak{D}_1$ represents the diffusivity of the unpenetrated polymer matrix in the limit $\phi \rightarrow 0$. After the relaxation has taken place an apparent diffusivity holds true $\mathfrak{D}_\infty$ given by

$$\mathfrak{D}_\infty = \mathfrak{D}_1 \; e^{K_d \phi} = \mathfrak{D}_F + \mathfrak{D}_R \tag{10}$$

Therefore $\mathfrak{D}_R$ is obtained from Eqs.(9) and (10).
By using Eqs.(6), (7), and (8) one obtains the following rate type constitutive equation, with a retardation term:

$$\tau \frac{\partial \underline{J}}{\partial t} + \underline{J} = - (\mathfrak{D}_F + \mathfrak{D}_R) \, \underline{\nabla} \phi - \tau \, \mathfrak{D}_F \frac{\partial}{\partial t}(\underline{\nabla} \phi) \tag{11}$$

After substitution into the species mass balance equation, the resulting problem is no longer hyperbolic but rather a higher order problem which gives rise to a hyperbolic-like behaviour whenever the ratio $\mathfrak{D}_{eq}/\mathfrak{D}_1$ is not unity and the diffusive Deborah number, Eq.(5), is neither too small nor too large, say $D_d \in (1,100)$. On the other side a parabolic limiting behaviour (Fickian like transport) is obtained for very large Deborah numbers, $D_d>100$, as well as for very low Deborah numbers, say $D_d<1$.
As it is discussed in ref.[22], the model above accounts for all the physical features actually observed as the Fickian limiting behaviours for very low and very high penetrant concentrations and in addition can represent Case II transport and anomalous diffusion behaviours, oscillations in the weight uptake, and is finally adequate to represent also the two stage sorption behaviour. To our knowledge the model above did not show physically unacceptable features as those encountered in the previous hyperbolic formulation.

THERMODYNAMIC CONSTRAINTS

In both the above rate type models, in the Cattaneo-Maxwell equation as well as in the viscoelastic equation with a retardation type term, the concentration dependence of the material parameters τ and $\mathfrak{D}$ or τ, $\mathfrak{D}_F$, and $\mathfrak{D}_R$ have been proposed based on steady state data or on the analogy with the relaxation behaviour of mechanical properties. No specific test of thermodynamic consistency was performed, based on the entropy inequality. On the other side under some specific conditions some physically unrealistic responses have been obtained which should be ruled out by the prior application of the second law of thermodynamics. In the present section we inspect the constraints which must be obeyed by the rate type equations for the diffusive mass flux, in order to give rise to non negative dissipations for all the processes.

Since thus far the model with the retardation time seems to accommodate a broader variety of the observed behaviours, let's consider first its compatibility with the second law of thermodynamics.

The latter is usually written in terms of an entropy volumetric density s and of an entropy flux $\underline{\psi}$ both of which are given through constitutive equations of the field variables. The dissipation inequality is

$$\frac{\partial s}{\partial t} + \underline{\nabla} \cdot \underline{\psi} \geq 0 \tag{12}$$

The system of Eqs.(4) and (11) may be rewritten in a quasi-linear first order form as:

$$\frac{\partial \phi}{\partial t} + \underline{\nabla} \cdot \underline{J} = 0 \tag{13}$$

$$\tau \left[\frac{\partial \underline{J}}{\partial t} + \mathfrak{D}_F \frac{\partial \underline{g}}{\partial t} \right] = - (\underline{J} + \mathfrak{D}_\infty \underline{g}) \tag{14}$$

$$\underline{\nabla} \phi = \underline{g} \tag{15}$$

where the field is ϕ, $\underline{J}$, and $\underline{g}$; the quantities τ, $\mathfrak{D}_F$, and $\mathfrak{D}_\infty$ are assumed to depend only on the concentration ϕ.

Let us impose the compatibility conditions for the constitutive equations following from the entropy principle (12). By using the Lagrange multiplier method indicated by Liu [30], from Eqs.(13)-(15) and inequality (12) we obtain:

$$\frac{\partial s}{\partial t} + \nabla \cdot \underline{\psi} - \xi \left[\frac{\partial \phi}{\partial t} + \nabla \cdot \underline{J} \right] - \underline{\lambda} \cdot \left\{ \tau \frac{\partial \underline{J}}{\partial t} + \mathfrak{D}_F \, \tau \frac{\partial \underline{g}}{\partial t} + \underline{J} + \mathfrak{D}_\infty \, \underline{g} \right\} -$$
$$- \underline{\varsigma} \cdot \left[\nabla \phi - \underline{g} \right] \geq 0 \qquad (16)$$

The quantities ξ, $\underline{\lambda}$, and $\underline{\varsigma}$ are the Lagrange multipliers associated with Eqs.(13)-(15), respectively.

Since inequality (16) must hold for all the processes, we have

$$ds = \xi \, d\phi + \tau \, \underline{\lambda} \cdot \left\{ d\underline{J} + \mathfrak{D}_F \, d\underline{g} \right\} \qquad (17)$$

$$d\underline{\psi} = \xi \, d\underline{J} + \underline{\varsigma} \, d\phi \qquad (18)$$

The residual inequality becomes:

$$- \underline{\lambda} \cdot \left[\underline{J} + \mathfrak{D}_\infty \, \underline{g} \right] + \underline{\varsigma} \cdot \underline{g} \geq 0 \qquad (19)$$

From Eq.(18) we obtain:

$$\frac{\partial \psi_i}{\partial J_j} = \xi(\phi, \underline{J}) \, \delta_{ij} \qquad \text{that implies} \qquad \underline{\psi} = \xi(\phi) \, \underline{J} \qquad (20)$$

As a consequence from Eq.(17)

$$\frac{\partial s}{\partial \phi} = \xi(\phi), \qquad (21)$$

so that

$$s = s_0(\phi) + k(\underline{J}, \underline{g}) \qquad (22)$$

$$\text{with} \quad \xi = s_0'(\phi) \qquad (23)$$

From Eq. (22), Eq.(17) implies

$$\tau \, \underline{\lambda} = \frac{\partial k}{\partial \underline{J}} \, ; \qquad \tau \, \underline{\lambda} \, \mathfrak{D}_F = \frac{\partial k}{\partial \underline{g}} \qquad (24)$$

From Eq.(24), in view of the constitutive assumption for $\mathfrak{D}_F$, we have soon:

$$\mathfrak{D}_F = \text{const} \, ; \qquad \tau \, \underline{\lambda} = \underline{\Gamma}(\underline{J}, \underline{g}) \qquad (25)$$

and

$$\mathfrak{D}_F \frac{\partial k}{\partial \underline{J}} = \frac{\partial k}{\partial \underline{g}} \qquad (26)$$

Taking into account that k is an objective scalar, the integration of Eq.(26) gives:

$$k = k(q^2) \qquad (27)$$

$$\text{where} \quad \underline{q} = \underline{J} + \mathfrak{D}_F \, \underline{g} \qquad (28)$$

thus Eq.(22) becomes

$$s = s_0(\phi) + k(q^2) \qquad (29)$$

Since in an equilibrium state $\underline{q}=\underline{0}$, we identify $s_0(\phi)$ with the equilibrium entropy function. If we confine our attention, as it is usual, to a linear perturbation from the equilibrium state, in view of stability considerations, we have:

$$s = s_0(\phi) - a\, q^2; \qquad a = \text{const} > 0 \tag{30}$$

At this stage all the conditions derived from Eqs.(17) and (18) are summarised as:

$$\xi(\phi) = s_0'(\phi); \qquad \underline{\zeta} = s_0''(\phi)\, \underline{J}$$

$$\underline{\lambda} = -\frac{2a}{\tau}(\underline{J} + \mathfrak{D}_F\, \underline{g}); \qquad \tau = \tau(\phi); \qquad \mathfrak{D}_F = \text{const} \tag{31}$$

and

$$s = s_0(\phi) - a(\underline{J} + \mathfrak{D}_F\, \underline{g})^2; \qquad \underline{\psi} = s_0'(\phi)\, \underline{J} \tag{32}$$

The residual dissipation inequality becomes:

$$\frac{2a}{\tau} J^2 + [s_0'' + \frac{2a}{\tau}(\mathfrak{D}_F + \mathfrak{D}_\infty)]\, \underline{J}\cdot\underline{g} + \frac{2a}{\tau}\mathfrak{D}_F\, \mathfrak{D}_\infty\, g^2 \geq 0 \tag{33}$$

The condition that the left hand side of inequality (33) is a quadratic form strictly positive for all the non equilibrium processes implies:

$$\tau > 0 \tag{34}$$

$$4\, \mathfrak{D}_F\, \mathfrak{D}_\infty > \left\{\frac{\tau\, s_0''}{2a} + (\mathfrak{D}_F + \mathfrak{D}_\infty)\right\}^2 \tag{35}$$

We remind that the present case is isothermal and completely stress free, so that only entropy plays a role instead of Gibbs free energy and chemical potentials which should enter the problem instead of s_0 and s_0' respectively. Noteworthily, inequality (35) imposes restrictions on the concentration dependence of the relaxation time τ, of $\mathfrak{D}_\infty$, and of the equilibrium entropy s_0. Indeed $\mathfrak{D}_\infty$ coincides with the apparently Fickian diffusivity which holds true during steady state diffusion; its concentration dependence can thus be derived from experimental data in the usual manner. On the other hand the entropy function $s_0(\phi)$ can be obtained through several molecular models [e.g. 31-33]. Without considering detailed analysis of the expressions for s_0 we simply notice that all of them are endowed with the following property:

$$\lim_{\phi \to 0} s_0'' = \lim_{\phi \to 1} s_0'' = -\infty \tag{36}$$

Since all terms in inequality (35) are finite, one also has:

$$\lim_{\phi \to 0} \tau = \lim_{\phi \to 1} \tau = 0 \tag{37}$$

Therefore, in view of the entropy inequality, the concentration dependence of the relaxation time may not be monotonous as it was on the contrary assumed by Camera-Roda and Sarti [22], based on the analogy with mechanical relaxations.

It is worth noting that Eq.(37) directly leads to Fickian behaviour both at very low and at very high penetrant concentrations, according to the experimental evidence. The second relevant requirement embodied by Eq.(25-1) was already accounted for in [22].

In passing we notice that Neogi's model [19] is equivalent to the previous model when the material properties are kept constant; in such a case however the conditions (37) may not be satisfied.

Let us now consider the thermodynamic constraints for the Cattaneo-Maxwell constitutive equation, that have been introduced also in [34].

The field equations are given by Eqs.(1) and (4). By associating the Lagrange multipliers $\underline{\chi}$ and ϑ to those equations, the second law of thermodynamics becomes:

$$\frac{\partial s}{\partial t} - \vartheta \frac{\partial \phi}{\partial t} - \underline{\chi}\, \tau \cdot \frac{\partial \underline{J}}{\partial t} + \underline{\nabla} \cdot \underline{\psi} - \vartheta\, \underline{\nabla} \cdot \underline{J} - \mathfrak{D}\, \underline{\chi} \cdot \underline{\nabla} \phi + \underline{\chi} \cdot \underline{J} \geq 0 \tag{38}$$

Following the same procedure as before we have:

$$ds = \vartheta\, d\phi + \tau\, \underline{\chi} \cdot d\underline{J} \tag{39}$$

$$d\underline{\psi} = \vartheta\, d\underline{J} + \mathfrak{D}\, \underline{\chi}\, d\phi \tag{40}$$

$$- \underline{\chi} \cdot \underline{J} \geq 0 \tag{41}$$

By applying the same arguments as in the previous case we finally obtain:

$$s = s_0(\phi) - \frac{1}{2} \eta\, J^2 \tag{42}$$

$$\underline{\psi} = s_0'(\phi)\, \underline{J} \tag{43}$$

where $s_0 = s_0(\phi)$ indicates the equilibrium entropy and with:

$$\vartheta = s_0'(\phi) \tag{44}$$

$$\underline{\chi} = \frac{s_0''}{\mathfrak{D}}\, \underline{J} \tag{45}$$

$$\frac{\tau(\phi)\, s_0''(\phi)}{\mathfrak{D}(\phi)} = - \eta \equiv \text{const} \tag{46}$$

From Eq.(45) the dissipation inequality (41) implies

$$- \frac{s_0''}{\mathfrak{D}}\, J^2 \geq 0 \tag{47}$$

so that, since stability of equilibrium states requires convexity of s_0 i.e. $s_0'' < 0$, we have

$$\mathfrak{D} \geq 0 \qquad (48)$$

On the other hand, stability considerations applied to Eq.(42) imply that the constant η be positive:

$$\eta > 0 \qquad (49)$$

therefore, from Eqs.(46) and (48) we conclude that the relaxation time is positive

$$\tau = -\frac{\eta\ \mathfrak{D}(\phi)}{s_0''(\phi)} = \tau(\phi) > 0 \qquad (50)$$

In the present model $\mathfrak{D}$ coincides with the steady state diffusion coefficient so that its dependence on concentration can be taken directly from steady state experimental data which are reasonably represented by Eq.(2).

It is worth noting, on the other hand, that the second law implies a precise relationship for the relaxation time, Eq.(46), in terms of diffusivity and equilibrium entropy. Such condition may no longer be given by Eq.(3) which was used based on an analogy with mechanical responses, giving rise to the unrealistic feature during desorption already discussed in the previous section.

Noteworthily, in view of Eq.(36), in both limits of low and high penetrant concentration the diffusional relaxation time τ becomes infinitesimally small:

$$\lim_{\substack{\phi \to 0 \\ \phi \to 1}} \tau(\phi) = \lim_{\substack{\phi \to 0 \\ \phi \to 1}} -\eta \frac{\mathfrak{D}}{s_0''} = 0 \qquad (51)$$

In other words the entropy inequality implies that the Maxwell-Cattaneo model reduces to the parabolic Fickian model for $\phi \to 0$ and $\phi \to 1$.

Remarkably, the latter property overcomes the difficulty which was found for the model by using the expression in Eq.(3).

The hyperbolic Maxwell-Cattaneo model allows for a concentration dependence of the relaxation time, which was on the contrary explicitly ruled out in the rate model with a retardation time. In this respect the hyperbolic model appears by far more satisfactory.

The concentration dependence of the relaxation time given by Eq.(50) is not monotonous and apparently this is not a common feature for the mechanical relaxation time.

CONCLUSIONS

Two rate type constitutive equations for the diffusive mass flux have been considered: one is the Maxwell-Cattaneo model while the other contains a retardation time. In both of the models the material properties, that is relaxation time and diffusivities, are concentration dependent. When the concentration dependence of the relaxation time is monotonous and taken simply based on the analogy with the relaxation of mechanical properties, Eq.(3), the model with a retardation term seems to accommodate a broad spectrum of the behaviours experimentally observed in the diffusion of low molecular weight penetrants through solid polymers, while the Maxwell-Cattaneo model leads to apparently unrealistic behaviours.
Compatibility with the second law of thermodynamics implies that i) $\mathfrak{D}_F$ is a constant value for the problem; ii) the dependence of the diffusional relaxation time τ on concentration has to obey the residual dissipation inequality (35), so that τ is not monotonous, in view of the properties of the equilibrium entropy function $s_0(\phi)$. In particular the value of the diffusional relaxation time vanishes in the limits of very high as well as of very small penetrant concentrations, thus recovering in both the cases the usual Fickian behaviour.
On the other side, when the Maxwell-Cattaneo model is made consistent with the second law, a concentration dependence of the relaxation time implied by Eq.(3). That expression, in addition, recovers the parabolic Fickian limit both for very small and very high penetrant concentrations, also for the Cattaneo-Maxwell model.

*Acknowledgements. This work was partially supported by C.N.R. grant No.*90.00871.*PF*72 *and by M.U.R.S.T.* 40%.

REFERENCES

1. C.CATTANEO, Sulla conduzione del calore. Atti Sem. Mat. Fis. Univ. Modena 3 (1948) 83.
2. I.MULLER, Thermodynamics (Pitman Publ. Co. London 1985).
3. P.VERNOTTE, Les Paradoxes de la Thorie Continue de l'équation de la chaleur. Compt. Rend. Acad. Sci. Paris. **246** (1958) 3154.

4. H.D.WEYMAN, Finite Speed of Propagation in Heat Conduction, Diffusion and Viscous Shear Motion. Am. J. Phys. **35** (1967) 488.
5. A.S.HARTLEY, Diffusion and Swelling in High Polymers. Part III. Trans. Faraday Soc. **45** (1946) 820-832.
6. F.A.LONG and R.J.KOKES, Diffusion of Benzene and Methylene Chloride Vapors into Polystyrene. J. Am. Chem. Soc. **75** (1953) 2232-2237.
7. S.PARK, An Experimental Study of the Influence of Various Factors on the Time Dependent Nature of Diffusion in Polymers. J. Polym. Sci. **11** (1953) 97-115.
8. A.C.NEWNS, The Sorption and Desorption Kinetics of Water in a Regenerated Cellulose. Trans. Faraday Soc. **52** (1956) 1533-1545.
9. F.A.LONG and D.J.RICHMAN, Concentration Gradients for Diffusion of Vapors and Their Relation to Time Dependent Diffusion Phenomena. J. Am. Chem. Soc. **82** (1960) 513-519.
10. E.F.GURNEE, Measurement of Concentration Gradients in Swelling Crosslinked Polymer Beads. J. Polym. Sci., A-2 **5** (1967) 799-816.
11. H.B.HOPFENBERG, Anomalous Transport of Penetrants in Polymeric Membranes. In Membrane Science and Tecnology, J. Flinn Ed. Plenum New York (1970) 16-32.
12. L.NICOLAIS, E.DRIOLI, H.B.HOPFENBERG, and D.TIDONE, Characterization and Effects of n-Alkane Swelling of Polystyrene Sheets. Polymer **18** (1977) 1137-1142.
13. J.H.PETROPOULOS and P.P.ROUSSIS, The Influence of Transverse Differential Swelling Stresses on the Kinetics of Sorption of Penetrants by Polymer Membranes. J. Membr. Sci. **3** (1978) 343-356.
14. J.H.PETROPOULOS, Application of Transverse Differential Swelling Stress Model to the Interpretation of Case II Diffusion Kinetics. J. Polym. Sci., Polym. Phys. Ed. **22** (1984) 183-189.
15. F.DOGHIERI, G.C.SARTI, and R.G.CARBONELL, Sorption of Penetrants in Rubbery Polymers: Coupled Effects of Concentration and Stress Field. In Diffusion in Polymers, P.R.I. Ed. Reading (1988) 18/1-18/30.
16. S.R.LUSTIG, J.M.CARHUTERS, and N.A.PEPPAS, Development and Implementation of a Model for Penetrant Transport in Glassy Polymers. paper 210f Annual AIChE Meeting S.Francisco (Nov. 1989).
17. K.N.MORMAN, Stress Assisted Diffusion of Perfect Fluids in Simple Solids with Fading Memory. paper 210d Annual AIChE Meeting

S.Francisco (Nov. 1989).

18. R.G.CARBONELL and G.C.SARTI, Coupled Deformation and Mass Transport Processes in Solid Polymers. I.E.C. Res. **29** (1990) 1194-1204.
19. P.NEOGI, Anomalous Diffusion of Vapors through Solid Polymers. II: Anomalous Sorption. AIChE J. **29** (1983) 833.
20. F.ABID and P.NEOGI, Sorption with Oscillation in Solid Polymers. AIChE J. **33** (1987) 164.
21. G.CAMERA-RODA and G.C.SARTI, Non Fickian Mass Transport through Polymers: A Viscoelastic Theory. Tran. Theory Stat. Physics. **15** (1986) 1023.
22. G.CAMERA-RODA and G.C.SARTI, Mass Transport with Relaxation in Polymers. AIChE J. **36** (1990) 851-860.
23. N.KALOSPIROS, R.OCONE, G.ASTARITA, and J.R.MELDON, Analysis of Anomalous Diffusion and Relaxation in Solid Polymers. I.E.C. Res. in Press 1991.
24. G.C.SARTI, G.GOSTOLI, G.RICCIOLI, and R.G.CARBONELL,Transport of Swelling Penetrants in Glassy Polymers: Influence of Convection. J. Appl. Polym. Sci. **32** (1986) 36.
25. C.J.DURNING and M.TABOR, Mutual Diffusion in Concentrated Polymer Solutions under a Small Driving Force. Macromol. **19** (1986) 2220.
26. H.FUJITA, Diffusion in Polymer-Diluent Systems. Fortschr. Hoch-polym.-Forsch. **3** (1961) 1.
27. J.D.FERRY, Viscoelastic Properties of Polymers, 2d Ed. (Wiley New York 1970).
28. M.FELS and R.Y.M.HUANG, Diffusion Coefficients of Liquids in Polymer Membranes by a Desorption Method. J. Appl. Polym. Sci. **14** (1970) 523.
29. S.M.FANG, S.A.STERN, and H.L.FRISCH, A 'Free Volume' Model of Permeation of Gas and Liquid Mixtures through Polymeric Membranes. Chem. Eng. Sci. **30** (1975) 773.
30. I-SHIH LIU, Method of Lagrange Multipliers for Explanation of the Entropy Principle. Arch. Rat. Mech. Anal. **46** (1972) 131.
31. P.J.FLORY, Principles of Polymer Chemistry (Cornell University Press Ithaca N.Y. 1953).
32. P.J.FLORY, Statistical Thermodynamics of Liquid Mixtures. J. Am. Chem. Soc. **87** (1965) 1833-1846.
33. R.H.LACOMBE and I.C.SANCHEZ, Statistical Thermodynamics of Fluid

Mixtures. J. Phys. Chem. **80** (1976) 2568-2580.

34. A.MORRO and T.RUGGERI, Non Equilibrium Properties of Solids obtained from Second Sound Measurements. J.Phys. C:Solid State Phys. **21** (1988) 1743-1752.

Giovanni Camera-Roda
Ferruccio Doghieri
Giulio C. Sarti

Dipartimento di Ingegneria Chimica e di Processo
Università di Bologna
Viale Risorgimento, 2
40136 BOLOGNA, ITALY

Tommaso A. Ruggeri

Dipartimento di Matematica
Università di Bologna
piazza Porta S.Donato, 5
40127 BOLOGNA, ITALY

V CIANCIO

Thermodynamic theory for heat radiation

Abstract.

Basing on the conclusions of a previous work [1] a simple thermodynamic model is established for radiating heat transfer through the air. The model is applied to the temperature distribution of the atmosphere and the radiating heat loss of the Earth.

1. Introduction.

In a recent paper [1] we shown out that the heat current can be spilt into $n+1$ parts, one of which is obeying Fourier's equation while the others are described by the following Cattaneo-Vernotte type equations.

$$\vec{\mathbf{J}}^{(k)} + \tau^{(k)} \frac{d\vec{\mathbf{J}}^{(k)}}{dt} = -\lambda^{(k)} gradT \quad (k = 1, 2, \ldots, n). \tag{1.1}$$

where $\tau^{(k)}$ and $\lambda^{(k)}$ are the relaxation times and the partial heat conductivities, respectively.

If insisting on the final velocity of signal propagations, one can not help but inferring that the heat conductivity in Fourier's equation vanishes, and in this case the heat propagation is purely radiation in the semi transparent medium.

In some practical applications, a sufficient accuracy is reached by assuming a uniform absorption coefficient, neglecting that it depends on the frequency of the electromagnetic wave. If applying this approximation, the heat flow becomes a unique process (n turns into 1) and our treatment reduces to Gyarmati's wave approach [3]. The simple model obtained this way includes also the hypothesis that the radiation field is in thermal equilibrium with the material. It seems to over-simplify the model for the phenomena of heat radiation in an atmosphere at rest we intend to investigate.

The purpose of this work is to establish a rather simple thermodynamic model for the heat emission of the Earth during a night when no convection is present in the atmosphere. The heat conduction in the air can be neglected with respect to the radiation according to the large (scale of) distances.

We suppose that the phenomenon is modeled well by two superposed continua, one of which does not conduct heat and the heat propagation follows Gyarmati's wave approach [3].

2. State variables and entropy.

We assume a medium consisting of two superposed continua the local state of which is given by the specific internal energy of the one u_m, the energy density of the other u_r and the heat flow $\vec{\mathbf{J}}$. The specific internal energy of the medium is given as the sum of the two energies

$$u = u_m + \frac{1}{\varrho} u_r , \tag{2.1}$$

where ϱ stands for the density of the material. The mass of the radiating field is neglected. The first law of thermodynamics reads

$$\varrho \frac{du}{dt} + div\, \vec{\mathbf{J}} = 0 . \tag{2.2}$$

This equation excludes the creation or diminishing of the energy but allows heat exchange between the two continua. The specific entropy function is supposed to consist of three terms

$$s = s_m(u_m) + \frac{1}{\varrho} s_r(u_r) - \frac{m}{2} \vec{\mathbf{J}}^2 . \tag{2.3}$$

The first term $s_m(u_m)$ is the specific entropy of the material (*air*) depending on the specific internal energy of the material, the second term $s_r(u_r)$ is the entropy density of the radiation while the last term is the *kinetic* part as proposed by Gyarmati [2]. The coefficient m is characteristic for the inertia of the radiating heat transfer.

The maximum of the entropy with the constraint condition is determined from the partial derivatives of the auxiliary function

$$F = s + \mu u = s_m(u_m) + \frac{1}{\varrho} s_r(u_r) + \\ - \frac{m}{2} \vec{\mathbf{J}}^2 + \mu \left(u_m + \frac{1}{\varrho} u_r \right), \qquad (2.4)$$

where μ is Lagrange's multiplier.

The equations for a local equilibrium state read

$$\frac{ds_m}{du_m} + \mu = 0, \qquad \frac{1}{\varrho} \frac{ds_r}{du_r} + \mu \frac{1}{\varrho} = 0, \qquad and \quad \vec{\mathbf{J}} = 0, \qquad (2.5)$$

from which the condition for an equilibrium is obtained in the form

$$\frac{1}{T_m} = \frac{ds_m}{du_m} = \frac{ds_r}{du_r} = \frac{1}{T_r} \qquad (2.6)$$

This equation says that the temperatures of the material and the radiation field are equal in equilibrium. It means that the energy density of the radiation field is determined by specific internal energy of the material if they are in equilibrium.

3. Entropy balance.

The entropy balance is sought in the usual form

$$\varrho \frac{ds}{dt} + div \frac{\vec{\mathbf{J}}}{T_r} = \sigma_s \geq 0 \qquad (3.1)$$

Here we presumed that the entropy flow $\vec{\mathbf{J}}/T_r$ is joined to the heat flow $\vec{\mathbf{J}}$ by the reciprocal temperature of the radiating field.

To determine the actual form of the entropy production density σ_s, combine the time derivative of eq. (2.1) with eq. (2.2)

$$\varrho \frac{du_m}{dt} + \frac{du_r}{dt} + div \, \vec{\mathbf{J}} = 0 \qquad (3.2)$$

This form of the internal energy balance together with the time derivative of the entropy (2.3) and the general form of the entropy balance (3.1) performs the actual form of the entropy production. It reads

$$\sigma_s = \varrho \frac{du_m}{dt} \left[\frac{1}{T_m} - \frac{1}{T_r} \right] + \vec{\mathbf{J}} \left[grad \frac{1}{T_r} - m \frac{d\vec{\mathbf{J}}}{dt} \right] . \tag{3.3}$$

It is seen, that the entropy production is due to two phenomena. The first term on the right hand side of (3.3) gives the contribution of heat exchange between the semitransparent material and the field of radiation and the second is the contribution of heat propagation. Each term is a product of the rate of a process and an affinity conjugate to it. The affinity conjugate to heat propagation contains a term proportional to the time derivative of the heat flow in accordance with Gyarmati's wave approach.

4. Phenomenological equations.

According to the usual procedure of non-equilibrium thermodynamics, we have the following phenomenological equations for an anisotropic medium by virtue of the form (3.3) for the entropy production

$$\varrho \frac{du_m}{dt} = L^{(e,e)} \frac{T_r - T_m}{T_r T_m} + \sum_{\gamma=1}^{3} L_\gamma^{(e,r)} \left[- \frac{1}{T_r^2} \frac{\partial T_r}{\partial x_\gamma} - m \frac{dJ_\gamma}{dt} \right] , \tag{4.1}$$

$$J_\alpha = L_\alpha^{(r,e)} \frac{T_r - T_m}{T_r T_m} + \sum_{\gamma=1}^{3} L_{\alpha\gamma}^{(r,r)} \left[- \frac{1}{T_r^2} \frac{\partial T_r}{\partial x_\gamma} - m \frac{dJ_\gamma}{dt} \right] , \tag{4.2}$$

The tensors L are called phenomenological tensors. If the material is isotropic the phenomenological tensors are invariants of the orthogonal group. Following the method of Smith and Rivlin we obtain for the components of the L's

$$L^{(e,e)} = k\, T_r\, T_m\, , \tag{4.3}$$

$$L_\gamma^{(e,r)} = 0\, , \tag{4.4}$$

$$L_{\alpha\gamma}^{(r,r)} = \lambda\, T_r^2\, \delta_{\alpha\gamma}\, . \tag{4.5}$$

The k and λ are new phenomenological quantities replacing the L's in an isotropic medium. Introducing the above relations in (4.1) and (4.2), moreover, replacing the inertial coefficient m by

$$m = \frac{\tau}{T_r^2 \lambda} \tag{4.6}$$

we obtain the phenomenological relations for an isotropic medium

$$\varrho \frac{du_m}{dt} = k (T_r - T_m), \tag{4.7}$$

$$\vec{\mathbf{J}} + \tau \frac{d\vec{\mathbf{J}}}{dt} = - \lambda \, grad \, T_r , \tag{4.8}$$

The first equation is the usual one for heat exchange between two media and the other coincides with the equation derived in [1] for radiating heat transfer.

5. Emission and absorption.

As it was derived in [1] the radiating energy density and the heat flow obey the equations

$$\frac{du_r}{dt} + div\vec{\mathbf{J}} = e - acu_r , \tag{5.1}$$

$$\frac{1}{ac} \frac{d\vec{\mathbf{J}}}{dt} = - \frac{c}{a} gradu_r , \tag{5.2}$$

which are the invariant forms of (A.7) and (A.8) in ref.[1]. Here c is the velocity of the wave, a is the absorption coefficient and e is the emission per unit length. In that paper we did not make distinction between the molecular absorption of radiation and the scattering processes due, e.g. to the repeated refraction on the surfaces of microscopic droplets. We simply supposed that the absorbed energy is remitted immediately.

In the present model the two process has to be distinguished. The absorption coefficient is split into two parts

$$a = a_m + s , \tag{5.3}$$

where a_m gives account on the molecular absorption and s does so on the scattering process. The scattered radiation can be assumed as if it was absorbed and emitted immediately so the emission e in (5.1) has also to be split into two parts

$$e = e_m + e_s \,, \tag{5.4}$$

The emission of the molecules are given an account on by e while e_s gives the scattered radiation

$$e_s = s\,c\,u_r \,. \tag{5.5}$$

The quantities applied are not independent. If analyzing the radiation in a sample of the material between two black bodies, we arrive at

$$c\,u_r = 2\,e_0(T_r)\,, \tag{5.6}$$

which is the actual form of Kirchhoff's law. The function $e_0(T_r)$ stands for the emission of the black body at temperature T_r.

From here

$$e_s = 2\,s\,e_0(T_r)\,, \tag{5.7}$$

and

$$e_m = 2\,a_m\,e_0(T_m) \tag{5.8}$$

follow.

Comparing the above equations with the equations (2.1), (2.2), (4.7) and (4.8), we detect complete agreement and find

$$k = 2\,a_m\,\frac{e_0(T_r) - e_0(T_m)}{T_r\,T_m} \approx 2\,a_m\,\frac{d\,e_0(T)}{dT}\,, \tag{5.9}$$

$$\tau = \frac{1}{a\,c} \tag{5.10}$$

$$\lambda = \frac{c}{a}\,\frac{d\,u_r}{dt} = \frac{2}{a}\,\frac{d\,e_0(T_r)}{dT_r} \tag{7.11}$$

for the phenomenological coefficients.

6. Cooling in the night.

When applying the equations derived for the heat emission of the Earth during a night the mathematical problem simplifies to one dimension. The set of the differential equations to be solved are

$$\varrho \frac{\partial u_m(T_m)}{\partial t} = 2 a_m \left[e_0(T_r) - e_0(T_m) \right] \tag{6.1}$$

$$J + \frac{1}{a c} \frac{\partial J}{\partial t} = -\frac{2}{a} \frac{\partial e_0(T_r)}{\partial x}, \tag{6.2}$$

$$\varrho \frac{\partial u_m(T_m)}{\partial t} + \frac{\partial u_r(T_r)}{\partial t} + \frac{\partial J}{\partial x} = 0, \tag{6.3}$$

where J is the heat flow upward and x is a vertical coordinate, say, the height above the surface. The boundary conditions are rather obvious, both temperatures are equal to that of the soil at the surface and to zero at infinity. It seems better to look for the functions $e_0(T_r)$ and $e_0(T_m)$ instead of T_r and T_m themselves.

The equations are easy to solve for a stationary state in an atmosphere the chemical composition of which is uniform. In this case the equations reduce to

$$e_0(T_r) - e_0(T_m) = 0, \tag{6.4}$$

$$J = -\frac{2}{a} \frac{d e_0(T_r)}{dx}, \tag{6.5}$$

$$\frac{dJ}{dx} = 0, \tag{6.6}$$

where the only complication is that the coefficient a depends on the place.

According to the spectroscopic facts, the coefficient a is supposed to be proportional with the density ϱ, so we write

$$\frac{a_0}{\varrho_0} \varrho J = -2 \frac{d e_0(T_r)}{dx} \tag{6.7}$$

with constant J. Here a_0 and ϱ_0 are the absorption coefficient and the density at the surface. The density can be replaced by the derivative of the atmosphere pressure with respect to x as

$$\frac{dp}{dx} = -\varrho g, \tag{6.8}$$

where g is the gravitational acceleration. So

$$\frac{a_0}{g\,\varrho_0} J \frac{dp}{dx} = 2\,\frac{d\,e_0(T_r)}{dx} \tag{6.9}$$

which is easy to solve

$$\frac{a_0}{g\,\varrho_0} J\,p = 2\,e_0(T_r) + C\,. \tag{6.10}$$

the integration constant C is obviously zero, as both the pressure and T_r is practically zero in the space. From here, we determine the intensity of the heat flow

$$J = \frac{2\,g\,\varrho_0}{a_0\,p_0}\,e_0(T_0)\,, \tag{6.11}$$

where p_0 and T_0 are the pressure and the temperature at the surface, respectively.

References.

[1] Ciancio, V., Verhás, J., A thermodynamic theory for radiating heat transfer, J. Non-Equilib. Thermodyn., 15 (1990), 33.

[2] Ciancio, V., Verhás, J., On heat conduction in media with isotropic microstructure, Atti Accad. Peloritana dei Pericolanti, Classe I di Sc.Mat., Fis. e Nat.,vol.LVIII (1990) (to be published).

[3] Gyarmati, I., On the wave approach of thermodynamics and some problems of non-linear theories, J. Non-Equilib. Thermodyn., 2 (1977), 233.

Vincenzo Ciancio
Department of Mathematics
University of Messina
contrada Papardo, salita Sperone 31
98166 Sant'Agata - Messina
ITALY

S GIAMBÓ AND G LEBON

Helium II: an extended thermodynamical approach

ABSTRACT

A phenomenological theory where liquid helium II is viewed as a one-component fluid is proposed. The model is constructed by following the line of thought of extended irreversible thermodynamics wherein the thermodynamical fluxes are taken as independent variables. Besides the classical variables that are mass and energy, or entropy, we consider as independent variables the heat flux **q** and a supplementary variable ξ describing the motion of elementary excitations, whose nature is similar to that of phonons in a cristal lattice. The most general linearized evolution equations for **q** and ξ are established. The equation for **q** is a generalization of Maxwell-Cattaneo relation while the equation for ξ is new and has no counterpart in the literature.

1. Introduction

Liquid helium II is usually modelled by Landau's two fluid model [1,2] : He II is considered as a mixture of a normal fluid with a non-zero viscosity and a superfluid with zero viscosity and zero entropy. Although the two-fluid model is able to interprete most of the striking effects observed in He II, it suffers from some important deficiencies reported by Landau himself [1] and several people [2]-[6]. That has motivated the search for another description based on a single fluid model. A first step towards this direction was done by Atkin and Fox [3]-[5]. They viewed He II as a single fluid and proposed a continuum theory based on rational thermodynamics. This work has been criticized by Lebon and CJou [7] who still considered He II as a one-component fluid but instead of rational thermodynamics, they used Onsager's classical theory of irreversible processes [8,9] was used. In addition to the classical variables, namely energy and volume, Lebon and Jou [7] introduced an internal variable describing the microscopic excitations of the atoms. When referred to Landau's two-fluid model, it is seen that internal variable is related to the relative velocity between the normal and the superfluid.

Our objective in this note is to describe the phenomenological behaviour of He II within the framework of extended irreversible thermodynamics. This theory has known an increasing interest during the last decade (for a review, see [10]) and it would be interesting to check whether such a formalism is able to cope with particular system like He II. The main field equations are derived in section 2. A particular case is considered in section 3.

2. Extended irreversible approach of He II

The main characteristic of extended thermodynamic is to raise the thermodynamic fluxes like the flux of matter, the heat flux, . . ., to the status of independent variables. Evolution equations for these extra variables are then derived on the basis of representation theorems of tensors; restriction on the forms of the evolution equations are placed by the second principle of thermodynamics.

We shall view He II as a one-component rather than a two-component fluid. In the present description, the classical variables, namely the internal energy u and the specific mass ρ, are augmented by two extra variables given by the heat flux vector $\mathbf{q}$ and by an internal vector valued variable ξ ; the latter takes into account for the motion of microscopic excitations known as phonons and rotons which are regarded here as a continuum. These excitations are known to be responsible for the particular phenomena observed im He II. More about the physical meaning of the variable ξ can be found in previous papers [5],[7],[13].

While in classical irreversible thermodynamics, the entropy of an ordinary fluid is function only of the internal energy u and of the specific mass ρ, in extended thermodynamics it is postulated that there exists a non-equilibrium entropy S with the following properties:

1) it is an analytic function of u, ρ, $\mathbf{q}$ and ξ;
2) it is a convex function of the whole set of variables;
3) the rate of entropy production per unit mass is positive definite.

As a consequence of property (1), one may write the generalised Gibbs equation under the form:

$$dS = T^{-1}du - pT^{-1}\rho^{-2}d\rho - T^{-1}\rho^{-1}q\,(\alpha_1 dq + \alpha_2 d\xi) \qquad (1)$$
$$-T^{-1}\rho^{-1}\xi\,(\gamma^1 dq + \gamma_2 d\xi)$$

where the coefficients α_1, α_2, γ_1 and γ_2 may depend on u and ρ. By writing (1), it was postulated that the absolute temperature T and the pressure p are given by their classical definitions

$$T^{-1} = \frac{\partial S}{\partial u} \qquad , \qquad p = -\rho^2 T \frac{\partial S}{\partial \rho} \tag{2}$$

In (1), it is also assumed that the derivatives of S with respect to the fluxes $\mathbf{q}$ and $\boldsymbol{\xi}$ are restricted to first order terms in the fluxes. From the equality of the crossed derivatives of S with respect to $\mathbf{q}$ and $\boldsymbol{\xi}$, it is inferred that

$$\alpha_2 = \gamma_1 \tag{3}$$

Combining the balance equations of mass and energy, namely

$$\frac{d\rho}{dt} = -\rho \nabla \cdot \mathbf{v} \, , \tag{4}$$

$$\rho \frac{du}{dt} = - \nabla \cdot \mathbf{q} - p \nabla \cdot \mathbf{v} \, , \tag{5}$$

with the time derivative of expression (1), one obtains

$$\begin{aligned} \rho \frac{dS}{dt} = &- \nabla \cdot \frac{\mathbf{q}}{T} + \mathbf{q} \cdot \left(\nabla T^{-1} - T^{-1} \alpha_1 \frac{d\mathbf{q}}{dt} - T^{-1} \gamma_1 \frac{d\boldsymbol{\xi}}{dt} \right) + \\ &- \boldsymbol{\xi} \cdot \left(T^{-1} \gamma_1 \frac{d\mathbf{q}}{dt} + T^{-1} \gamma_2 \frac{d\boldsymbol{\xi}}{dt} \right) \geq 0 \, . \end{aligned} \tag{6}$$

where ∇ denotes the nabla operator and viscous effects are ignored.

Relation (6) may be considered as a balance equation for the entropy on condition of identifying the entropy flux $\mathbf{J}^S$ and the entropy production σ^S respectively with

$$\mathbf{J}^S = \frac{1}{T} \mathbf{q} \, , \tag{7}$$

$$\begin{aligned} \sigma^S = &\, \mathbf{q} \cdot \left(\nabla T^{-1} - T^{-1} \alpha_1 \frac{d\mathbf{q}}{dt} - T^{-1} \gamma_1 \frac{d\boldsymbol{\xi}}{dt} \right) + \\ &- \boldsymbol{\xi} \cdot \left(T^{-1} \gamma_1 \frac{d\mathbf{q}}{dt} + T^{-1} \gamma_2 \frac{d\boldsymbol{\xi}}{dt} \right) \geq 0 \end{aligned} \tag{8}$$

The entropy production (8) has the structure of a bilinear form. The simplest linearized expressions of the evolution equations for $\mathbf{q}$ and $\boldsymbol{\xi}$ compatible with the positiveness property of σ^S are readily obtained. Following the usual procedure of extended thermodynamics [10], the evolution equations for the fluxes can be cast into the form

$$\begin{bmatrix} \frac{dq}{dt} \\ \\ \frac{d\xi}{dt} \end{bmatrix} = \frac{T}{\Delta} \nabla \begin{bmatrix} \frac{\gamma_2}{T} \\ \\ -\frac{\gamma_1}{T} \end{bmatrix} + \underset{\approx}{B} \cdot \begin{bmatrix} q \\ \\ \xi \end{bmatrix} \tag{9}$$

where Δ and $\underset{\approx}{B}$ stand, respectively, for

$$\Delta = \alpha_1 \gamma_2 - \gamma_1^2 , \tag{10}$$

$$\underset{\approx}{B} = -\frac{T}{\Delta} \begin{bmatrix} \gamma_2 \mu_{11} - \gamma_1 \mu_{21} & \gamma_2 \mu_{12} - \gamma_1 \mu_{22} \\ \\ \alpha_1 \mu_{21} - \gamma_1 \mu_{11} & \alpha_1 \mu_{22} - \gamma_1 \mu_{12} \end{bmatrix} \tag{11}$$

Clearly, equations (9) have the structure of conservation laws.

Substitution of equations (9) into the expressions (8) of σ^S gives rise to

$$\sigma^S = \mu_{11} q \cdot q + (\mu_{12} + \mu_{21}) q \cdot \xi + \mu_{22} \xi \cdot \xi \geq 0 \tag{12}$$

whereupon the following constraints are obtained

$$\mu_{11} \geq 0 , \quad \mu_{22} \geq 0 , \quad \mu_{11} \mu_{22} > \frac{1}{4} (\mu_{12} + \mu_{21})^2 . \tag{13}$$

Further restrictions on the coefficients μ_{12} and μ_{21} are derived by invoking Onsager's reciprocity relations. To derive them, let us write the derivatives of S with respect to the fluxes in matricial form; from Gibbs' equation (1), one has

$$\begin{bmatrix} \frac{\partial S}{\partial q} \\ \\ \frac{\partial S}{\partial \xi} \end{bmatrix} = \underset{\approx}{g} \cdot \begin{bmatrix} q \\ \\ \xi \end{bmatrix} \tag{14}$$

where $\underset{\approx}{g}$ denotes the simmetrix matrix

$$\underset{\approx}{g} = -\frac{1}{\rho T} \begin{bmatrix} \alpha_1 & \gamma_1 \\ \gamma_1 & \gamma_2 \end{bmatrix} . \tag{15}$$

On a time-scale at which the temperature remains constant, the evolution equations (9) are given by

$$\begin{bmatrix} \dfrac{dq}{dT} \\ \\ \dfrac{d\xi}{dT} \end{bmatrix} = \underset{\approx}{\mathbf{M}} \cdot \begin{bmatrix} \dfrac{\partial S}{\partial q} \\ \\ \dfrac{\partial S}{\partial \xi} \end{bmatrix} \tag{16}$$

with

$$\underset{\approx}{\mathbf{M}} = \underset{\approx}{\mathbf{B}} \cdot \underset{\approx}{\mathbf{g}}^{-1} \quad . \tag{17}$$

Expression (16) is similar to the relation satisfied by the α variables in Onsager's theory [8],[9]. According to Onsager, the matrix M is simmetric so that it follows

$$(\alpha_1\gamma_2 - \gamma_1^{2})\mu_{12} = (\alpha_1\gamma_2 - \gamma_1^{2})\mu_{21}$$

or

$$\mu_{12} = \mu_{21} \tag{18}$$

In short, we have modelled liquid helium by means of four variables ρ, u, **q**, ξ. The time evolution in the course of time of ρ and u is given by the classical conservation laws of mass (4) and energy (5); as far as the fluxes are concerned, their evolution obeys equations (9) which can be written in the form

$$\frac{d\mathbf{q}}{dt} = -\frac{T}{\Delta}\Big((\gamma_2\mu_{11} - \mu_{12}\gamma_1)\mathbf{q} + (\gamma_2\mu_{12} - \mu_{22}\gamma_1)\boldsymbol{\xi} \; - \gamma_2\nabla T^{-1}\Big), \tag{19}$$

$$\frac{d\boldsymbol{\xi}}{dt} = -\frac{T}{\Delta}\Big((\alpha_1\mu_{12} - \gamma_1\mu_{11})\mathbf{q} + (\alpha_1\mu_{22} - \gamma_1\mu_{12})\boldsymbol{\xi} \; + \gamma_1\nabla T^{-1}\Big), \tag{20}$$

where the coefficients μ_{ij} (i, j = 1, 2) are subject to the constraints (13) and (18). Supplementary informations about the coefficients α_1, γ_1 and γ_2 can be derived from the convexity property of S :

$$\frac{\partial^2 S}{\partial\mathbf{q}\partial\mathbf{q}} < 0 \;, \quad \frac{\partial^2 S}{\partial\boldsymbol{\xi}\partial\boldsymbol{\xi}} < 0 \;,$$

or, equivalently,

$$\alpha_1 > 0 \qquad\qquad \gamma_2 > 0 \,. \tag{21}$$

The results (13), (18) and (21) provide all the informations that can be drawn from our macroscopic description.

3. A particular case

In order to simplify somewhat the model, we make the following assumption

$$\frac{\gamma_2}{\gamma_1} = \frac{\mu_{22}}{\mu_{12}} = \text{const.} \tag{22}$$

so that the pair of equations (20) and (21) reduces to

$$\frac{d}{dt}\left(q + \frac{\gamma_2}{\gamma_1}\,\xi\right) = -\,T\frac{\mu_{22}}{\gamma_2}\left(q + \frac{\gamma_2}{\gamma_1}\,\xi\right) \tag{23}$$

It is straightforward to ascertain that the equation (23) admits the particular solution

$$q = -\,\frac{\mu_{22}}{\mu_{12}}\,\xi \tag{24}$$

which means that in He II, the heat flux is mainly driven by excitation motions. The property that, in He II, heat can flow even in absence of a temperature gradient was explicitly pointed out by Kapitza [12].

Substitution of the relation (23) into equation (19), yields

$$\frac{d\xi}{dt} = \frac{\gamma_1}{\Delta}\,\frac{\nabla T}{T} - \frac{\gamma_1 T}{\mu_{12}\Delta}\,(\mu_{11}\mu_{22} - \mu_{12}{}^2)\xi \tag{25}$$

Bearing in mind the approximation (22), Gibbs equation (1) reads as

$$du = TdS - p\rho^{-2}d\rho + \alpha\,\xi\cdot d\xi\,, \tag{26}$$

where we have set

$$\rho\alpha = \frac{\gamma_2}{\gamma_1{}^2}\,\Delta\,. \tag{27}$$

By comparing with Landau's two fluid theory, it has been shown earlier [11] that the quantities ξ, α and $\frac{\gamma_1}{\Delta}$ can be identified as

$$\xi = \frac{\rho_n}{\rho}\left(v_n - v_s\right), \tag{28}$$

$$\frac{\gamma_1}{\Delta} = - S, \tag{29}$$

$$\alpha = \frac{\rho_s}{\rho_n} . \tag{30}$$

where the subscript n refers to the normal fluid and index s to the superfluid.

ACKNOWLEDGEMENTS

Part of this work was done during the stay of one of us (G.Lebon) at the Dipartimento di Matematica dell'Universita' di Messina. It is a pleasure for him to thank all the members of the Department and, more particularly, Professors A. PALUMBO and G. VALENTI for their warm hospitality and for useful discussions.

REFERENCES

[1] L. Landau, J. Phys. U.S.S.R. , 5, (1941), 71.

[2] L. Landau and E. Liftschitz, *Fluid Mechanics*, Pergamon Press, 1966.

[3] R. Atkin and N. Fox, Acta Mech., 21, (1978), 221.

[4] R. Atkin and N. Fox, J. Sound Vibr., 42, (1975), 13.

[5] R. Atkin and N. Fox, Rheol. Acta, 16, (1977), 213.

[6] K. Mendelsohn, *Liquid Helium*,Hd der Physic, XV, Springer Berlin (1956).

[7] G. Lebon and D. Jou, J. Non-Equilib. Thermo., 4, (1979), 259.

[8] L. Onsager, Phys. Res., 37, (1931), 405.

[9] S. De Groot and P. Mazur, *Non-equilibrium thermodynamics*, North Holland, Amsterdam (1962).

[10] D. Jou, J. Casas-Vazquez and G. Lebon, Rep. Prog. Phys., 51, (1988), 1105.

[11] G. Lebon and D. Jou, J. Phys. C., 16, (1983), 6199.

[12] P. Kapitza, *Collected papers*, Ter Haar ed., Pergamon, New York (1967).

S GIAMBÓ AND G VALENTI

On the anisotropic relativistic magnetohydrodynamics of heat conducting fluid

ABSTRACT

A scheme for the anisotropic relativistic magnetohydrodynamics of a heat conducting fluid of infinite conductivity is proposed by using an extended irreversible thermodynamics approach. It is deduced the fundamental system of anisotropic magnetohydrodynamics, which is able to describe a collisionless anisotropic plasma embedded in a strong magnetic field.

1. Introduction

The study of a diluted and collisionless plasma embedded in a strong magnetic field has received a considerable attention during the last twenty years [1]-[7].

In the Chew, Goldberger and Low (C.G.L.) approximation the adiabatic motion of such a plasma is studied by means of field equations that are formally the same as those of the usual Magnetohydrodynamics (MHD) with an anisotropic pressure distribution. Furthermore all the extensive thermodynamic quantities, such as the internal energy ϵ, the entropy S etc.... , are decomposed with respect to the magnetic field because the transversal and longitudinal components of each quantity are different one to another, while for the intensive ones we distinguish the longitudinal and the transversal quantities .

In order to study several astrophysical problems, for example the state of the matter outside the immediate vicinity of planets or sun, it is necessary to consider such a theory within the relativistic context.

Therefore, in 1974-75 Cissoko [8]-[10] has proposed a relativistic theory of anisotropic MHD, limiting his attention only to the adiabatic motions.

Nevertheless, the afore-mentioned theoretical studies of Astrophysics point out that, in the most of real cases, it is necessary to take into consideration also phenomena of viscosity and of heat conduction.

In this first approach, within the irreversible estended thermodynamic context (see [11] and references quoted there), we propose a relativistic model for the anisotropic MHD of a heat conducting fluid, which is supposed to be non viscous and of infinite conductivity.

Notations

Space-time is a four dimensional manifold whose normal hyperbolic metric ds^2, (with signature + - - -), can be express in local coordinates in the usual form $ds^2 = g_{\alpha\beta}\, dx^\alpha\, dx^\beta$; the metric tensor is assumed to be of class C^1; the four velocity is defined as $u^\alpha = \frac{dx^\alpha}{ds}$, which implies its unitary character $u^\alpha u_\alpha = 1$; ∇_α is the operator of covariant differentiation with respect to the assigned metric; everywhere the units are such that the velocity of light is equal to one.

2. Thermodynamical considerations

Along the lines indicated by Cissoko in [10], for generalizing to the relativistic framework some results obtained by Abraham-Shauner about the anisotropic MHD under the C.G.L. approximation, the thermodynamic properties of anisotropic MHD are completely determined assigning the explicit dependence of the specific free energy $\psi^{(0)}$ on the longitudinal $T_{||}$ and transversal $T_\perp$ temperatures, the proper mass density r and $h^2 = - h^\alpha h_\alpha$, h^α being the intensity magnetic space-like vector ($h^\alpha u_\alpha = 0$), that is :

$$\psi^{(0)} = \psi^{(0)}\,(T_{||}\,,\, T_\perp\,,\, r,\, h^2) \tag{1}$$

But, as well known, the relation (1) is valid for situations near equilibrium. Far from equilibrium, we must generalize this relation by including among the field variables $q^2 = - q^\alpha q_\alpha$, q^α being the heat flux spce-like vector ($q^\alpha u_\alpha = 0$). So the free energy becomes a function also of q^2, namely $\psi = \psi(T_{||}, T_\perp, r, h^2, q^2)$ and we assume :

$$\psi(T_{||}, T_\perp, r, h^2, q^2) = \psi^{(0)}(T_{||}, T_\perp, r, h^2) + \tfrac{1}{2}(\alpha_{||} + 2\alpha_\perp)q^2 \tag{2}$$

with $\alpha_{||} = \alpha_{||}(T_{||}, r, h^2)$ and $\alpha_\perp = \alpha_\perp(T_\perp, r, h^2)$.

Furthermore, by analogy with the classical relations, we define the longitudinal and transversal components of the specific total entropy $S = 2S_\perp + S_{||}$ and longitudinal and transversal pressures as follows :

$$S_{||} = - \frac{\partial\psi}{\partial T_{||}} \quad ; \quad S_\perp = - \frac{1}{2}\frac{\partial\psi}{\partial T_\perp} \tag{3}$$

$$p_{||} = r^2\frac{\partial\psi}{\partial r} \quad ; \quad p_\perp = r\left(2h^2\frac{\partial}{\partial h^2} + r\frac{\partial}{\partial r}\right)\psi \tag{4}$$

which, by virtue of (2), yield :

$$S_{||}(T_{||}, r, h^2, q^2) = S_{||}^{(0)}(T_{||}, r, h^2) - \frac{1}{2}\frac{\partial \alpha_{||}}{\partial T_{||}} q^2 \tag{5}$$

$$S_{\perp}(T_{\perp}, h^2, q^2) = S_{\perp}^{(0)}(T_{\perp}, h^2) - \frac{1}{2}\frac{\partial \alpha_{\perp}}{\partial T_{\perp}} q^2 \tag{6}$$

$$p_{||}(T_{||}, r, h^2, q^2) = p_{||}^{(0)}(T_{||}, r, h^2) + \frac{r^2}{2}\frac{\partial \alpha_{||}}{\partial r} q^2 \tag{7}$$

$$p_{\perp}(T_{\perp}, r, h^2, q^2) = p_{\perp}^{(0)}(T_{\perp}, r, h^2) + 2rh^2 \frac{\partial \alpha_{\perp}}{\partial h^2} q^2 \tag{8}$$

with the conditions :

$$\frac{\partial \alpha_{\perp}}{\partial r} = 0 \qquad ; \qquad \frac{\partial \alpha_{||}}{\partial r} = -\frac{2h^2}{r}\frac{\partial \alpha_{||}}{\partial h^2} . \tag{9}$$

where here and in the follows the superscript $^{(0)}$ stands for the quantity evalueted at equilibrium.

Hence, writing (2) in the differential form, we have :

$$d\psi = -S_{||}dT_{||} - 2S_{\perp}dT_{\perp} + \frac{p_{||}}{r^2}dr - \frac{(p_{||} - p_{\perp})}{2rh^2}dh^2 + \frac{1}{2}(\alpha_{||} + 2\alpha_{\perp})dq^2 \tag{10}$$

Next, by introducing the total specific internal energy :

$$\epsilon = \epsilon_{||} + 2\epsilon_{\perp} \tag{11}$$

so that :

$$\psi = \epsilon - T_{||}S_{||} - 2\,T_{\perp}S_{\perp} \tag{12}$$

we obtain :

$$\epsilon_{||}(T_{||}, r, h^2, q^2) = \epsilon_{||}^{(0)}(T_{||}, r, h^2) + \frac{1}{2}\left(\alpha_{||} - T_{||}\frac{\partial \alpha_{||}}{\partial T_{||}}\right) q^2 \tag{13}$$

$$\epsilon_{\perp}(T_{\perp}, r, h^2, q^2) = \epsilon_{\perp}^{(0)}(T_{\perp}, r, h^2) + \frac{1}{2}\left(\alpha_{\perp} - T_{\perp}\frac{\partial \alpha_{\perp}}{\partial T_{\perp}}\right) q^2 \tag{14}$$

where we have taken into consideration the relation (2) and the expressions (5), (6).

Therefore, using (11)-(12) and taking into account the independence of the longitudinal components by the transversal ones, the differential relation (10) gives rise the following two generalized Gibbs equations :

$$
\begin{aligned}
T_{||}dS_{||} &= d\epsilon_{||} - \frac{p_{||}}{r^2}dr + \frac{p_{||}}{2rh^2}dh^2 - \frac{1}{2}\,\alpha_{||}dq^2 \\
T_{\perp}dS_{\perp} &= d\epsilon_{\perp} - \frac{p_{\perp}}{4rh^2}dh^2 - \frac{1}{2}\,\alpha_{\perp}dq^2 \,.
\end{aligned}
\tag{15}
$$

3. Field equations

Following the physical intuitions developed by Landau [12], for studying the evolution of a relativistic inviscid heat-conducting fluid, we assume the energy tensor to have the same structure as that for an inviscid non heat-conducting fluid and also we modify the conservation equation for the proper material density .

Therefore the field equations for the anisotropic relativistic MHD are:

$$\nabla_\alpha T^{\alpha\beta} = 0 \;;\; T^{\alpha\beta} = wu^\alpha u^\beta - (p_{\perp} + \mu\,\frac{h^2}{2})g^{\alpha\beta} - \overline{\mu}\; h^\alpha h^\beta; \tag{16}$$

$$\nabla_\alpha(ru^\alpha + q^\alpha) = 0 \,; \tag{17}$$

$$\nabla_\alpha(u^\alpha h^\beta - h^\alpha u^\beta) = 0 \,; \tag{18}$$

where $w = r(1 + \epsilon) + p_{\perp} + \mu h^2$, $\overline{\mu} = \mu\Big(1 - \dfrac{p_{||} - p_{\perp}}{\mu h^2}\Big)$, with μ the constant magnetic permeability.

By projecting $(16)_1$ on the time direction u^α, by virtue of the equations (17) and (18), we obtain the energy equation :

$$ru^\alpha\Big(\partial_\alpha\epsilon - \frac{p_{||}}{r^2}\partial_\alpha r + \frac{p_{||} - p_{\perp}}{2rh^2}\,\partial_\alpha h^2\Big) = (1 + \epsilon + \frac{p_{||}}{r})\;\nabla_\alpha q^\alpha. \tag{19}$$

Therefore, owing to eq. (19), the eqs. (15) assume the following form :

$$
\begin{aligned}
T_{||}DS_{||} &= \Big(\frac{\gamma_1 + \epsilon_{||}}{r} + \frac{p_{||}}{r^2}\Big)\;\nabla_\alpha q^\alpha + \alpha_{||}q_\alpha Dq^\alpha \\
T_{\perp}DS_{\perp} &= \frac{(\gamma_2 + \epsilon_{\perp})}{r}\;\nabla_\alpha q^\alpha + \alpha_{\perp}q_\alpha Dq^\alpha
\end{aligned}
\tag{20}
$$

where $D = u^\alpha \nabla_\alpha$; γ_1 and γ_2 are two adimensional factors subjected to the restriction $\gamma_1 + 2\gamma_2 = 1$.

Now we may consider the most general form of the entropy balance equation, that is :

$$\nabla_\alpha(rSu^\alpha + J^\alpha) = \sigma \qquad ; \qquad \sigma \geq 0 \tag{21}$$

where σ is the entropy production and J^α the space-like vector entropy flux, whose general expression is given by :

$$J^\alpha = \Big(\beta_{||}(T_{||}, r, h^2, q^2) + 2\beta_\perp(T_\perp, r, h^2, q^2)\Big)q^\alpha. \tag{22}$$

In order to derive the phenomenological law governing the evolution of q^α, first it is necessary to obtain the entropy production σ in terms of such a variable. The equations (20) present the standard form (21) provided that the entropy flux J^α and the entropy production are defined by :

$$J^\alpha = -\left(\frac{g_{||}}{T_{||}} + 2\frac{g_\perp}{T_\perp}\right)q^\alpha \tag{23}$$

$$\sigma = \frac{q^\alpha}{T_{||}^{\,2}T_\perp^{\,2}}\Big(rT_{||}T_\perp(\alpha_{||}T_\perp + 2\alpha_\perp T_{||})Dq_\alpha + g_{||}T_\perp^{\,2}\partial_\alpha T_{||} + $$
$$+ 2g_\perp T_{||}^{\,2}\partial_\alpha T_\perp - T_{||}T_\perp(T_\perp\partial_\alpha g_{||} + 2T_{||}\partial_\alpha g_\perp)\Big) \tag{24}$$

where $g_{||} = \gamma_1 + \epsilon_{||} + \frac{p_{||}}{r} - T_{||}S_{||}$; $g_\perp = \gamma_2 + \epsilon_\perp - T_\perp S_\perp$.

Therefore the requirement that the entropy production σ must be a positive semidefinte quantity implies the following simplest assumption for the evolution equation of the heat flux :

$$q_\alpha = -\chi\,\gamma_\alpha^\beta\Big(rT_{||}T_\perp(\alpha_{||}T_\perp + 2\alpha_\perp T_{||})Dq_\beta + g_{||}T_\perp^{\,2}\partial_\beta T_{||} + $$
$$+ 2g_\perp T_{||}^{\,2}\partial_\beta T_\perp - T_{||}T_\perp(T_\perp\partial_\beta g_{||} + 2T_{||}\partial_\beta g_\perp)\Big), \tag{25}$$

where $\chi \geq 0$ is the thermal conductivity.

Finally, taking into account the relations (3)-(4) and imposing the equality of the second order mixed derivative of ψ, from (20) we obtain respectively :

$$\begin{aligned} & rC_{||}DT_{||} - \frac{T_{||}}{r}\frac{\partial p_{||}}{\partial T_{||}} Dr + \frac{T_{||}}{2h^2}\frac{\partial p_{||}}{\partial T_{||}} Dh^2 + r\Big(T_{||}\frac{\partial \alpha_{||}}{\partial T_{||}} + \\ & - \alpha_{||}\Big) q_\alpha D q^\alpha - (\gamma_1 + \epsilon_{||} + \frac{p_{||}}{r}) \nabla_\alpha q^\alpha = 0 \end{aligned} \tag{26}$$

$$\begin{aligned} & rC_{\perp}DT_{\perp} - \frac{T_{\perp}}{4h^2}\frac{\partial p_{\perp}}{\partial T_{\perp}} Dh^2 + r\Big(T_{\perp}\frac{\partial \alpha_{\perp}}{\partial T_{\perp}} - \alpha_{\perp}\Big) q_\alpha D q^\alpha + \\ & - (\gamma_2 + \epsilon_{\perp}) \nabla_\alpha q^\alpha = 0 \end{aligned} \tag{27}$$

where :

$$C_{||} = T_{||}\frac{\partial S_{||}}{\partial T_{||}} \quad ; \quad C_{\perp} = T_{\perp}\frac{\partial S_{\perp}}{\partial T_{\perp}} . \tag{28}$$

Summarizing, our system is completely determined by the twelve equations $(16)_1$, (17), (18), (25), (26) and (27) for the twelve variables

$$r,\ u_\alpha,\ T_{||},\ T_{\perp},\ h_\alpha,\ q_\alpha .$$

REFERENCES

[1] G.F. Chew, M.L. Goldberg and F.E. Low, Proc. Roy. Soc., **A236**, (1956), 112-118.

[2] B. Abraham-Shrauner, J. Plasma Phys., 1, (1967), 361-381.

[3] F.M. Neubauer, Zs. Phys., 237, (1970), 205-208.

[4] V.B. Baranov and M.D. Kartalev, Izv. Akad. Nauk SSSR, Mekhan. Zhidk. i Gaza, 6, (1970), 3-10.

[5] M.D. Kartalev,Soviet Phys. Doklady, 17, (1973), 744-746.

[6] Y.Kato, M. Tajiri and T.Taniuti, J. Phys. Soc. Japan, 21, (1967), 765-777.

[7] A.V. Gopalakrishna, J. Math. Anal. Appl., 35, (1971), 349-360.

[8] M. Cissoko, Comptes rendus, 278 A, (1974), 463-467; 641-644; 1233-1236.

[9] M. Cissoko, Ann. Inst. Henri Poincare', 22, (1975), 1-27.

[10] M. Cissoko, Ann. Mat. Pura Appl., 111, (1976), 331-368.

[11] D. Jou, J. Casas-Vazquez and G. Lebon, Rep. Prog. Phys.,51,

(1988), 1105-1179.

[12] L.D. Landau and E.M. Lifshitz, *Fluid Mechanics*, Pergamon Press, 1959.

Sebastiano Giambò

Dipartimento di Matematica Contrada Papardo - Salita Sperone 31, 98166 Sant'Agata - Messina

Georgy Lebon

Institut de Physique B5 - Universitè de Liegè - B-4000 Sart Tilman - Liegè - Belgium

I-SHIH LIU

Viscoelasticity in extended thermodynamics

Abstract

A theory of isotropic viscoelastic materials without heat conduction is formulated within the general scheme of extended thermodynamics. An additional equation of balance is needed to account for the viscous stress as an independent field variable. It is interesting to point out that if the stress tensor itself is taken to be the density tensor field for the additional balance equation the theory will turn out to be not only inadequate but may also be inconsistent with the stability criterion of the maximal entropy in equilibrium.

1. Introduction

The governing equations of extended thermodynamics consist of the usual conservation laws of mass, momentum and energy and a set of additional equations of balance. For monatomic gases [1], such equations could be well-motivated by the moment equations of kinetic theory of gases. The basic structure of such equations can also be established from the requirement of Galilean invariance of the system [2], and has been applied to the formulation of extended thermodynamics of fluids [3,4] as well as viscoelastic solids [5,6].

It is presented here, a theory of isotropic viscoelastic materials without heat conduction, for simplicity, but with effects of bulk viscosity, which has not been considered in [5,6]. Viscous fluids and linear viscoelastic solids are treated as special cases. For the later, the usual rate-type stress-strain relations are obtained. However the general results are by no means restricted to small deformations, thus they provide a generalization of linear theory to viscoelasticity of finite deformations.

2. Governing Equations

2.1. Equations of Balance in Lagrangian Form

The conservation laws of mass, momentum and energy in a spatial coordinate system (x_i, t), relative to an inertial frame, can be written as

$$
\begin{aligned}
&\frac{\partial \varrho}{\partial t} + \frac{\partial \varrho v_k}{\partial x_k} = 0, \\
&\frac{\partial \varrho v_i}{\partial t} + \frac{\partial}{\partial x_k}(\varrho v_i v_k - T_{ik}) = 0, \\
&\frac{\partial \varrho e}{\partial t} + \frac{\partial}{\partial x_k}(\varrho e v_k - v_i T_{ik} + q_k) = 0,
\end{aligned} \tag{2.1}
$$

where ϱ is the mass density, v_i the velocity, T_{ik} the Cauchy stress tensor, q_k the energy flux. $e = (v^2/2 + \varepsilon)$ is the total specific energy and ε is the specific internal energy.

In order to obtain a hyperbolic mathematical model for viscoelastic materials without heat conduction, we propose in addition, the following equation of balance:

$$
\frac{\partial u_{ij}}{\partial t} + \frac{\partial}{\partial x_k}(u_{ij} v_k + G_{ijk}) = P_{ij} \tag{2.2}
$$

for a tensorial density u_{ij}, a quantity which is related to the momentum flux. The most suggestive choice of u_{ij} would be the momentum flux itself. However, we shall see later that our theory will not allow such a choice.

According to the tenet of extended thermodynamics, the requirement of Galilean invariance of the system of balance equations (2.1) and (2.2) implies the following explicit dependence of the velocity v_i, [2]

$$
\begin{aligned}
&u_{ij} = \varrho v_i v_j + \varrho m_{ij}, \\
&G_{ijk} = -v_i T_{jk} - v_j T_{ik} + M_{ijk}, \\
&P_{ij} = \varrho \ell_{ij},
\end{aligned} \tag{2.3}
$$

and that m_{ij}, M_{ijk}, ℓ_{ij} like ϱ, ε, T_{ik}, q_k are objective quantities.

For solids, it is more convenient to rewrite this system in Lagrangian form. Let (X_α, t) be a material coordinate system in the reference configuration of the material body and $F_{i\alpha}$ be the deformation gradient $\partial x_i / \partial X_\alpha$. We introduce the following quantities:

$$
\widehat{T}_{i\alpha} = \frac{\varrho_\kappa}{\varrho} T_{ik} F^{-1}_{\alpha k}, \qquad \widehat{q}_\alpha = \frac{\varrho_\kappa}{\varrho} q_k F^{-1}_{\alpha k}, \qquad \widehat{M}_{ij\alpha} = \frac{\varrho_\kappa}{\varrho} M_{ij\alpha} F^{-1}_{\alpha k}, \tag{2.4}
$$

where ϱ_κ is the mass density in the reference configuration. The system (2.1) and (2.2) in Lagrangian form read

$$
\begin{aligned}
&\dot{\varrho}_\kappa = 0, \\
&\varrho_\kappa \dot{v}_i - \frac{\partial \widehat{T}_{i\alpha}}{\partial X_\alpha} = 0, \\
&\varrho_\kappa \dot{e} + \frac{\partial \widehat{G}_\alpha}{\partial X_\alpha} = 0, \\
&\dot{\widehat{u}}_{ij} + \frac{\partial \widehat{G}_{ij\alpha}}{\partial X_\alpha} = \varrho_\kappa \ell_{ij},
\end{aligned} \tag{2.5}
$$

where

$$\begin{aligned}\widehat{u}_{ij} &= \varrho_\kappa v_i v_j + \varrho_\kappa m_{ij},\\ \widehat{G}_\alpha &= -v_i \widehat{T}_{i\alpha} + \widehat{q}_\alpha,\\ \widehat{G}_{ij\alpha} &= -v_i \widehat{T}_{j\alpha} - v_j \widehat{T}_{i\alpha} + \widehat{M}_{ij\alpha}.\end{aligned} \tag{2.6}$$

2.2 Thermodynamic Process

For viscoelastic materials, a state can be characterized by the following thermodynamic fields:

$$\begin{array}{ll} F_{i\alpha} & \text{deformation gradient,}\\ v_i & \text{velocity,}\\ \varepsilon & \text{specific internal energy,}\\ T_{ij} & \text{stress tensor.}\end{array} \tag{2.7}$$

Obviously, we have regarded the deformation gradient, in stead of the density, as a thermodynamic field so that solids can also be included in the theory.

The system (2.5) must be supplemented by constitutive relations which in extended thermodynamics are assumed to be local and instantaneous,

$$\mathcal{C} = \widetilde{\mathcal{C}}(F_{i\alpha}, v_i, \varepsilon, T_{ij}). \tag{2.8}$$

Here $\mathcal{C}$ stands for the constitutive quantities $\{m_{ij}, \widehat{q}_\alpha, \widehat{M}_{ij\alpha}, \ell_{ij}\}$.

Any fields $(F_{i\alpha}, v_i, \varepsilon, T_{ij})$ that satisfies the balance equations (2.5) together with the constitutive relations of the form (2.8) will be called a *thermodynamic process* in viscoelasticity.

3. General Constitutive Restrictions

3.1. Entropy Principle

The entropy principle states that for every thermodynamic process the entropy inequality must hold

$$\varrho_\kappa \dot{\eta} + \frac{\partial \widehat{\Phi}_\alpha}{\partial X_\alpha} = s \geq 0. \tag{3.1}$$

This is also written in the material coordinate system and similarly we have introduced $\widehat{\Phi}_\alpha = (\varrho_\kappa/\varrho)\Phi_k F^{-1}_{\alpha k}$, where Φ_k is the entropy flux. Both the specific entropy density η and $\widehat{\Phi}_\alpha$ are also given by constitutive relations of the form (2.8). The entropy production s is a non-negative quantity.

This constraint on thermodynamic processes can be taken into account by use of Lagrange multipliers [7]. Before doing this, we must first observe that the fields $F_{i\alpha}$ and v_i are not entirely independent. They are related by the identity

$$\dot{F}_{i\alpha} - \frac{\partial v_i}{\partial X_\alpha} = 0. \tag{3.2}$$

This identity must also be taken into account by a Lagrange multiplier. On the other hand the first equation of (2.5), *viz.* $\dot{\varrho}_\kappa = 0$, merely asserts that ϱ_κ is a time-independent field, a result which has been made explicitly already. Therefore by use of Lagrange multiplier: The inequality

$$\begin{aligned}\varrho_\kappa \dot{\eta} + \frac{\partial \widehat{\Phi}_\alpha}{\partial X_\alpha} - \lambda_{i\alpha}\, \varrho_\kappa \big(\dot{F}_{i\alpha} - \frac{\partial v_i}{\partial X_\alpha}\big) - \Lambda_i \,\big(\varrho_\kappa \dot{v}_i - \frac{\partial \widehat{T}_{i\alpha}}{\partial X_\alpha}\big) \\ - \Lambda \,\big(\varrho_\kappa \dot{e} + \frac{\partial \widehat{G}_\alpha}{\partial X_\alpha}\big) - \Lambda_{ij}\big(\dot{\widehat{u}}_{ij} + \frac{\partial \widehat{G}_{ij\alpha}}{\partial X_\alpha} - \varrho_\kappa \ell_{ij}\big) \geq 0\end{aligned} \tag{3.3}$$

must hold for all fields $(F_{i\alpha}, v_i, \varepsilon, T_{ij})$. The Lagrange multipliers $\lambda_{i\alpha}$, Λ_i, Λ and Λ_{ij} are functions of $(F_{i\alpha}, v_i, \varepsilon, T_{ij})$ in general.

This requirement implies the following relations

$$\begin{aligned}\varrho_\kappa d\eta &= \varrho_\kappa \lambda_{i\alpha} dF_{i\alpha} + \varrho_\kappa \Lambda_i dv_i + \varrho_\kappa \Lambda\, de + \Lambda_{ij} d\widehat{u}_{ij}, \\ d\widehat{\Phi}_\alpha &= \varrho_\kappa \lambda_{i\alpha} dv_i - \Lambda_i d\widehat{T}_{i\alpha} + \Lambda\, d\widehat{G}_\alpha + \Lambda_{ij} d\widehat{G}_{ij\alpha},\end{aligned} \tag{3.4}$$

and the inequality

$$s = \varrho_\kappa \Lambda_{ij} \ell_{ij} \geq 0. \tag{3.5}$$

3.2. Material Objectivity and Isotropy

According to the principle of material objectivity the constitutive equation of an objective quantity must be independent of the observer and thus it must be independent of the velocity. In particular, this must hold for η and $\widehat{\Phi}_\alpha$. Therefore, by (2.6), from (3.4) we must have

$$\begin{aligned}\frac{\partial \eta}{\partial v_i} &= \Lambda_i + \Lambda\, v_i + 2\Lambda_{ij} v_j = 0, \\ \frac{\partial \widehat{\Phi}_\alpha}{\partial v_i} &= -\varrho_\kappa \lambda_{i\alpha} - \Lambda\, \widehat{T}_{i\alpha} - 2\Lambda_{ij} \widehat{T}_{j\alpha} = 0,\end{aligned} \tag{3.6}$$

and (3.4) reduces to

$$\begin{aligned}d\eta &= \lambda_{i\alpha} dF_{i\alpha} + \Lambda\, d\varepsilon + \Lambda_{ij} dm_{ij}, \\ d\widehat{\Phi}_\alpha &= \Lambda\, d\widehat{q}_\alpha + \Lambda_{ij} d\widehat{M}_{ij\alpha}.\end{aligned} \tag{3.7}$$

These relations also show that the Lagrange multipliers Λ, Λ_{ij} and $\lambda_{i\alpha}$ are independent of the velocity, and Λ_i can be eliminated by use of $(3.6)_1$.

We shall consider isotropic materials only. For isotropic materials, the entropy flux Φ_k must vanish. Since it is an isotropic vector-valued function of scalars and tensors only and it is well-known that such a function is a zero function. Similarly q_k and M_{ijk} must also vanish and hence the relation $(3.7)_2$ is trivially satisfied.

For further evaluation, we introduce

$$\eta' = \eta - \Lambda_{ij} m_{ij}, \tag{3.8}$$

so that $(3.7)_1$ becomes

$$d\eta' = \lambda_{i\alpha} dF_{i\alpha} + \Lambda\, d\varepsilon - m_{ij}\, d\Lambda_{ij}. \tag{3.9}$$

By use of $(3.6)_2$ and (3.9), for isotropic materials one can show that

$$\Lambda_{ij}(m_{jk} - \frac{1}{\varrho} T_{jk}) = \Lambda_{kj}(m_{ji} - \frac{1}{\varrho} T_{ji}). \tag{3.10}$$

This symmetry relation* implies that (see the proof in Appendix of [5])

$$m_{jk} - \frac{1}{\varrho} T_{jk} = m_1 \delta_{jk} + m_2 \Lambda_{jk} + m_3 \Lambda_{jl} \Lambda_{lk}. \tag{3.11}$$

This determines m_{ij} in terms of T_{ij} to within three scalar functions $m_i(F_{k\alpha}, \varepsilon, \Lambda_{kl})$.

3.3. *Equilibrium and Temperature*

The relation (3.9) suggests that η' can be considered as a function of $(F_{i\alpha}, \varepsilon, \Lambda_{ij})$. Although the Lagrange multiplier Λ_{ij} does not have a clear physical meaning, it do have an important property, namely, it vanishes in equilibrium.

Indeed, inspection of (3.5) shows that the entropy production s has a minimum, namely zero, when Λ_{ij} vanishes. Consequently it also implies

$$\left.\frac{\partial s}{\partial \Lambda_{ij}}\right|_{\Lambda_{ij}=0} = \varrho_\kappa \ell_{ij}\Big|_{\Lambda_{ij}=0} = 0, \tag{3.12}$$

a necessary condition for a minimum. We shall call a process $(F_{i\alpha}, \varepsilon, \Lambda_{ij} = 0)$ an *equilibrium* since it is a process with no productions.

In equilibrium, by $(3.6)_2$ the relation (3.7) reduces to

$$d\eta_o = \Lambda|_E \big(d\varepsilon - \frac{1}{\varrho_\kappa} \widehat{T}^o_{i\alpha}\, dF_{i\alpha}\big). \tag{3.13}$$

Here the suffix o denotes the value in equilibrium. Hence, by comparison with the Gibbs equation in equilibrium thermodynamics, we conclude that

$$\Lambda|_E = \Lambda(F_{i\alpha}, \varepsilon, 0) = \frac{1}{T}, \tag{3.14}$$

---

* This relation has been derived in [5,6] from the assumption that the entropy production must be independent of rotation of the reference frame.

where T is the absolute temperature, and we can write

$$\varepsilon = \varepsilon(F_{i\alpha}, T). \tag{3.15}$$

Henceforth T can be chosen as a variable in the place of ε.

Another necessary condition for the entropy production s to be minimum at equilibrium is that

$$\frac{\partial^2 s}{\partial \Lambda_{ij} \partial \Lambda_{kj}}\bigg|_E = \varrho_\kappa \Big(\frac{\partial \ell_{ij}}{\partial \Lambda_{kl}} + \frac{\partial \ell_{kl}}{\partial \Lambda_{ij}} \Big) \bigg|_E \quad \text{be positive semi-definite} \tag{3.16}$$

or for arbitrary symmetric tensor M

$$\sigma_{ijkl} M_{ij} M_{kl} \geq 0, \quad \text{where} \quad \sigma_{ijkl} = \frac{\partial \ell_{ij}}{\partial \Lambda_{kl}}\bigg|_E. \tag{3.17}$$

4. Constitutive Equations

4.1. Representations

From the Gibbs equation, we obtain

$$\begin{aligned} T^o_{ij} &= \varrho \frac{\partial \psi_o}{\partial F_{i\alpha}} F_{j\alpha} = 2\varrho \frac{\partial \psi_o}{\partial B_{im}} B_{jm}, \\ \varepsilon &= \psi_o - T \frac{\partial \psi_o}{\partial T}, \qquad \eta_o = -\frac{\partial \psi_o}{\partial T}, \end{aligned} \tag{4.1}$$

where $\psi_o = \varepsilon - T\eta_o$ is the equilibrium free energy function. These are well-known relations in thermoelasticity. For isotropic materials, ψ_o is an isotropic scalar function of (B_{kl}, T), where $B_{kl} = F_{k\alpha} F_{l\alpha}$ is the left Cauchy-Green tensor.

For convenience, we introduce the elasticity tensors,

$$\begin{aligned} A_{i\alpha k\beta} &= \varrho \frac{\partial^2 \psi_o}{\partial F_{i\alpha} F_{k\beta}}, \\ C_{ijkl} &= A_{i\alpha k\beta} F_{j\alpha} F_{l\beta} - T^o_{jl} \delta_{ik}. \end{aligned} \tag{4.2}$$

From these definitions, one can easily express C_{ijkl} in terms of $\psi_o(B_{mn}, T)$. Moreover one can show that C_{ijkl} is symmetric in (ij) and in (kl).

For isotropic materials, $\eta' = \eta'(B_{kl}, T, \Lambda_{kl})$ is also an isotropic scalar function. Since we are mainly interested in processes near equilibrium we shall give a quadratic representation for η' in Λ_{kl},

$$\eta' = \eta_o + g_{ij} \Lambda_{ij} + \frac{1}{2} h_{ijkl} \Lambda_{ij} \Lambda_{kl} \tag{4.3}$$

where η_o, g_{ij} and h_{ijkl} are themselves isotropic functions of (B_{mn}, T). η_o is the equilibrium entropy.

Similarly the relation (3.11) to the first order in Λ_{kl} can be written as

$$m_{jk} - \frac{1}{\varrho}T_{jk} = -g_o\delta_{jk} - \gamma_{mn}\Lambda_{mn}\delta_{jk} - \gamma_2\Lambda_{jk}, \tag{4.4}$$

where g_o, γ_{mn} and γ_2 are also isotropic functions of (B_{kl}, T).

Further evaluation of the relations (3.9) and $(3.6)_2$ by use of (4.3) and (4.4) is now straightforward. We obtain after some calculations the following representations,

$$\begin{aligned} S_{ij} &= T_{ij} - T^o_{ij} = D_{ijkl}\Lambda_{kl}, \\ m_{ij} &= -g_{ij} - h_{ijkl}\Lambda_{kl}, \\ \eta &= \eta_o - \frac{1}{2}h_{ijkl}\Lambda_{ij}\Lambda_{kl}, \end{aligned} \tag{4.5}$$

where coefficients satisfy the following relations,

$$\begin{aligned} g_{ij} &= g_o\delta_{ij} - \frac{1}{\varrho}T^o_{ij}, \\ \gamma_{ij} &= \gamma_1\delta_{ij} + 2T\frac{\partial g_o}{\partial B_{im}}B_{jm} + \frac{T^2}{\varrho c_v}\frac{\partial g_o}{\partial T}\frac{\partial T^o_{ij}}{\partial T}, \end{aligned} \tag{4.6}$$

and

$$\begin{aligned} D_{ijkl} &= TC_{ijkl} - 2\varrho T\frac{\partial g_o}{\partial B_{im}}B_{jm}\delta_{kl} - \frac{T^2}{c_v}\frac{\partial T^o_{ij}}{\partial T}\frac{\partial g_{kl}}{\partial T}, \\ h_{ijkl} &= \delta_{ij}\gamma_{kl} + \frac{1}{2}\gamma_2(\delta_{ik}\delta_{jl} + \delta_{il}\delta_{jk}) - \frac{1}{\varrho}D_{ijkl}. \end{aligned} \tag{4.7}$$

and $c_v = \partial\varepsilon/\partial T$ is the specific heat at constant volume.

We shall call the equilibrium stress T^o_{ij} the *elastic stress* and S_{ij} the *viscous stress*. From $(4.5)_1$, it is clear that the viscous stress must also vanish in equilibrium. Suppose that we can invert this relation so that we have

$$\Lambda_{ij} = D^{-1}_{ijkl}S_{kl}, \tag{4.8}$$

where D^{-1}_{ijkl} is the inverse of D_{ijkl} or

$$D^{-1}_{ijkl}D_{klmn} = \frac{1}{2}(\delta_{im}\delta_{jn} + \delta_{in}\delta_{jm}). \tag{4.9}$$

With (4.8) we can eliminate the Lagrange multiplier Λ_{ij} and express constitutive equations in terms of viscous stress. We get from (4.5) and (3.17)

$$\begin{aligned} T_{ij} &= T^o_{ij} + S_{ij}, \\ m_{ij} &= -g_o\delta_{ij} + \frac{1}{\varrho}T^o_{ij} + \frac{1}{\varrho}S_{ij} - (\gamma_2 D^{-1}_{ijkl} + \delta_{ij}\gamma_{mn}D^{-1}_{mnkl})S_{kl}, \\ \eta &= \eta_o - \frac{1}{2}h_{ijkl}D^{-1}_{ijmn}D^{-1}_{klpq}S_{mn}S_{pq}, \\ \ell_{ij} &= \sigma_{ijmn}D^{-1}_{mnkl}S_{kl}. \end{aligned} \tag{4.10}$$

4.2. Hyperbolicity

In order that the theory predict finite speed of propagation for small disturbances, we shall also require the system to be hyperbolic near equilibrium. This leads us to assume that the entropy $\eta = \widetilde{\eta}(F_{i\alpha}, v_i, e, \widehat{u}_{ij})$ is a concave function near equilibrium [2], i.e.,

$$Q = \frac{\partial^2 \widetilde{\eta}}{\partial X_A \partial X_B}\bigg|_E \delta X_A \delta X_B < 0 \tag{4.11}$$

for any variations δX_A, where X_A stands for $(v_i, e, \widehat{u}_{ij})$.*

With the help of $(3.4)_1$, after some calculations we obtain

$$Q = -\frac{1}{T^2}\frac{\partial \varepsilon}{\partial T}(\delta T)^2 - \frac{1}{T}\delta v_i \delta v_i - \frac{\partial^2 \eta'}{\partial \Lambda_{ij} \partial \Lambda_{kl}}\bigg|_E \delta \Lambda_{ij} \delta \Lambda_{kl} < 0 \tag{4.12}$$

for any variations δT, δv_i and $\delta \Lambda_{ij}$. It follows immediately that

$$c_v > 0, \qquad h_{ijkl} M_{ij} M_{kl} > 0 \tag{4.13}$$

for any symmetric tensor M.

The first inequality of (4.13) implies that the specific heat is positive, while the second inequality together with $(4.5)_3$ gives the usual thermodynamic stability criterion – the entropy attains its maximum in equilibrium.

5. Field Equations of Viscoelasticity

For isotropic viscoelastic materials, the field equations for B_{ij}, v_i, T and S_{ij} consist of the usual conservation laws (2.1) and the additional equations of balance (2.2) which can be written as

$$\varrho \dot{m}_{ij} + \frac{\partial M_{ijk}}{\partial x_k} - \frac{\partial v_i}{\partial x_k} T_{jk} - \frac{\partial v_j}{\partial x_k} T_{ik} = \varrho \ell_{ij}. \tag{5.1}$$

With the knowledge of constitutive equations derived in the previous sections, if nonlinear terms which are products of S_{ij} with the gradients of B_{kl}, T and v_k are left out, it becomes

$$h_{ijkl} D^{-1}_{klmn} \dot{S}_{mn} + \sigma_{ijkl} D^{-1}_{klmn} S_{mn} = \frac{1}{\varrho T} D_{mnij} \frac{\partial v_m}{\partial x_n}. \tag{5.2}$$

---

* It has been pointed out that the concavity with respect to $F_{i\alpha}$ is not compatible with the material objectivity, unless it is restricted to the case of small displacements.

Although the equation (5.2) is valid only for slow processes near equilibrium, it is by no means restricted to small deformations in general.

Two special cases for which (5.2) is particularly simple will be considered below.

5.1. Viscous Fluid

For fluids, the four material functions ψ_o, g_o, γ_1 and γ_2 are functions of (ϱ, T). The elastic stress reduces to a hydrostatic pressure,

$$T^o_{ij} = -p\,\delta_{ij}, \qquad p = \varrho^2 \frac{\partial \psi_o}{\partial \varrho}, \tag{5.3}$$

and by (4.7)

$$\begin{aligned} D_{ijkl} &= T(d_1 \delta_{ij}\delta_{kl} + 2p\,\delta_{i\langle k}\delta_{l\rangle j}), \\ h_{ijkl} &= h_1 \delta_{ij}\delta_{kl} + h_2 \delta_{i\langle k}\delta_{l\rangle j}, \end{aligned} \tag{5.4}$$

where angular brackets denote traceless symmetrization and

$$\begin{aligned} d_1 &= -\frac{p}{3} + \varrho\frac{\partial p}{\partial \varrho} + \varrho^2 \frac{\partial g_o}{\partial \varrho} + \frac{T}{c_v}\frac{\partial p}{\partial T}\Big(\frac{\partial g_o}{\partial T} + \frac{1}{\varrho}\frac{\partial p}{\partial T}\Big), \\ h_1 &= \gamma_1 + \frac{\gamma_2}{3} - \frac{d_1 T}{\varrho}, \\ h_2 &= \gamma_2 - \frac{2pT}{\varrho}. \end{aligned} \tag{5.5}$$

σ_{ijkl} also reduces to contain only two material functions of (ϱ, T),

$$\sigma_{ijkl} = \sigma_1 \delta_{ij}\delta_{kl} + \sigma_2 \delta_{i\langle k}\delta_{l\rangle j}. \tag{5.6}$$

Therefore we obtain the following constitutive relations,

$$\begin{aligned} T_{ij} &= -p\,\delta_{ij} + S_{ij}, \\ m_{ij} &= -(g_o + \frac{p}{\varrho}) - \frac{h_1}{3d_1 T} S_{kk}\delta_{ij} - \frac{h_2}{2pT} S_{\langle ij\rangle}, \\ \eta &= \eta_o - \frac{h_1}{18 d_1^2 T^2} S_{kk}^2 - \frac{h_2}{8p^2T^2} S_{\langle kl\rangle}S_{\langle kl\rangle}, \\ \ell_{ij} &= \frac{\sigma_1}{3d_1 T} S_{kk}\delta_{ij} + \frac{\sigma_2}{2pT} S_{\langle ij\rangle}. \end{aligned} \tag{5.7}$$

Moreover, by (3.17) and $(4.13)_3$ we have

$$h_1 > 0, \quad h_2 > 0, \quad \sigma_1 > 0, \quad \sigma_2 > 0. \tag{5.8}$$

Here we have assumed that σ_1 and σ_2 do not equal to zero.

The equation (5.2) now takes the form

$$\tau_s \dot{S}_{\langle ij\rangle} + S_{\langle ij\rangle} = 2\mu \frac{\partial v_{\langle i}}{\partial x_{j\rangle}},$$
$$\tau_\pi \dot{\Pi} + \Pi = -\nu \frac{\partial v_i}{\partial x_i}, \tag{5.9}$$

where $\Pi = -S_{kk}/3$ is the *viscous pressure* and

$$\tau_s = \frac{h_2}{\sigma_2}, \qquad \mu = \frac{2p^2 T}{\varrho\, \sigma_2},$$
$$\tau_\pi = \frac{h_1}{\sigma_1}, \qquad \nu = \frac{d_1^2 T}{\varrho\, \sigma_1}. \tag{5.10}$$

By the inequalities (5.8) the relaxation times τ_s and τ_π as well as the shear and bulk viscosities μ and ν are all positive quantities. By letting τ_s and τ_π go to zero in (5.9) one recovers the usual Navier-Stokes relation.

5.2. Linear Viscoelastic Solids

We assume that the displacement gradient is small and the reference configuration is stress-free. Let ε_{ij} be the *infinitesimal strain tensor* defined as the symmetric part of displacement gradient, then

$$B_{ij} \approx \delta_{ij} + 2\varepsilon_{ij}, \tag{5.11}$$

and the linear elastic stress is given by

$$T^o_{ij} = \lambda\, \varepsilon_{kk}\delta_{ij} + 2\mu\, \varepsilon_{ij}, \tag{5.12}$$

where the Lamé constants λ and μ are functions of T only. Let $\nu = \lambda + \frac{2}{3}\mu$. In the linear theory we shall also neglect products of S_{ij} with ε_{ij} in the equation (5.2), therefore we can express

$$D_{ijkl} = T(d_1\delta_{ij}\delta_{kl} + d_2\delta_{i\langle k}\delta_{l\rangle j}),$$
$$h_{ijkl} = h_1\delta_{ij}\delta_{kl} + h_2\delta_{i\langle k}\delta_{l\rangle j}, \tag{5.13}$$
$$\sigma_{ijkl} = \sigma_1\delta_{ij}\delta_{kl} + \sigma_2\delta_{i\langle k}\delta_{l\rangle j},$$

where the coefficients are functions of T only. They can be related to λ and μ as well as the other material functions of Section 4.1. In particular, we have

$$d_2 = 2\mu. \tag{5.14}$$

Moreover the inequalities (5.8) also hold for this case.

The equation (5.2) can now be written as

$$\begin{aligned}\tau_s\dot{S}_{\langle ij\rangle} + S_{\langle ij\rangle} &= 2\widetilde{\mu}\,\dot{\varepsilon}_{\langle ij\rangle},\\ \tau_\pi\dot{S}_{kk} + S_{kk} &= 3\widetilde{\nu}\,\dot{\varepsilon}_{kk},\end{aligned} \tag{5.15}$$

where

$$\begin{aligned}\tau_s &= \frac{h_2}{\sigma_2}, & \widetilde{\mu} &= \frac{2\mu^2 T}{\varrho\,\sigma_2},\\ \tau_\pi &= \frac{h_1}{\sigma_1}, & \widetilde{\nu} &= \frac{d_1^2 T}{\varrho\,\sigma_1}.\end{aligned} \tag{5.16}$$

Again, the relaxation times τ_s and τ_π as well as the shear and bulk viscosities $\widetilde{\mu}$ and $\widetilde{\nu}$ are positive quantities according to (5.8).

More frequently the equations (5.15) are written in terms of the total deviatoric stress and the total pressure,

$$T_{\langle ij\rangle} = 2\mu\,\varepsilon_{\langle ij\rangle} + S_{\langle ij\rangle}, \quad P = -\nu\,\varepsilon_{kk} - \frac{S_{kk}}{3}. \tag{5.17}$$

They read

$$\begin{aligned}\tau_s\dot{T}_{\langle ij\rangle} + T_{\langle ij\rangle} &= 2\mu\,\varepsilon_{\langle ij\rangle} + 2(\tau_s\mu + \widetilde{\mu})\dot{\varepsilon}_{\langle ij\rangle},\\ \tau_\pi\dot{P} + P &= -\nu\,\varepsilon_{kk} - (\tau_\pi\nu + \widetilde{\nu})\dot{\varepsilon}_{kk}.\end{aligned} \tag{5.18}$$

These are the usual rate-type stress-strain relations in linear viscoelasticity.

6. Remarks on the Physical Interpretation of u_{ij}

The postulate for the existence of additional equations of balance such as (2.2) is fundamental to the formulation of extended thermodynamics. Even though no immediate physical interpretations of tensorial densities like u_{ij} are available, they are shown to be related to usual non-equilibrium physical quantities in quite an explicit manner in the theory.

For ideal gases such equations are well-motivated by the moment equations of kinetic theory of gases, in which the tensorial density u_{ij} is simply the momentum flux, i.e.,

$$u_{ij} = \varrho v_i v_j - T_{ij}, \tag{6.1}$$

and its trace is twice the total energy density. This gives a clear meaning to the tensorial density u_{ij}, but at the same time it puts a severe restriction on the theory, namely, the internal energy and the pressure are related by $2\varrho\varepsilon = 3p$ which is valid for monatomic ideal gases only.

In the present theory, with an independent energy equation postulated, the trace of u_{ij} is not longer identified as the total energy. With the above mentioned restriction removed, one is tempted to adopt (6.1) again for simplicity and clarity. However we have objected to make such a choice for the following two reasons.

Firstly, compared with (2.3) such a choice leads to

$$\varrho\, m_{ij} = -T_{ij}, \tag{6.2}$$

which implies that in equilibrium

$$\varrho\, m^o_{\langle ij\rangle} = -T^o_{\langle ij\rangle}. \tag{6.3}$$

On the other hand, from the relation (3.11), one has

$$\varrho\, m^o_{\langle ij\rangle} = T^o_{\langle ij\rangle}. \tag{6.4}$$

Therefore the deviatoric part of elastic stress must vanish. Obviously such a conclusion can not be true for solids in general.

Secondly, if one adopts (6.2) for fluids, from $(5.7)_1$ and $(5.7)_2$, one has

$$g_o = -\frac{2p}{\varrho}, \quad h_1 = \frac{d_1 T}{\varrho}, \quad h_2 = \frac{2pT}{\varrho} \tag{6.5}$$

and from $(5.5)_1$

$$d_1 = \frac{5}{3}p - \varrho\frac{\partial p}{\partial \varrho} - \frac{T}{\varrho c_v}\left(\frac{\partial p}{\partial T}\right)^2. \tag{6.6}$$

Therefore once the equation of state $p = p(\varrho, T)$ is known, all these material functions can be determined explicitly. Recall that the stability condition requires that both h_1 and h_2 be positive according to (5.8). This implies that d_1 must be positive. Unfortunately, from (6.6) this may fail to hold. Indeed, if the equation of state is given by virial expansion

$$p = R\varrho T(1 + A\varrho), \tag{6.7}$$

where A is the second virial coefficient and R the gas constant, then

$$d_1 = -\frac{5}{3}R\varrho^2 T(A + \frac{4}{3}TA' + \frac{4}{15}T^2 A''), \tag{6.8}$$

which is normally a negative quantity.

References

[1] Liu, I-Shih; Müller, I.: Extended thermodynamics of classical and degenerate ideal gases. Arch. Ration. Mech. Anal. 83 (1983) 285-332

[2] Ruggeri, T.: Galilean invariance and entropy principle for system of balance laws. Continuum Mech. Thermodyn. 1 (1989) 3-20

[3] Kremer, G. M.: Extended thermodynamics of molecular ideal gases. Continuum Mech. Thermodyn. 1 (1989) 21-45

[4] Müller, I.; Ruggeri, T.: Eds. ISIMM Symposium on kinetic theory and extended thermodynamics. Tecnoprint, Bologna (1987)

[5] Liu, I-Shih: An extended field theory of viscoelastic materials. Int. J. Engng. Sci. 26 (1988) 331-342

[6] Liu, I-Shih: Extended thermodynamics of viscoelastic materials. Continuum Mech. Thermodyn. 1 (1989) 143-164

[7] Liu, I-Shih: Method of Lagrange multipliers for exploitation of the entropy principle. Arch. Rational Mech. Anal. 46 (1972) 131-148

I-Shih Liu

Instituto de Matemática
Universidade Federal do Rio de Janeiro
Caixa Postal 68530
21944 Rio de Janeiro, Brazil

I MÜLLER

Light scattering and extended thermodynamics

ABSTRACT

The spectrum of light scattered at large angles is not well-described at all by the Navier-Stokes-Fourier theory of noble gases. There is, however, an alternative theory for the description of such gases, viz. the kinetic theory of gases and the moment theories based on the Boltzmann equation. The classical 13-moment method by Grad [1] does only a little better than the Navier-Stokes-Fourier theory. Therefore more moments are needed. Following the dissertation by W. Weiss [2] the paper shows how to determine the number of moments necessary for the representation of the scattering spectrum at a given scattering angle.

1. BASICS OF LIGHT SCATTERING

Light scattering in a gas is the result of fluctuations of the density due to thermal motion of the molecules. When the density fluctuates so does the dielectric constant and we denote its fluctuation by

$$\delta\epsilon(\underline{r},t) = \epsilon(\underline{r},t) - \epsilon_0$$

where ϵ_0 is the mean value of the fluctuating field. Figure 1 gives a schematic view of the scattering process.

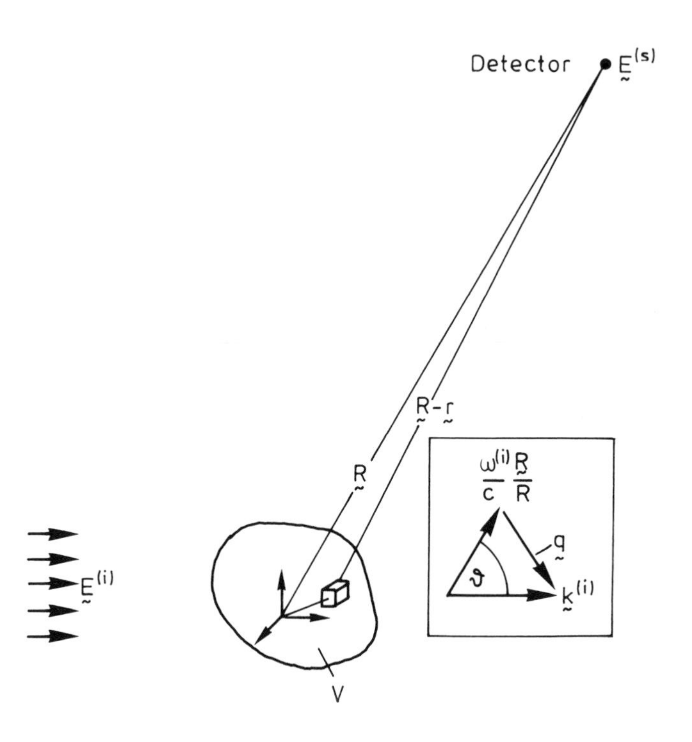

FIGURE 1 Schematic view of light scattering.
$\underset{\sim}{E}^{(i)}$ - incident electric field,
$\underset{\sim}{E}^{(s)}$ - scattered electric field, V - scattering volume.

In electrodynamics we calculate the scattered field, e.g. see Jackson [3], and in the far-field approximation we obtain for the field at the detector

$$E_i^{(s)}(\underset{\sim}{R},t)=\frac{1}{4\pi\epsilon_0 c^2}\frac{1}{R}\left(\frac{R_i R_n}{R^2}-\delta_{in}\right)\int_V d\underset{\sim}{r}\;\delta\epsilon(\underset{\sim}{r},t)\;\frac{\partial^2 E_n^{(i)}(r,t')}{\partial t'^2} \tag{1.1}$$

If the incident field is harmonic we may write

$$E_n^{(i)}(\underset{\sim}{r},t') = n_n^{(i)}\;\hat{E}_0^{(i)} e^{i(k_l^{(i)} r_l - \omega^{(i)} t')} \tag{1.2}$$

$\underset{\sim}{n}^{(i)}$ is the polarisation of the incident wave and $\hat{E}_0^{(i)}$ is its complex amplitude. $\underset{\sim}{k}^{(i)}$ and $\omega^{(i)}$ are the wave vector and angular frequency respectively. t' is the time of scattering of the wave that is detected at time t and we have

$$t' \approx t - \frac{1}{c}\left(R - \frac{R_l}{R} r_l\right) \tag{1.3}$$

Insertion into (1.1) provides

$$E_i^{(s)}(\underset{\sim}{R},t) = -\frac{\hat{E}_0^{(i)}}{4\pi\epsilon_0 c^2}\frac{1}{R}\,\omega^{(i)^2}\,e^{i\left(\frac{\omega^{(i)}}{c}R-\omega^{(i)}t\right)}\left(\frac{R_i\,R_n}{R^2} - \delta_{in}\right) n_n^{(i)}\delta\epsilon(\underset{\sim}{q},t);$$

$$\delta\epsilon(\underset{\sim}{q},t) = \int_V d\underset{\sim}{r}\; e^{i\left(k_l^{(i)} - \frac{\omega^{(i)}}{c}\frac{R_l}{R}\right) r_l}\;\delta\epsilon(\underset{\sim}{r},t) \tag{1.4}$$

We denote $k_l^{(i)} - \frac{\omega^{(i)}}{c}\frac{R_l}{R}$ by q_l and conclude that the integral in (1.4) is the spatial Fourier harmonic appropriate to a wavelength $\lambda = 2\pi/q$. In the inset of Figure 1 we see that forward scattering, i.e. small scattering angle, corresponds to a small value of q and therefore to a large-scale fluctuation, i.e. a fluctuation with a Fourier harmonic of large wavelength. Conversely, backward scattering corresponds to large values of q and therefore to small-scale fluctuations.

2. THE DYNAMIC FORM FACTOR $S(q,\omega)$

Since $\delta\epsilon(\underset{\sim}{r},t)$ is a fluctuating quantity, so is $E^{(s)}(t)$ the scattered field at the detector. Figure 2a shows a typical fluctuation as a function of time. Irregular as that function may seem, it reveals an intrinsic regularity, if we construct the mean regression of given fluctuations E_1, or E_2, or E_3, see Figure 2b. The autocorrelation function also represents that regularity. One may say that the autocorrelation function is the mean of all mean regressions for different fluctuations. see Figure 2c.

The dynamic form factor $S(q,\omega)$ is defined as the temporal Fourier transform of the autocorrelation function. ω is the Fourier frequency and the dependence on q is determined by the position of the detector where $E^{(s)}(t)$ is registered. The importance of the dynamic form factor $S(q,\omega)$ is due to the fact that <u>it is this function that is measured in a spectrometer.</u>

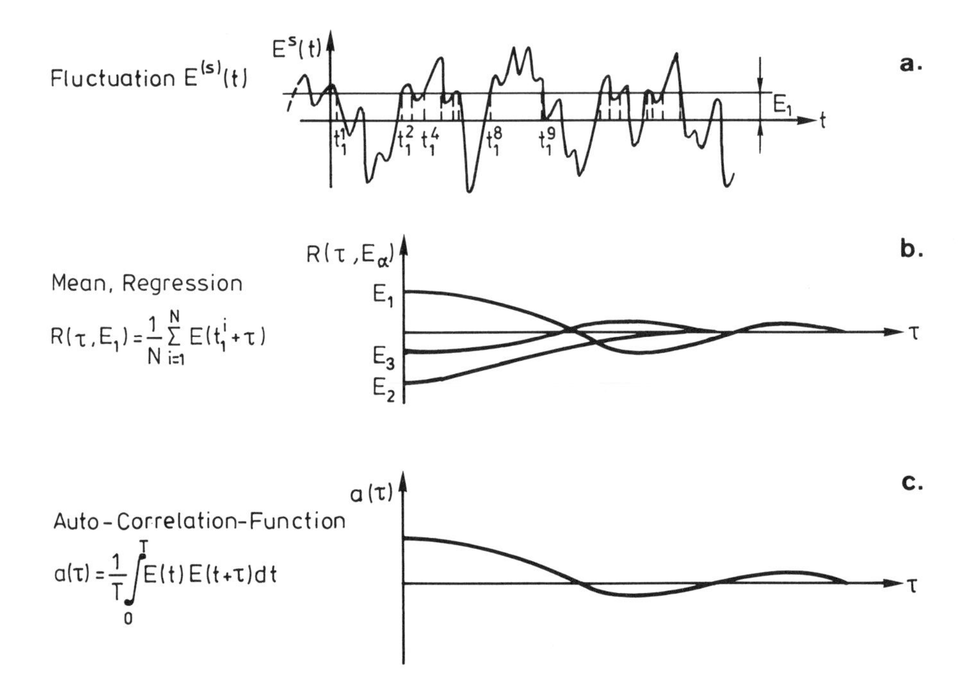

FIGURE 2 a. Fluctuation, b. Mean regression, c. Auto-correlation

The central importance of the auto-correlation function of the scattered field for thermodynamics is due to the validity of the Onsager hypothesis which states that

A mean$^{+)}$ macroscopic deviation of the spatial Fourier q-harmonic of the density field from its equilibrium value decays in the same manner as the auto-correlation function of the q-harmonic of the fluctuating density field.

Alternatively we may say that

The dynamic form factor $S(q,\omega)$ of a fluctuation is equal to the mean$^{+)}$ spatial and temporal Fourier transform $\rho(q,\omega)$ of the macroscopic density field.

+) The mean is taken over all initial values of the other fields like velocity, temperature, etc.

3. TWO POSSIBLE APPLICATIONS

Since $S(q,\omega)$ is measured and $\rho(q,\omega)$ can be calculated for a given macroscopic theory, there are two possible applications:

i.) If the two functions of ω agree qualitatively for a fixed q we may adjust the constitutive coefficients to obtain quantitative agreement. If our theory is a Navier-Stokes-Fourier theory this amounts to a (very indirect) measurement of such coefficients as heat conductivity, viscosities, compressibility, thermal expansion and specific heat.

ii.) If the two functions disagree qualitatively we must conclude that the macroscopic theory was not good enough.

In experiments with noble gases it turns out that situation i.) prevails for small scattering angles, while situation ii.) prevails for large scattering angles. Thus we conclude that for large scale fluctuations the Navier-Stokes-Fourier theory is good, while for small-scale fluctuations we need a better theory, viz. Extended Thermodynamics.

Actually there are many types of Extended Thermodynamics depending on the number of moments we take into account for the characterization of the state. Roughly speaking we expect that the theory is better, if we choose more moments. This expectation is confirmed by the kinetic theory of mon-atomic gases where we know all field equations explicitly, independent of the number of moments and also the Navier-Stokes-Fourier theory. We proceed to show this.

4. $S(q,\omega)$ FOR THE NAVIER-STOKES-FOURIER THEORY AND FOR EXTENDED THERMODYNAMICS OF 13 AND 14 MOMENTS

In this chapter we rely heavily on Weiss' dissertation [2]. On the basis of the kinetic theory of gases he has calculated the explicit forms of the field equations for gases of Maxwellian molecules up to systems of many hundreds of moments. Once that system is known, we take its double Fourier transform, spatial and temporal, and solve for the mean value $\rho(q,\omega)$. The actual calculations are formidable and can only be performed numerically. The result should agree with measured values of $S(q,\omega)$, if the theory is good. Figure 3 shows some results for the Navier-Stokes-Fourier theory and for Extended Thermodynamics of 13 and 14 moments. The diagrams of Figure 3 differ by the scattering angle, or the wave length $\lambda=2\pi/q$ of the Fourier harmonic. Actually the parameter in Figure 3 is

$$y = \frac{1}{3} \frac{\rho \, (2p_\rho)^{1/2}}{\eta} \frac{1}{q} \tag{4.1}$$

where η is the shear viscosity and p_ρ stands for $(\partial p/\partial \rho)_T$. Large values of y correspond to small scattering angles, i.e. large scale fluctuations while small values of y correspond to large scatteringangles, i.e. small-scale fluctuations.

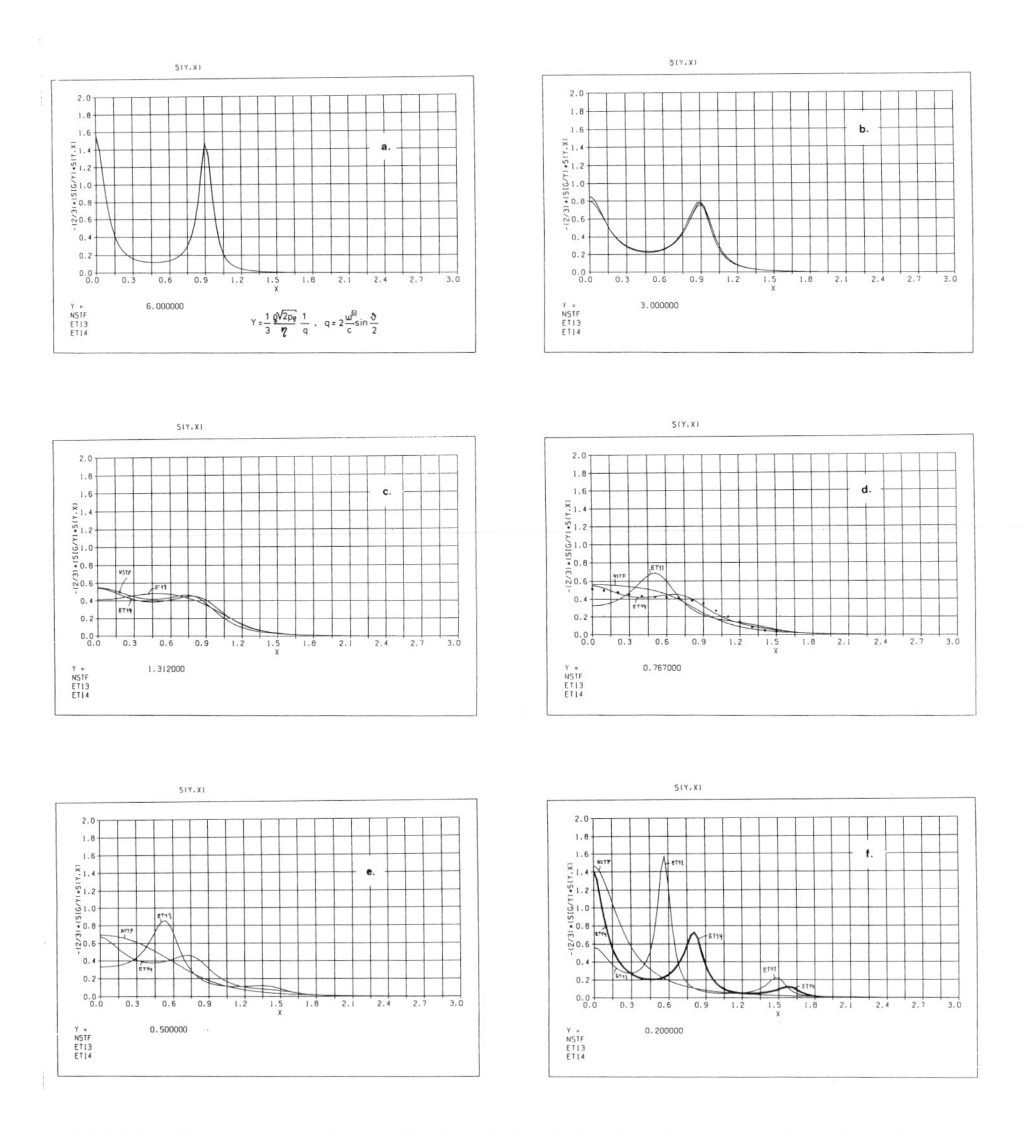

FIGURE 3 Dynamic form factors for the Navier-Stokes-Fouriertheory and for 13 and 14 moments at different scattering angles.

Inspection shows that for large-scale fluctuations i.e. y=6 all three theories – Navier-Stokes-Fourier and 13 and 14 moment – provide the same results, because there is only one curve in Figure 3a. This is still so for y=3 in Figure 3b even though small differences begin to appear. Actually these curves are also well confirmed by experiments but experimental points are not shown in Figures 3a and 3b. For values y=1.3 and y=0.767 we detect strong disagreement between the three theories, see Figures 3c and 3d. We expect the 14 moment theory to be best as it takes most moments into account. And, indeed, the experimental points of Figure 3d, which are taken from a paper by Clarke [4] lie approximately on the 14-moment curve.

For still smaller values of y, namely y=0.5 and y=0.2 we will then definitely ignore Navier-Stokes-Fourier and 13 moments.The only theory that might still be valid is the 14–moment theory .The corresponding curve for y=0.2 shows three peaks indicating the emergence of a second sound speed. Unfortunately the apparent quality of the 14-moment theory for y=0.767 and the suspected validity of that theory for y=0.5 and y=0.2 is an illusion. We find this out by investigating theories with more moments. Let us consider:

5. S(q,ω) FOR THEORIES WITH MORE MOMENTS

If, as it seemed, 14 moments are good for y=0.767, more moments should give the same result for that value of y. But now let us look at Figure 4a which represents the curves S(q,ω) for y=0.767 for 20, 35, 56 and 84 moment theories. Even those theories differ widely. Thus if any one of them is to be trusted, it is the 84-moment theory. Therefore certainly the agreement of the 14-moment theory with the experimental results was fortuitous. It is only from 120 moments onward that the theories agree as illustrated by Figure 4b which shows curves for 120, 165, 220 and 286 all coinciding. And indeed those curves roughly also coincide - by sheer accident - with the 14-moment theory of Figure 3d.

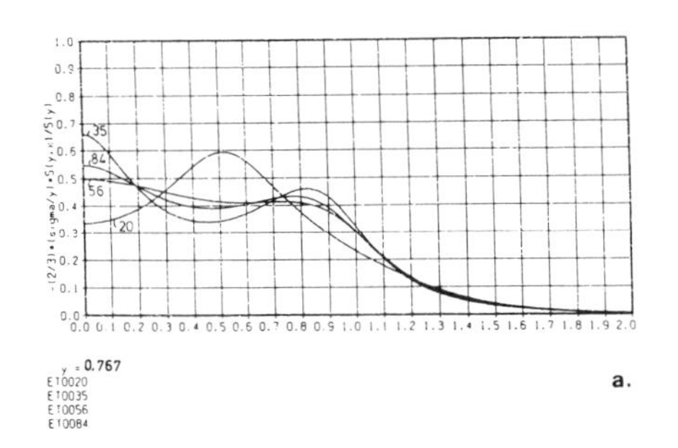

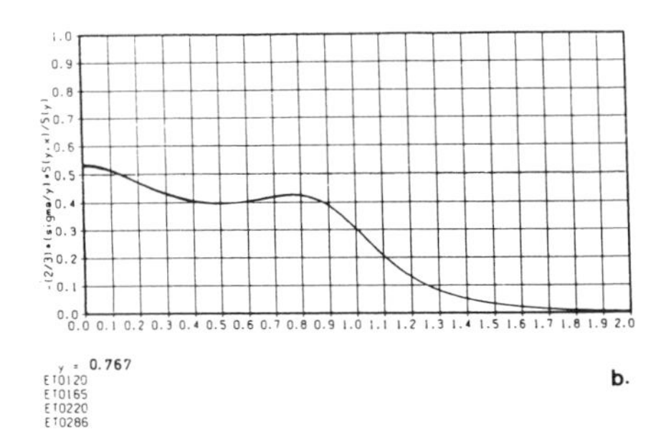

FIGURE 4: S(q,w) for y=0.767

a. Number of moments 20, 35 56, 84

b. Number of moments 120, 165, 220, 286.

We can never win definitely though! Indeed, for y=0.767 the number of moments between 200 and 300 is good enough to calculate $S(q,\omega)$, but, if we set y=0.1 we need more moments again.This is illustrated by Figure 5 where the theories for 210, 225, 240 and 256 disagree so that none of them can be good, except possibly the one with most moments.

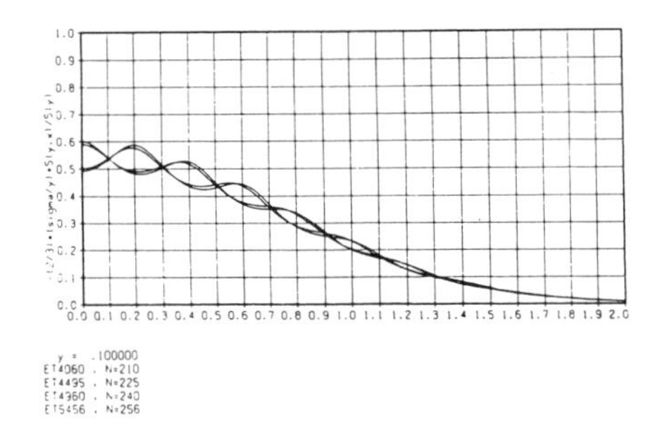

FIGURE 5: $S(q,\omega)$ for y=0.1 and 210, 225, 240 and 256 as number of moments.

6. CONCLUSION

The main conclusion to be drawn from all this is as follows: If we have a process whose spatial Fourier spectrum contains harmonics of small wavelengths we need Extended Thermodynamics of many moments. Let us consider Helium at 1 bar and T=300 K with

$\eta=0.45\cdot10^{-5}$ Ns/m^2. In that case equation (4.1) with $q=2\pi/\lambda$ reads

$$\lambda = 6\pi \frac{\eta}{\rho\,(2p_\rho)^{1/2}}\, y \approx 1.15\cdot10^{-7}\text{m} \qquad \text{for } y=0.767$$

and this is a wave length of the order of magnitude of the mean free path. Thus in processes where the density field (say) changes considerably over the length of a mean free path, as it might in a shock wave for instance, ordinary thermodynamics is not appropriate.

REFERENCES

1. H. GRAD, On the Kinetic Theory of Rarefied Gases. Comm. Pure Appl. Math. 2 (1949).
2. W. WEISS, Zur Hierarchie der Erweiterten Thermodynamik. Dissertation TU Berlin (1990) .
3. J.D. JACKSON, Classical Electrodynamics. John Wiley & Sons. Inc. NY, London, Sydney (1960).
4. N.A. CLARKE, Light Scattering from Density Fluctuations in Dilute Gases. The Kinetic Hydrodynamic Transition in a Monatomic Gas. Phys. Rev. A 12 (1975).

Ingo Müller
TU Berlin
Sekr. HF 2
Strasse des 17 . Juni 135
1000 BERLIN 12
GERMANY

A MURACCHINI, T RUGGERI AND L SECCIA

Heat shocks in a rigid conductor

Abstract

In this paper a generalized nonlinear Maxwell-Cattaneo equation is used to study shock waves propagating in a rigid heat conductor at low temperature.

The existence of a critical temperature $\tilde{\theta}$ characteristic of the materials and separating two families of shocks, the "hot" and the "cold" ones, is proved both numerically and analitically. In all the pure crystals taken up the critical temperatures calculated are very close to the values at which the second sound was identified experimentally. Finally a possible explanation of the distortion of the initial thermal pulse during its propagation is proposed.

1 Introduction

In 1947 Peshkow [1] suggested that heat could propagate in pure crystals as a true temperature wave, called second sound. In the following years a great work has been developed to understand the theoretical bases of this idea (see, in particular, the papers of Guyer and Krumhansl [2], [3]) and for finding experimentally the new wave. At the first time, second sound was observed in pure crystals of ^4He (1966) and then in high-purity crystals of ^3He (1969), NaF (1970) and Bi (1972).

To study the heat pulses in very pure crystals at low temperatures the starting point lies in considering the crystal as a phonon system. Here the normal processes (N-processes) in which phonon momentum is conserved are stronger, in certain temperature ranges, than the R-processes (dissipative processes not conserving momentum) and so the second sound can be identified. Two interesting features must be underlined: the first is the existence of a *critical temperature* such that the second sound is most clearly seen (for example, about 15^oK in NaF and 3.5^oK in Bi) and the second one concerns the modifications of the initial square wave form during its propagation according to the different temperatures of the crystal.

2 Generalized Maxwell-Cattaneo equation and second sound propagation

The phenomenology previously illustrated cannot be interpreted by Fourier's theory because of the "paradox of instantaneous propagation" and so it is necessary to find a suitable set of hyperbolic field equations.

In the spirit of Extended Thermodynamics [4], [5], [6], let us now consider a general system of two balance laws writing, in correspondence to the state pair (θ, $\mathbf{q}$)

$$\rho\dot{e} + \operatorname{div}\mathbf{q} = 0 \tag{2.1}$$

$$\dot{\mathbf{w}} + \operatorname{div}\mathbf{T} = -\mathbf{b}. \tag{2.2}$$

The first equation is the usual balance law of energy; ρ, $e \equiv e(\theta)$, $\mathbf{q}$, are respectively the (constant) mass density, the internal energy and the heat flux vector. Moreover the superposed dot indicates the time derivative. Using representation theorems for the tensors $\mathbf{w}$, $\mathbf{T}$, $\mathbf{b}$ and supposing to be near the equilibrium state, the system (2.1)-(2.2) becomes [7], [8], [9], [10]

$$\rho\dot{e} + \operatorname{div}\mathbf{q} = 0 \tag{2.3}$$

$$(\alpha\mathbf{q})^{\bullet} + \nabla\nu = -\frac{\nu'}{\kappa}\mathbf{q} \tag{2.4}$$

Here $\kappa \equiv \kappa(\theta)$ represents the heat conductivity and the remaining functions α, ν are constitutive quantities depending on the absolute temperature θ (the apex denotes the derivative with respect to θ and ∇ is the gradient operator). When α is equal to a constant, the Maxwell-Cattaneo equation

$$\tau\dot{\mathbf{q}} + \mathbf{q} = -\kappa\nabla\theta \tag{2.5}$$

is obtained ($\tau = \alpha\kappa/\nu'$) while, if $\alpha = 0$, we have the Fourier law.

The most important feature of (2.4) lies in the presence of the not constant factor α playing the role of thermal inertia. In fact, if $\alpha \equiv \alpha(\theta)$, the entropy principle as well as the hyperbolicity of the differential system (2.3)-(2.4) are satisfied without requiring the dependence of e on $\mathbf{q}$ in addition to temperature [11]. Besides the great generality due to the function $\alpha(\theta)$ allows us to recover the stability criterion of the maximum of entropy at equilibrium.

Let us impose now the compatibility of eqs. (2.3)-(2.4) with the entropy principle taken in the form

$$\dot{h}^{o} + \operatorname{div}\mathbf{h} \leq 0 \tag{2.6}$$

with

$$h^{o} = -\rho S, \qquad \mathbf{h} = -\frac{\mathbf{q}}{\theta} \tag{2.7}$$

(S is the specific entropy). Then we obtain [10]

$$\alpha = \gamma/(\nu'\theta^2), \quad \gamma = \text{const.}, \qquad \kappa > 0, \tag{2.8}$$

$$h^{o} = -\rho S = -\rho S_E(\theta) + \frac{\gamma q^2}{2(\nu'\theta^2)^2} \tag{2.9}$$

where S_E is the equilibrium entropy density.

Also the convexity condition for h^o, with respect to the field $\mathbf{u} \equiv (\rho e, \alpha \mathbf{q})^T$ is imposed and this implies our system is symmetric-hyperbolic (in the sense of Friedrichs) [1] if

$$\gamma > 0, \qquad c(\theta) = e'(\theta) > 0 \tag{2.10}$$

where c is the equilibrium specific heat.

Taking into account that $e = e(\theta)$ is known (for example, in the case of crystals at low temperature $e = \epsilon\theta^4/4$) and also $\kappa(\theta)$ is found through experimental data, we have at this step that the only arbitrary quantities are $\nu(\theta)$ and the constant γ. Besides the second sound velocity at equilibrium, $U_E \equiv U_E(\theta)$, can be identified with the characteristic velocities of the system (2.3)-(2.4) evaluated in an equilibrium state ($\mathbf{q} = 0$). The characteristic velocities in a generic state are given by the roots of the characteristic polynomial

$$\rho c \alpha \lambda^2 + \lambda \alpha' q_n - \nu' = 0, \tag{2.11}$$

where $q_n \equiv \mathbf{q} \cdot \mathbf{n}$ and $\mathbf{n}$ is the unit normal to the shock wave front.

Therefore from $(2.8)_1$ and (2.11), when $\mathbf{q} = 0$, the constitutive function ν in terms of U_E is obtained

$$\frac{\nu}{\sqrt{\rho\gamma}} = \int \frac{U_E(\theta)}{\theta} \sqrt{c(\theta)} d\theta. \tag{2.12}$$

Since it is possible to verify that γ is an inessential common factor we have no more free parameters: in other words all the constitutive functions are univocally determined knowing the equilibrium quantities $e \equiv e(\theta)$, $\kappa \equiv \kappa(\theta)$, $U_E \equiv U_E(\theta)$ [10].

3 Shock waves in high purity crystals

As (2.3), (2.4) represent a system of balance laws (i.e. the first member is in the form of space-time divergence), it is possible to write it in an integral form and to study weak solutions and, in particular, shock waves [14]; then the Rankine-Hugoniot compatibility conditions across the shock front allow us to evaluate the shock velocity s in terms of the temperature θ_o, θ_1 respectively ahead and behind the shock surface.

In order to pick out the physically relevant shocks among all the mathematical solutions of the Rankine-Hugoniot equations, two selection rules are often used: i) *the entropy growth criterion* [15], [16] and ii) *the Lax shock conditions* [17], [18], [19], [20].

The first one consists in accepting only the shock wave solutions for which the entropy production η across the shock front is non-negative; the second one states

[1]For such systems a general theorem on the well-position of the Cauchy problem (locally) holds and it follows, with regard to the shock waves, that: a) we have an entropy production across the shock wave front; b) a generating function of the shocks exists; c) the shock propagation speeds are bounded. (See, for more details, [12], [13], [7], [8]).

that the admissible shocks are those satisfying the condition $U_1 > s > U_o$ (U_o and U_1 are the values of the characteristic velocities λ_o and λ_1 evaluated respectively in the states ahead and behind the shock front). From the mathematical theory of shock waves it is well known that the two criteria are equivalent for weak shocks, i.e. in a neighbourhood of the null shock, but we underline that, in general, this is not true for strong shocks as it will be showed in the following.

Let us apply now the present approach to the case of NaF and Bi crystals specifying in accordance with previous results, only the functions $U_E(\theta)$ and $e(\theta)$.

The values of U_E obtained from experiments by Jackson et al. (for NaF) [21] and by Narayanamurti-Dynes (for Bi) [22] are well described by the empirical equation [23]

$$U_E^{-2} = A + B\theta^n \tag{3.1}$$

in the temperature range $10^oK \leq \theta \leq 18.5^oK$ (for NaF) and $1.4^oK \leq \theta \leq 4^oK$ (for Bi), where heat pulses were observed with properties expected of second sound. Values of the parameters A, B, n giving an excellent fit are [23]

$$n = 3.10, \quad A = 9.09 \cdot 10^{-12}, \quad B = 2.22 \cdot 10^{-15} \quad \text{(NaF)}$$

$$n = 3.75, \quad A = 9.07 \cdot 10^{-11}, \quad B = 7.58 \cdot 10^{-13} \quad \text{(Bi)}$$

for U_E in centimeters per second and θ in Kelvin degrees. Furthermore we take the equilibrium specific heat $c = \epsilon\theta^3$, with $\epsilon = 23$ erg cm^{-3} $^oK^{-4}$ for NaF and $\epsilon = 550$ erg cm^{-3} $^oK^{-4}$ for Bi.

Considering a plane shock wave propagating in the x-direction ($\mathbf{n} = (1,0,0)$), from the Rankine-Hugoniot equations it is possible to obtain $s = s(\theta_o, \theta_1)$ and $U_1 = U_1(\theta_o, \theta_1)$ where θ_o is the unperturbed temperature while θ_1 is the perturbed one (shock parameter).

Then, a numerical evaluation [14] allows us to plot s (cm sec^{-1}) and U_1 (cm sec^{-1}) vs. temperature θ_1 (^oK), for a fixed value of θ_o, in both cases of NaF and Bi. Figures $1 \div 3$ refer to NaF case. We observe that when the temperature θ_o increases in the range $10^oK \sim 18.5^oK$ the plots are, at first, of type displayed in fig. 1 (for $\theta_o < \tilde{\theta}$) and then as in fig. 2 ($\theta_o > \tilde{\theta}$).

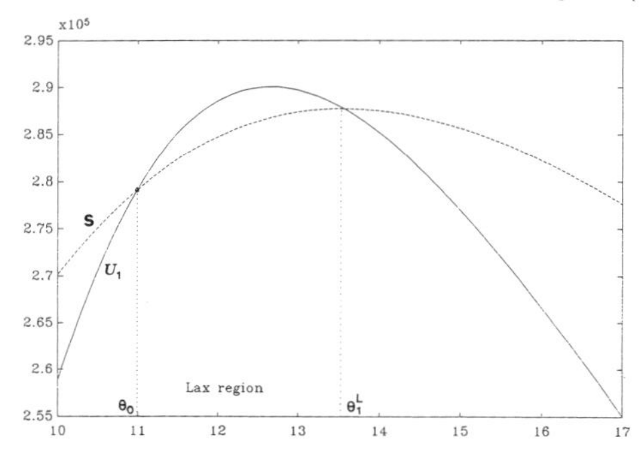

The shock wave speed s (cm sec^{-1}) and the characteristic velocity U_1 (cm sec^{-1}) behind the shock front vs. temperature θ_1 (^oK), for $\theta_o < \tilde{\theta}$, in NaF case.

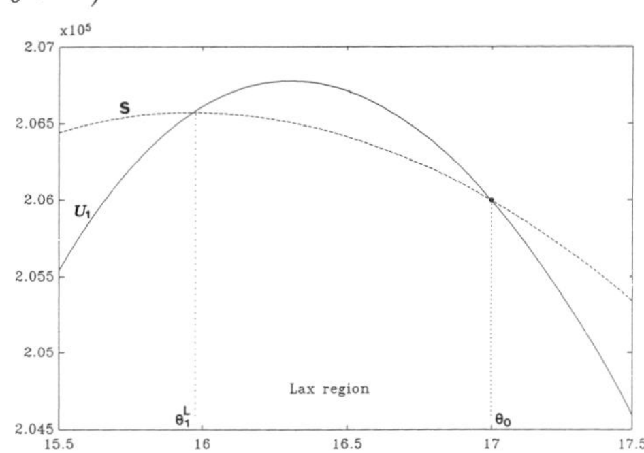

The shock wave speed s (cm sec^{-1}) and the characteristic velocity U_1 (cm sec^{-1}) behind the shock front vs. temperature θ_1 (^oK), for $\theta_o > \tilde{\theta}$, in NaF case.

Note that, in both figures, the Lax conditions impose that the possible shocks there exist only if $\mid \theta_1 - \theta_o \mid$ is bounded (unlike the usual shocks which occur, for example, in fluid dynamics). In particular in fig. 1 it is clearly seen that the Lax conditions are verified in the range $\theta_o < \theta_1 < {\theta_1}^L$ (${\theta_1}^L$ depending on θ_o). The shock wave can then propagate through the material only if we generate a heat pulse with a positive jump of temperature not exceeding the maximum value of ${\theta_1}^L - \theta_o$. Let us call this shock a *hot shock*. In fig. 2 we note that there is a very different physical situation since the Lax conditions are verified in the range ${\theta_1}^L < \theta_1 < \theta_o$. The shock propagation takes place now if the initial temperature jump is negative and does not exceed in absolute value $\mid {\theta_1}^L - \theta_o \mid$ (*cold shock*).

The transition from a situation to the other one is shown in fig. 3 where it is pointed out the existence of a *critical temperature* $\tilde{\theta} = 15.36^o K$ such that $\theta_o = \tilde{\theta} = {\theta_1}^L$. In this particular case, the Lax conditions are not satisfied and no shock is possible.

It turns out that $\tilde{\theta}$ is a structural temperature, i.e. characteristic of NaF, defining the boundary between two very different phenomena: for $\theta_o < \tilde{\theta}$ a hot shock is generated while the cold shock appears for $\theta_o > \tilde{\theta}$ and we point out that $\tilde{\theta}$ is the temperature for which the heat flux behind the front changes sign [14].

The same qualitative behaviour is observed in Bi in the range $1.4^o K \sim 4^o K$. In this case the critical temperature $\tilde{\theta} = 3.38^o K$ is found.

The non usual *cold shock* might appear inconsistent with thermodynamics but the study of the function η characterizing the entropy growth across the shock surface shows that $\eta > 0$ in the Lax region. Furthermore note that the temperature range for which $\eta > 0$ is larger than the previous one: ${\theta_1}^\eta < {\theta_1}^L < \theta_1 < \theta_o$ (in fig. 4 η/ρ (ergs cm^3 $^oK^{-1}$ gr^{-1}) vs. θ_1 with a fixed $\theta_o > \tilde{\theta}$, i.e. in the case of cold shock, is plotted for NaF).

The explanation of this fact is that the density of entropy at non equilibrium depends not only on the temperature but also on the heat flux [10] and in the function η the heat flux q_1 plays a very important role. Therefore cold shocks are compatible with the thermodynamics principles (in a different context a similar situation was already noted by Nielsen and Shklovskii [24]).

The condition $\eta > 0$ provides the same qualitative results of the Lax conditions also for hot shocks with a ${\theta_1}^\eta > {\theta_1}^L$. Observing as the plot of the function η/ρ is modified changing the unperturbed temperature θ_o, it results that the value of the critical temperature $\tilde{\theta}$ remains unchanged.

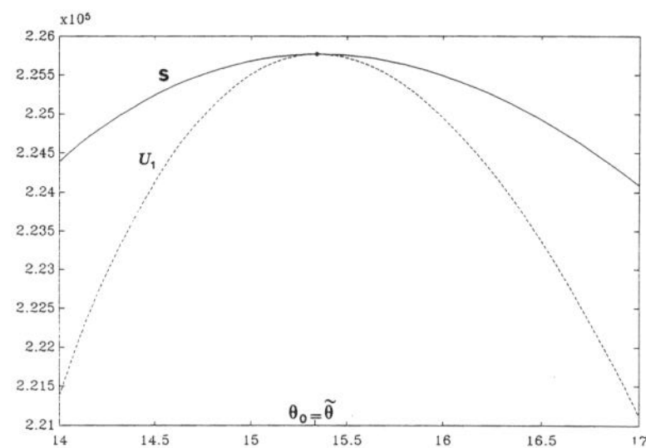

The shock wave speed s (cm sec^{-1})and the characteristic velocity U_1 (cm sec^{-1}) behind the shock front vs. temperature θ_1 (^oK), for $\theta_o = \tilde{\theta}$, in NaF case.

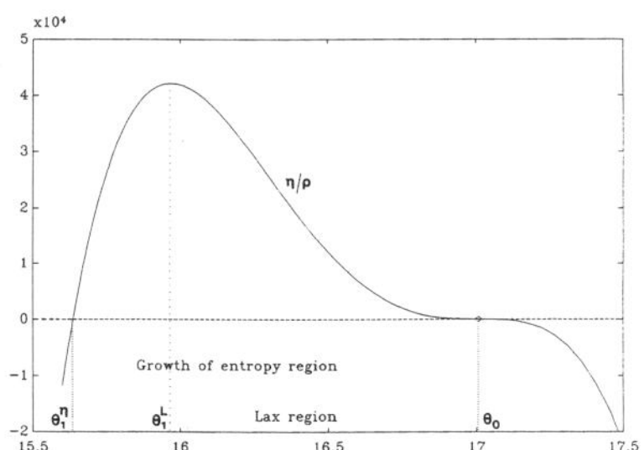

The function η/ρ (ergs cm^3 $^oK^{-1}$ gr^{-1}) vs. temperature θ_1 (^oK) in NaF case. Note that the temperature range for which $\eta > 0$ (i.e. ${\theta_1}^\eta < \theta_1 < \theta_o$) is larger than Lax region (i.e. ${\theta_1}^L < \theta_1 < \theta_o$).

So we want to remark that the present model shows unusual shocks characterized, from a macroscopic point of view, by the existence of a critical structural temperature $\tilde{\theta}$ for which the "state" of the material changes in a very unexpected way. In particular the value $\tilde{\theta} = 15.36^oK$ is found very close to the value ($\sim 15^oK$) at which a new pulse, identified as second sound, is clearly seen in a highly pure dielectric crystal of NaF. Also the value $\tilde{\theta} = 3.38^oK$ is practically coincident with the value ($\sim 3.5^oK$) at which the saturation of the velocity in the second sound regime has been observed in a pure crystal of the semimetal bismuth.

The value $\tilde{\theta}$ of the critical temperature was found by the plots of s and U_1 changing the value of θ_o. In fact, the presence of two very different shocks (hot and cold shocks) enables us to find numerically the transition temperature for which any shock at all is forbidden by the Lax conditions.

However, it is possible for a generic system to prove [25] that there exists an analytical condition for the existence of a particular unperturbed state $\tilde{\mathbf{u}}_o$, such that the Lax conditions are violated also for weak shocks, i.e.

$$(\nabla\lambda \cdot \mathbf{d})_{\tilde{\mathbf{u}}_o} = 0 \tag{3.2}$$

where $\nabla \equiv \partial/\partial\mathbf{u}$, and λ, $\mathbf{d}$ are respectively an eigenvalue and the corresponding right eigenvector of the characteristic eigenvalue problem. Therefore for the existence of $\tilde{\mathbf{u}}_o$ it is necessary that the system is not genuinely non linear.

In the present case (3.2) implies that $\tilde{\theta}$ is the value for which the function

$$\Phi(\theta) = U_E(\theta)\theta^{5/6} \tag{3.3}$$

has a maximum ($\Phi'(\tilde{\theta}) = 0$, $\Phi''(\tilde{\theta}) < 0$). Using for U_E the empirical relationship (3.1) it follows

$$\tilde{\theta} = \left\{ \frac{5A}{B(3n-5)} \right\}^{1/n} \tag{3.4}$$

and then also $\tilde{U}_E$ can be found analitically i.e.

$$\tilde{U}_E = \sqrt{\frac{3n-5}{3nA}}. \tag{3.5}$$

The relationships (3.4), (3.5) give

$$\tilde{\theta} = 15.36^oK, \qquad \tilde{U}_E = 2.26 \cdot 10^5 \text{cm/sec} \qquad (\text{for NaF}) \tag{3.6}$$

$$\tilde{\theta} = 3.38^oK, \qquad \tilde{U}_E = 7.83 \cdot 10^4 \text{cm/sec} \qquad (\text{for Bi}) \tag{3.7}$$

coincident with the values obtained numerically in our previous paper [14].

It is interesting to underline that using the function (3.3) it is possible to find $\tilde{\theta}$ also for the cases of ^3He and ^4He; as it will be reported in [25], the values so obtained are again very close to the values for which the second sound is clearly picked out in these crystals.

4 Changes of shape on second sound wave: a possible explanation

To conclude we present a possible explanation, based on the previous general results, of the distortion of the initial thermal pulse during its propagation in a rigid heat conductor. The results obtained could be a check for verifying experimentally the limits of validity of our model.

Suppose we generate an heat pulse by some type of heater: usually, with a good approximation, the schematic shape of the initial pulse is rectangular. In our opinion, the front part of the pulse is quite similar to a hot shock while the back part to a cold one. Then it should be possible to verify the expected cold shock propagation also without particular experiments.

In fact, the following three cases are possible.

For $\theta_o < \tilde{\theta}$ a hot shock can propagate (the Lax conditions $U_o < s < U_1$ are satisfied if $\theta_o < \theta_1 < {\theta_1}^L$) while a cold shock not (the Lax conditions are never satisfied). Therefore we expect that a regularisation takes place in the back part of the signal.

For $\theta_o > \tilde{\theta}$ we have a similar situation but now only the cold shock can propagate and so a regularisation of the front part of the signal occurs.

For $\theta_o = \tilde{\theta}$ the Lax conditions are never satisfied and both the cold shock and the hot shock are regularised (all these cases are sketched in fig. 5).

We note that the situation predicted for $\theta_o \leq \tilde{\theta}$ can be identified enough clearly on the ground of the experimental data, in particular in the case of ^3He [26], while the case with $\theta_o > \tilde{\theta}$ is not so well seen because in the front part the signal appears regularized, as we expect it, but in the back part a cold shock is not present. However, when θ_o is greater than $\tilde{\theta}$ diffusion, not included in our theory, comes into play and so we think it prevents us from observing the shock phenomenon.

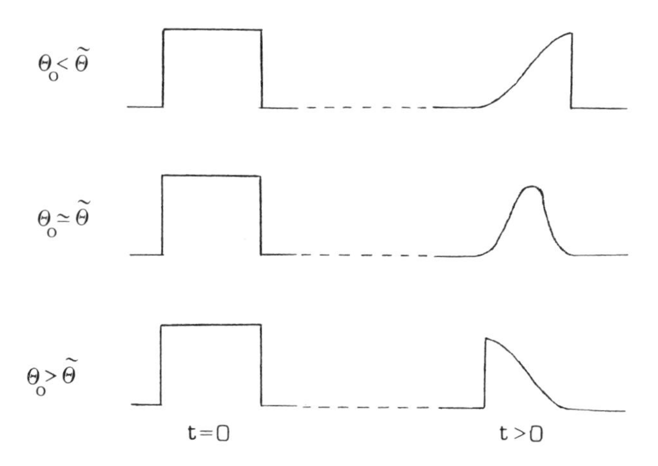

Changes in the wave-form of the initial square heat pulse.

References

[1] **V. Peshkov**, in: *Report on an International Conference on Fundamental Particles and Low Temperature Physics* Vol. II , The Physical Society of London, (1947).

[2] **R. A. Guyer & J. A. Krumhansl**, Phys. Rev. *148*, 766 (1966).

[3] **R. A. Guyer & J. A. Krumhansl**, Phys. Rev. *148*, 778 (1966).

[4] **I. Müller**, *Thermodynamics*, Pitman, Boston (1985).

[5] **I. Müller & T. Ruggeri**, Eds. *Kinetic Theory and Extended Thermodynamics*, Pitagora, Bologna (1987).

[6] **I-Shih Liu, I. Müller & T. Ruggeri**, Ann. Phys. *169*, 191 (1986).

[7] **T. Ruggeri**, Suppl. BUMI - Fisica Matematica, *4*, 261 (1985).

[8] **T. Ruggeri**, Rend. Sem. Mat. Univ. Torino. Fascicolo speciale *Hyperbolic Equations*, *167* (1987).

[9] **A. Morro & T. Ruggeri**, Int. J. Non-Linear Mech., *22*, 27 (1987).

[10] **A. Morro & T. Ruggeri**, J. Phys. C: Solid State Phys., *21*, 1743 (1988).

[11] **B. D. Coleman, M. Fabrizio & D. R. Owen**, Arch. Rat. Mech. Anal. *80*, 135 (1982).

[12] **G. Boillat & T. Ruggeri**, Compt. Rend. Acad. Sci. Paris, *289* A, 257 (1979).

[13] **T. Ruggeri & A. Strumia**, Ann. Inst. H. Poincaré, *34*, 65 (1981).

[14] **T. Ruggeri, A. Muracchini & L. Seccia**, Phys. Rev. Lett. *64*, 2640, (1990).

[15] **P. D. Lax**, in: *Contributions to Nonlinear Functional Analysis* , E. H. Zarantonello Ed. Academic Press, New York (1971).

[16] **G. Boillat** , C.R. Acad. Hebd. Seances Sci. A *283*, 409 (1976).

[17] **P. D. Lax**, Comm. Pure Appl. Math. *10*, 537 (1957).

[18] **A. Jeffrey**, *Quasilinear Hyperbolic Systems and Waves*, Pitman, London (1976).

[19] **L. Brun**, in: *Mechanical Waves in Solids*, J. Mandel and L. Brun Eds. Springer-Verlag, Wien (1975).

[20] **G. Boillat & T. Ruggeri**, Proc. Royal Soc. of Edinburgh *83 A*, 17 (1979).

[21] **H. E. Jackson, C. T. Walker & T. F. McNelly,** Phys. Rev. Lett. *25*, 26 (1970).

[22] **V. Narayanamurti & R. C. Dynes,** Phys. Rev. Lett. *28*, 1461 (1972).

[23] **B. D. Coleman & D. C. Newman,** Phys. Rev. B *37*, 1492 (1988).

[24] **H. Nielsen & B. I. Shklovskii,** Soviet Phys. JETP *29*, 386 (1969).

[25] **A. Muracchini, T. Ruggeri & L. Seccia,** *Shock Waves Theory and Changes of Shape on Second Sound Wave in Low Temperature Physics.* To appear.

[26] **C. C. Ackerman & R. A. Guyer,** Ann. of Phys. *50*, 128 (1968).

A. Muracchini, T. Ruggeri, L. Seccia

Department of Mathematics and Research Center

of Applied Mathematics - C.I.R.A.M.

University of Bologna - Via Saragozza 8, 40123 Bologna, Italy.

A PALUMBO

A model of relativistic magnetizable fluid

Abstract. In the framework of extended thermodynamics we examine the influence of the magnetic field on the internal structure of a relativistic magnetizable fluid.

Introduction.

In the last years, several attempts have been made in order to generalize the M.H.D.' equations to the case of a magnetic fluid represented by a magnetically polarizable medium, for which the magnetic permeability μ is not a constant , but depends on the temperature, the mass density and the modulus of the magnetic field [1] - [10].

These fluid models have been studied expecially in connection with problems of nonlinear wave propagation.

The different analyses carried on point out that, owing to the constitutive assumption

$$\mu = \mu(T,r,h^2) \ , \tag{1}$$

where T is the absolute temperature, r is the proper material density and $h^2 = - h_\alpha h^\alpha$, h_α being the magnetic intensity field, the thermodynamic state variables must be suitably modified in order to take into account the contribution of the magnetic field to the internal structure of the fluid.

Within such a framework there have been proposed several theories which, usually, are not in total agreement one to another.

Then, we want to propose a model of relativistic magnetizable fluid in the context of extended thermodynamics. In this environment, in order to describe the supposed thermodynamical effects acting on the fluid, we need to introduce a thermodynamic extravariable in addition to those usually considered.

Thermodynamical hypotheses.

In the classical description of thermodynamics of a simple fluid, the thermodynamical state is completely determined by assuming the specific free energy ψ_0 as a function of two thermodynamic variables, for istance the absolute temperature T and the proper material density r [11]:

$$\psi_0 = \psi_0(T,r) \quad . \tag{2}$$

By differentiating ψ_0 with respect to T and r respectively, the specific entropy S_0 and the thermodynamical pressure p_0 are defined as

$$\left(\frac{\partial \psi_0}{\partial T}\right)_r = - S_0(T,r) \quad , \quad \left(\frac{\partial \psi_0}{\partial r}\right)_T = \frac{p_0(T,r)}{r^2} \quad . \tag{3}$$

Since the equations (3) express S_0 and p_0 in terms of T and r they represent the so-called state equations.

The fundamental relation (2), by virtue of (3) , can be written in a differential form as follows:

$$d\psi_0 = - S_0 \, dT + \frac{p_0}{r^2} \, dr \, , \tag{4}$$

which, introducing the specific internal energy ϵ_0 related to ψ_0 by the relation: $\psi_0 = \epsilon_0 - T \, S_0$, yields the following Gibbs equation:

$$T \, dS_0 = d\epsilon_0 - \frac{p_0}{r^2} \, dr \, . \tag{5}$$

However equation (2) is too restrictive for a wide class of phenomena. In particular, it is not able to describe the thermodynamical effects acting on a magnetizable medium. Then it is necessary to introduce one extra variable in order to take into account the contribution of the magnetic field to the internal structure of the fluid.

Hence, in the framework of extended thermodynamics, we are led to consider the following fundamental relation for a relativistic magnetizable medium

$$\psi(T,r,\xi^2) = \psi_0(T,r) + \frac{1}{2} \int_0^{\xi^2} \Lambda(T,r,\xi^2) \, d\xi^2 \quad , \tag{6}$$

where $\xi^2 = - \xi_\alpha \xi^\alpha$, ξ^α being a new space-like four-vector which plays the role of internal extra variable whose expression will be determined later on.

Moreover , in analogy with the classical description, we define the specific entropy S and the thermodynamical pressure p by means of analogous relations:

$$\left(\frac{\partial \psi}{\partial T}\right)_{r,\xi^2} = - S(T,r,\xi^2) \quad , \quad \left(\frac{\partial \psi}{\partial r}\right)_{T,\xi^2} = \frac{p(T,r,\xi^2)}{r^2} \quad , \tag{7}$$

which , through (6), lead to :

$$S(T,r,\xi^2) = S_0(T,r) - \frac{1}{2} \int_0^{\xi^2} \Lambda'_T \, d\xi^2 \quad , \tag{8}$$

$$p(T,r,\xi^2) = p_0(T,r) + \frac{r^2}{2} \int_0^{\xi^2} \Lambda'_r \, d\xi^2 \quad . \tag{9}$$

In (8) and (9), as in the following, the subscript $_0$ means that the quantity is evalued when the magnetic field is zero and the prime denotes partial differentiation with respect to the subscripted variable.

Furthermore, by introducing the specific internal energy ϵ so that $\psi = \epsilon - TS$,the equation of state for ϵ, owing to (6), and (8), is

$$\epsilon(T,r,\xi^2) = \epsilon_0(T,r) + \frac{1}{2}\int_0^{\xi^2} (\Lambda - T\,\Lambda'_T)\, d\xi^2 \quad , \tag{10}$$

while the Gibbs equation (5) is replaced by the following generalized Gibbs equation :

$$T\, dS = d\epsilon - \frac{p}{r^2}\, dr + \Lambda\, \xi_\alpha\, d\xi^\alpha \quad . \tag{11}$$

Field equations.

The fundamental system of equations governing the evolution of a relativistic isotropic magnetizable fluid (inviscid, with infinite conductivity, embedded into a electromagnetic field), is constitued by the set of covariant laws:

- energy-momentum conservation:

$$\nabla_\alpha T^{\alpha\beta} = 0 \tag{12}$$

with the following energy -momentum tensor

$$T_{\alpha\beta} = r\, f\, u_\alpha u_\beta - p\, g_{\alpha\beta} + \sigma_{\alpha\beta} \quad ,$$

where

$$f = 1 + \epsilon + \frac{p}{r} = f_0 + \frac{1}{2}\int_0^{\xi^2} (\Lambda - T\Lambda'_T + r\Lambda'_r)\, d\xi^2$$

is the index of the magnetofluid and $\sigma_{\alpha\beta}$ is the simmetric tensor defined in terms of the total stress tensor $\tau_{\alpha\beta}$ by the relation

$$\sigma_{\alpha\beta} = \tau_{\alpha\beta} + p\, g_{\alpha\beta} \quad ,$$

satisfying the orthogonality condition with respect to u^α: $\sigma_{\alpha\beta}u^\alpha=0$;

- matter conservation:

$$\nabla_\alpha(r u^\alpha) = 0 \quad ; \tag{13}$$

- Maxwell equations:

$$\nabla_\alpha(u^\alpha b^\beta - b^\alpha u^\beta) = 0 \quad , \tag{14}$$

$b_\alpha = \mu(T,r,h^2)h_\alpha$ being the magnetic induction vector;

- constitutive relations for $\sigma_{\alpha\beta}$ and ξ_γ :

$$\sigma_{\alpha\beta} = \sigma_{\alpha\beta}(T,r,\xi_\gamma) \quad , \quad \xi_\gamma=\xi_\gamma(T,r,h_\nu) \quad , \tag{15}$$

together with the generalized Gibbs relation (11).

By contracting eq. (12) with u_β, taking into account the equation of conservation of the matter and the Gibbs relation, we obtain the the continuity equation (or energy equation):

$$r\,T\,u^\alpha\,\partial_\alpha S = r\,\Lambda\,u^\alpha \xi_\beta\,\nabla_\alpha \xi^\beta + \sigma^{\alpha\beta}\theta_{\alpha\beta} \quad , \tag{16}$$

where

$$\theta_{\alpha\beta} = \tfrac{1}{2}\,(\gamma^\lambda_\alpha\,\nabla_\lambda u_\beta + \gamma^\lambda_\beta\,\nabla_\lambda u_\alpha)$$

is the spatial rate of strain and $\gamma_{\alpha\beta} = g_{\alpha\beta} - u_\alpha u_\beta$.

Since the process is adiabatic, the energy equation (16) reduces to:

$$u^\alpha\,\partial_\alpha S = 0 \quad \Leftrightarrow \quad r\,\Lambda\,u^\alpha\,\xi_\beta\,\nabla_\alpha\xi^\beta + \sigma^{\alpha\beta}\theta_{\alpha\beta} = 0 \quad . \tag{17}$$

Therefore from (17) we get for ξ_α the most general form for the evolution equations along the current lines:

$$u^\alpha\,\nabla_\alpha\xi_\beta = \lambda_1\,\xi^\alpha\,\nabla_\alpha u_\beta + \lambda_2\,\xi^\alpha\,\nabla_\beta u_\alpha + \lambda_3\,\xi_\beta\,\nabla_\alpha u^\alpha + \lambda_4\,\partial_\beta r + \lambda_3\,\partial_\beta T + \\ +\eta_\beta \quad (\,\eta_\beta\xi^\beta = 0\,) \quad , \tag{18}$$

where λ_i (i=1,2,3,4) are coefficients whose expressions will be determined later.

Firstly, substitution of (18) into (17) yields $\lambda_4 = \lambda_5 = 0$ and the following constitutive equation for $\sigma_{\alpha\beta}$:

$$\sigma_{\alpha\beta} = -\,r\,\Lambda\,(\lambda_1+\lambda_2)\,\xi_\alpha\,\xi_\beta + r\,\Lambda\,\lambda_3\,\xi^2\,\gamma_{\alpha\beta} \quad . \tag{19}$$

The next step is to obtain the constitutive equation for ξ_α.

In order to accomplish this goal, we observe that the equation (18) must be a consequence of the field equation, hence there exist the multipliers g_α , m , l_α , n so that we have identically:

$$u^\alpha\,\nabla_\alpha\xi_\beta - \lambda_1\,\xi^\alpha\,\nabla_\alpha u_\beta - \lambda_2\,\xi^\alpha\,\nabla_\beta u_\alpha - \lambda_3\,\xi_\beta\,\nabla_\alpha u^\alpha - \eta_\beta \equiv$$

$$\equiv g_\beta\,\nabla_\alpha(r u^\alpha) + m\,(r f\,u^\alpha\,\nabla_\alpha u_\beta - \gamma^\alpha_\beta\,\partial_\alpha p + \gamma_{\rho\beta}\,\nabla_\alpha\sigma^{\alpha\rho}) + \tag{20}$$

$$+\,l_\beta\,(r\,u^\alpha\,\partial_\alpha\epsilon - \tfrac{p}{r}\,u^\alpha\,\partial_\alpha r - \sigma^{\alpha\rho}\theta_{\alpha\rho}) + n\,\nabla_\alpha(u^\alpha b_\beta - b^\alpha u_\beta) \quad .$$

So, from (20) we have respectively:

$$\lambda_1 \ \xi^\alpha = n \ b^\alpha \quad , \tag{21}$$

$$\lambda_2 = m = l_\beta = 0 \quad , \tag{22}$$

$$\eta_\beta = n \ u_\beta \ \nabla_\alpha b^\alpha \ , \tag{23}$$

$$\lambda_3 \ \xi_\beta + r \ g_\beta + n \ b_\beta = 0 \quad , \tag{24}$$

$$u^\alpha \ \nabla_\alpha \xi_\beta = g_\beta \ u^\alpha \ \partial_\alpha r + n \ u^\alpha \ \nabla_\alpha b_\beta \quad . \tag{25}$$

Relations (21), (24) and (25) are compatible iff along the current lines the following conditions hold:

$$\lambda_1 = 1 \quad , \quad n \ (1+\lambda_3) + r \ n'_r = 0 \ . \tag{26}$$

Choising $\lambda_3=0$, $(31)_2$ yields

$$n = \frac{k}{r} \qquad (\text{k=const.}). \tag{27}$$

Consequently, the relation (21) gives the constitutive equation for ξ^α :

$$\xi^\alpha = \frac{k}{r} \ b^\alpha \quad , \tag{28}$$

which, by virtue of (23), satisfies the orthogonality condition $\xi^\alpha \eta_\alpha=0$, and, furthermore, transforms (19) into the following:

$$\sigma_{\alpha\beta} = - \frac{\Lambda}{r} \ k^2 \ b_\alpha b_\beta \quad . \tag{29}$$

Now, we note that we obtain the Maxwell equations (14) when (28) is inserted into the evolution equations (18).

Therefore the motion of our fluid is completely described by the following set of equations:

$$\nabla_\alpha \ (r u^\alpha) = 0 \quad ,$$

$$\begin{aligned} & rf \ u^\alpha \ \nabla_\alpha u^\beta - \gamma^{\alpha\beta} \ \partial_\alpha p - \frac{k^2}{r} \ b^\alpha b^\beta \ (\ \partial_\alpha \Lambda - \frac{\Lambda}{r} \ \partial_\alpha r \ ) + \\ & \quad - \frac{\Lambda k^2}{r} \ (\ b^\beta \ \nabla_\alpha b^\alpha + \gamma^\beta_\lambda \ b^\alpha \ \nabla_\alpha b^\lambda \ ) = 0 \quad , \end{aligned} \tag{30}$$

$$u^\alpha \ \partial_\alpha S = 0 \quad ,$$

$$\nabla_\alpha \ (\ u^\alpha \ b^\beta - b^\alpha \ u^\beta \ ) = 0 \quad ,$$

together with the algebraic constraints

$$u^\alpha u_\alpha = 1 \quad , \qquad u^\alpha b_\alpha = 0 \quad , \tag{31}$$

and the fundamental thermodynamic differential relation (11), which, by means of (28), takes the form :

$$T\, dS = d\epsilon - \frac{p}{r^2}\, dr + \frac{\Lambda k^2}{r}\, b_\alpha\, d\left(\frac{b^\alpha}{r}\right) . \tag{32}$$

Comparison with other theories.

By setting

$$k = 1 \quad , \qquad \Lambda = \frac{r}{\mu} \quad , \tag{33}$$

the relations (6), (8), (9), (10) become respectively

$$\psi(T,r,h^2) = \psi_0(T,r) - \frac{1}{r}\left(h_\alpha b^\alpha - \int_0^{h^\alpha} \mu\, h_\alpha\, dh^\alpha \right) , \tag{34}$$

$$S(T,r,h^2) = S_0(T,r) - \frac{1}{r} \int_0^{h^\alpha} \mu'_T\, h_\alpha\, dh^\alpha \quad , \tag{35}$$

$$p(T,r,h^2) = p_0(T,r) - \int_0^{h^\alpha} (\, \mu - r\, \mu'_r \,)\, h_\alpha\, dh^\alpha \quad , \tag{36}$$

$$\epsilon(T,r,h^2) = \epsilon_0(T,r) - \frac{1}{r}\left(h_\alpha b^\alpha + \int_0^{h^\alpha} (\, T\, \mu'_T - \mu \,)\, h_\alpha\, dh^\alpha \right), \tag{37}$$

which are the relativistic generalisation of the corresponding classical expressions introduced by Tarapov.

Obviously, in addition to (33), by setting μ = const. we recover the usual relativistic M.H.D. [12] , [13].

REFERENCES

[1] De Groot S.R. and Suttorp L.G., *Foundation of Electrodynamics*, North-Holland, Amsterdam, 1972.

[2] Tarapov I.E., *On the Basic Equations and Problems of the Hydrodynamics of Polarizable and Magnetizable Media*, in book: *Theory of Functional Analysis and Their Applications*, Ed. 17, Khar'lov, Izd-vo KhGU, 1973.

[3] Sedova G.L., Izv. Akad. Nauk SSSR MZhG, **6**, (1974), 114-120.

[4] Gorskii V.B., PMM, **50**, 3, (1986), 388-391.

[5] Tarapov I.E., PMM, **37**, 5, (1973), 770-778.

[6] Sutyrin G.G. and Taktarov N.G., PMM, **39**, 3, (1975), 547-550.

[7] Patsegon N.F., Polovin R.V. and Tarapov I.E., *PMM*, **43**, 1, (1979), 57-64.

[8] Tarapov I.E., PMM, **48**, 3, (1984), 275-279.

[9] Cissoko M., Phys. Rev. A, **36**, 4, (1987), 1786-1794.

[10] Barilaro M. and Giambó S., Rend. Matem., Serie VII, **9**,(1989), 681-687.

[11] Callen H., *Thermodynamics*, J. Willey, New York, 1960.

[12] Choquet-Bruhat Y., Astronauta Acta, VI, (1960), 354-365.

[13] Lichnerowicz A., *Relativistic Fluid Dynamics*, C.Cattaneo Ed. Cremonese, Roma, 1971.

ANNUNZIATA PALUMBO
Istituto del Biennio della Facoltá di Ingegneria
Via E.Cuzzocrea, 48 - 89128 Reggio Calabria - Italy

T RUGGERI

Shock waves in hyperbolic dissipative systems: non equilibrium gases

Abstract

We study shock waves for the hyperbolic dissipative systems of the Extended Thermodynamics. Using the selection rules for physical shocks, we prove the existence of two families of shocks and we compare the results with the classical ones. The main goal is to verify the existence of a small set of Mach numbers for which there exists a shock such that just behind the front the temperature is smaller than the unperturbed one (cold shock). Moreover we prove that there exists also an upper bound for Mach number (*umbilic point*), such that beyond this the theory is no longer valid.

1 Introduction

The Extended Thermodynamics is a physical example of an hyperbolic dissipative system of balance laws. In this framework it is possible to define weak solutions and, in particular, to study shock waves just like in a non dissipative system.

In this paper we study shock waves propagating through an equilibrium state. A qualitative analysis of the Rankine-Hugoniot equations and of the selection rules for admissible shocks (Lax conditions, entropy growth) gives different results with respect to the classical approach.

In fact, in the parabolic case, the dissipation variables influence only the thickness of the shock front and behind and ahead it there are equilibrium states. Consequently we have only one shock moving in the $x > 0$ region with an unperturbed Mach number greater than one.

In Extended Thermodynamics we prove that just behind the shock front a non equilibrium state always exists and two fronts are present: a *hot shock*, in which the temperature behind the shock is greater than the unperturbed temperature and a *cold shock* in which the perturbed temperature is smaller than the unperturbed one. In these cases there exists a small set of Mach numbers less than one for which the admissibility criteria are satisfied.

The consequences of these results are discussed, in particular pointing out that the non equilibrium temperature loses the usual role since it is no longer a *measure* of the disorder as in the equilibrium case.

Moreover we find the limit values of the Mach numbers corresponding to the so called *umbilic points*. After these values the differential system is not hyperbolic and so the theory cannot be applied.

2 The Hyperbolic System of the Extended Thermodynamics

The equations of the Extended Thermodynamics (E.T.) for a non equilibrium mono-atomic gas in one space dimension represent a system of balance laws [1,2,3]:

$$\frac{\partial}{\partial t}\rho + \frac{\partial}{\partial x}(\rho v) = 0 \tag{2.1}$$

$$\frac{\partial}{\partial t}(\rho v) + \frac{\partial}{\partial x}(\rho v^2 + p - \sigma) = 0 \tag{2.2}$$

$$\frac{\partial}{\partial t}(\rho v^2 + 3p) + \frac{\partial}{\partial x}(\rho v^3 + 5pv - 2\sigma v + 2q) = 0 \tag{2.3}$$

$$\frac{\partial}{\partial t}(\frac{2}{3}\rho v^2 - \sigma) + \frac{\partial}{\partial x}\left(\frac{2}{3}\rho v^3 + \frac{4}{3}pv - \frac{7}{3}\sigma v + \frac{8}{15}q\right) = \tau_o\sigma \tag{2.4}$$

$$\frac{\partial}{\partial t}\left(2q + 5pv - 2\sigma v + \rho v^3\right) + \frac{\partial}{\partial x}\left(\rho v^4 + 5\frac{p^2}{\rho} - 7\frac{\sigma p}{\rho} + \right.$$
$$\left. + \frac{32}{5}qv + v^2(8p - 5\sigma)\right) = 2\tau_o v\sigma - \tau_1 q, \tag{2.5}$$

where $\rho, v, e, p = 2/3\rho e = k\rho\theta, \sigma, q$ are respectively the mass density, velocity, internal energy, pressure, shear stress and heat flux. Moreover τ_o, τ_1 are parameters related to the heat conductivity and the viscosity coefficient and θ is the absolute temperature.

3 Shocks

Let us now consider a plane shock front with velocity s. Introducing the relative velocity

$$u = v - s, \tag{3.1}$$

the Rankine-Hugoniot (R. H.) compatibility conditions across the shock front read

$$[\![\rho u]\!] = 0 \tag{3.2}$$

$$\left[\!\left[\rho u^2 + p - \sigma\right]\!\right] = 0 \tag{3.3}$$

$$\left[\!\left[2q + 5pu - 2\sigma u + \rho u^3\right]\!\right] = 0 \tag{3.4}$$

$$\left[\!\left[\frac{8}{15}q + \frac{4}{3}pu - \frac{7}{3}\sigma u + \frac{2}{3}\rho u^3\right]\!\right] = 0 \tag{3.5}$$

$$\left[\!\left[\rho u^4 + 5\frac{p^2}{\rho} - 7\frac{p\sigma}{\rho} + \frac{32}{5}uq + u^2(8p - 5\sigma)\right]\!\right] = 0. \tag{3.6}$$

where $[\![f]\!] = f_1 - f_o$ indicates the jump and f_o, f_1 are the limit values of f across the shock front evaluated respectively in the unperturbed and perturbed state.

We consider now the propagation in an equilibrium state, i.e. when the unperturbed state is characterized by $\rho_o, p_o, v_o = q_o = \sigma_o = 0$. Introducing the adimensional variables (c_o is the sound velocity):

$$M_o = \frac{s - v_o}{c_o} = \frac{s}{c_o};\ w = \frac{u_1}{u_o};\ \pi = \frac{p_1}{p_o};\ r = \frac{\rho_1}{\rho_o};\ \chi = \frac{q_1}{p_o c_o};\ \tau = \frac{\sigma_1}{p_o}, \tag{3.7}$$

it is possible, after straightforward calculations, to give the solution of the R. H. equations (for non characteristic shocks) in the form:

$$r = \frac{1}{w} \tag{3.8}$$

$$\tau = \frac{10}{27}M_o{}^2\frac{w^2 - 1}{w} \tag{3.9}$$

$$\pi = 1 - \frac{5}{27}M_o{}^2\frac{(w-1)(7w-2)}{w} \tag{3.10}$$

$$\chi = -\frac{5}{18}M_o(w-1)\left(10M_o{}^2w - 5M_o{}^2 - 9\right) \tag{3.11}$$

where w and M_o are related by the following polynomial relation:

$$5M_o{}^4\left(694w^3 - 710w^2 + 143w + 8\right) - 2106M_o{}^2w(2w-1) + 729w = 0. \tag{3.12}$$

Therefore for a fixed value M_o of the unperturbed Mach number (characterizing the shock parameter), we obtain w from (3.12) and then r, τ, π and χ from (3.8)-(3.11). From (3.8) and (3.7) we note that w is the volume ratio, π is the pressure ratio, τ and χ are the adimensional shear stress and heat flux in the non equilibrium perturbed region. Moreover taking into account (3.1), (3.7) and the state function $p = k\rho\theta$, it is easy to evaluate also the velocity and the temperature just behind the shock front:

$$\bar{v}_1 = \frac{v_1}{c_o} = M_o(1 - w) \qquad \Theta = \frac{\theta_1}{\theta_o} = \pi w. \tag{3.13}$$

In accordance with the general theory of shock waves the shocks are one parameter families corresponding to the bifurcated branches of the trivial solution (null shock), occurring when s approaches to a characteristic eigenvalue λ of the differential system (2.1)-(2.5). In fact, observing that from (3.8)-(3.11) the null shock corresponds to $w \to 1$, (3.12) reduces to

$$25M_o{}^4 - 78M_o{}^2 + 27 = 0, \tag{3.14}$$

i.e. the characteristic polynomial of the differential system at equilibrium [4]. The roots of (3.14) are

$$M_o \simeq \pm 0.6297; \quad \pm 1.65028. \tag{3.15}$$

Therefore considering the shocks propagating in the $x > 0$ region, we have two families of shocks bifurcating from the null solution, when M_o approaches respectively to ${M_o}^{(S)} = 0.6297$ (slow shock) and ${M_o}^{(F)} = 1.65028$ (fast shock).

First of all it is easy to verify the following general properties:

Statement 3.1 *Just behind the shock front we have always a non equilibrium state.*

In fact from (3.9), (3.11) the only possibility for which the heat flux and the shear stress are both zero is $w = 1$, i.e. the null shock.

This result is substantially in contrast with the classical one. In fact in the Fourier Navier-Stokes theory and also in the Irreversible Extended Thermodynamics the approach to the shock theory is concentrated in the solutions called *shock structure solutions* in which the two states (perturbed and unperturbed) are always of equilibrium and they are connected by a continuous very rapid non equilibrium process.

Another general result is the following one:

Statement 3.2 *For both the non characteristic shocks there exists a lower bound for the volume ratio w:*

$$w = \frac{V_1}{V_o} > w_{\text{crit}} \simeq 0.709949 \tag{3.16}$$

instead of $w_{\text{crit}} = 0.25$ of the classic case (see e.g. [5]).

The proof of the statement follows by an analysis of the coefficients of (3.12) such that the biquadratic equation (3.12) may admit real roots M_o for a fixed value of w. When $w \to w_{\text{crit}}$ in the fast shock, M_o approaches to $+\infty$. Therefore in E.T. the volume ratio is always greater with respect to the classical expectation.

3.1 Selection Rules for Shocks

As it is well known, we have for a generic hyperbolic system of balance laws two criteria for selecting the physical shocks among all the mathematical solutions of the R.H. equations.

The first one is to satisfy the Lax conditions [6], i.e. $\forall k = 1, 2, \ldots, N$, we have:

$${\lambda_0}^{(1)} < \ldots < {\lambda_0}^{(k)} < s < {\lambda_0}^{(k+1)} < \ldots < {\lambda_0}^{(N)} \tag{3.17}$$

$${\lambda_1}^{(1)} < \ldots < {\lambda_1}^{(k-1)} < s < {\lambda_1}^{(k)} < \ldots < {\lambda_1}^{(N)} \tag{3.18}$$

where N is the number of equations and λ_o and λ_1 are the characteristic velocities evaluated respectively in the unperturbed and perturbed state.

In particular (3.17) and (3.18) imply that for each family of shocks we have:

$$\lambda_o < s < \lambda_1. \tag{3.19}$$

In the classical fluid dynamics the physical meaning of (3.19) lies in the well known condition that the shock is supersonic in one side and subsonic in the other one.

An other selection rule for shocks arises from a thermodynamics requirement. In fact in the non linear case the shock produces a source of entropy $\eta \neq 0$ and for the *entropy growth* criterion the physical shocks are those for which

$$\eta > 0. \tag{3.20}$$

Under a suitable condition of concavity for the entropy density it is well known that the two criteria are equivalent at least for weak shocks.

In the next section we discuss the consequences of the selection conditions on the solutions of the R.H. equations for both the non characteristic shocks (fast and slow) occurring in E.T.

3.2 The Fast Shock

We choose, according to the Lax conditions, $s > \lambda_o$, i.e. $M_o > M_o{}^{(F)}$ in (3.12) and therefore for a fixed M_o we obtain w from (3.12), and then from (3.8)-(3.11) r, τ, π and χ, and from (3.13) $\overline{v}_1, \Theta$.

A simple analysis of (3.8)-(3.13) permits to prove the following statement:

Statement 3.3 *Behind the fast shock front the volume is less than the corresponding unperturbed one (compression wave):*

$$w_{\text{crit}} < w < 1, \tag{3.21}$$

the pressure ratio π and the temperature ratio Θ are greater than one (hot shock), the shear stress is negative and the heat flux, together with the velocity, is positive.

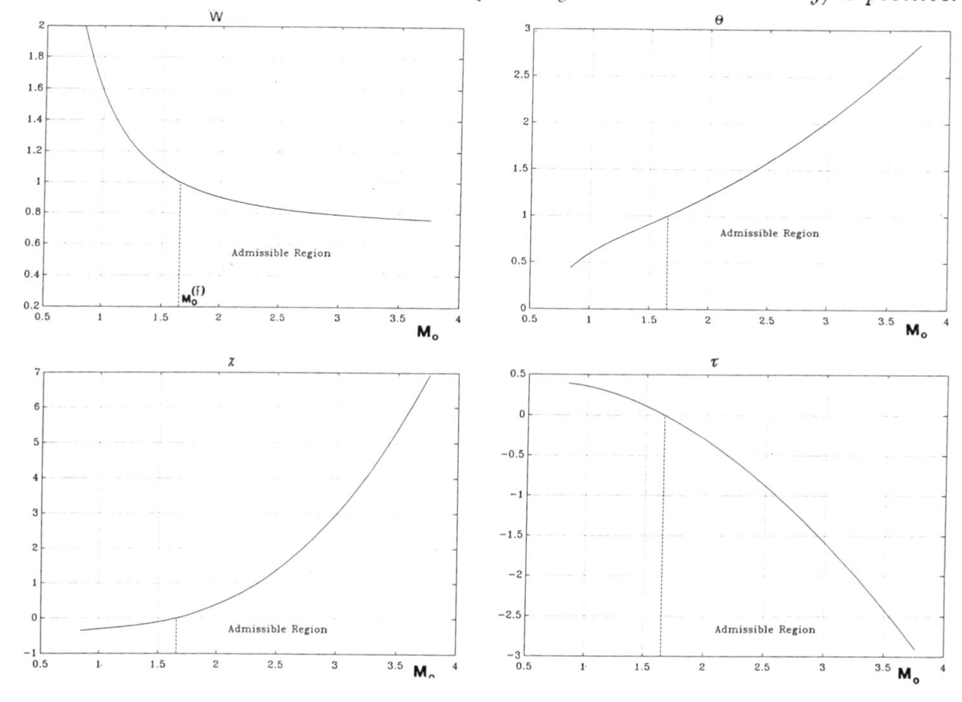

For what concerns volume, pressure, temperature and velocity, these results are qualitatively quite similar to the ones of the classical case. The figg. 1-4 give the behavior of w, Θ, χ and τ as functions of the Mach number M_o.

The previous results are obtained by using only a part of the Lax condition, i.e. $s > \lambda_o$. To verify the complete Lax conditions (3.17), (3.18) it is necessary to evaluate the characteristic velocities λ_1 in the perturbed state. The characteristic polynomial was deduced in [4] for a generic non equilibrium state. Introducing the adimensional perturbed characteristic velocity $\hat{\lambda} = \lambda_1/c_o$, it is possible to rewrite the characteristic polynomial in the form:

$$\overline{\lambda}^4 - \frac{2}{25}\overline{\lambda}^2 w(39\pi - 31\tau) - \frac{288}{125} w\chi\overline{\lambda} + \frac{27}{25}\pi w^2(\pi - 2\tau) = 0, \qquad (3.22)$$

where

$$\overline{\lambda} = \hat{\lambda} - \overline{v}_1 = \hat{\lambda} - M_o(1 - w). \qquad (3.23)$$

The fig. 5 gives the adimensional characteristic velocities as function of the Mach number. Two very interesting properties arise from fig. 5: the first is that the Lax conditions (3.17), (3.18) are completely satisfied, when $M_o > M_o{}^{(F)}$, the second one is that there exists a critical value of the Mach number $M_{oc}{}^{(F)} \simeq 2.7$ for which two eigenvalues are coincident. This case is called in the literature an *umbilic point.* After this point the two eigenvalues become complex and the system is no longer hyperbolic.

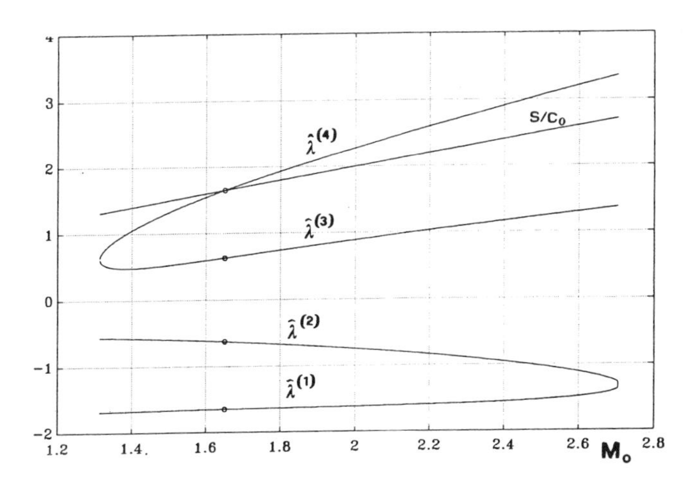

This situation does not surprise, because after the shock passage we have a non equilibrium region and then when the Mach number increases, also the absolute values of the heat flux and shear stress increase and as consequence the theory of the E.T. is no longer applicable since the process becomes far from an equilibrium state. The interesting feature is that this analysis gives a precise evaluation of the critical Mach number for which the theory is true. From fig. 5 it is also possible to see that another umbilic point occurs in the non admissible region ($M_o \simeq 1.3$).

We observe that these questions are not related to the critical Mach numbers observed in [7,8] for shock structure solutions of the irreversible thermodynamics.

We summarize these results as a statement:

Statement 3.4 *The fast hot shock is admissible and the differential system of the E.T. is hyperbolic in the set of Mach numbers*

$$\sim 1.6509 < M_o <\sim 2.7. \tag{3.24}$$

3.3 The Slow Shock

Proceeding as in the previous case, we consider the shock bifurcating from $M_o = M_o^{(S)}$. The main results are:

Statement 3.5 *The slow shock there exists only in a small set of Mach numbers*

$$\sim 0.629727 < M_o <\sim 0.907948. \tag{3.25}$$

In this range the slow shock has the following properties: the volume ratio is less than one (compression wave): $w_{\text{crit}} < w < 1$. *When* $w \to 1$, $M_o \to M_o^{(S)}$, *while when* $w \to w_{\text{crit}}$, $M_o \to 0.907948$. *The pressure ratio* π *is greater than one, but the temperature ratio* Θ *is less than one (cold shock) and according to this fact the heat flux just behind the shock front is negative. The velocity is positive and the shear stress is negative.*

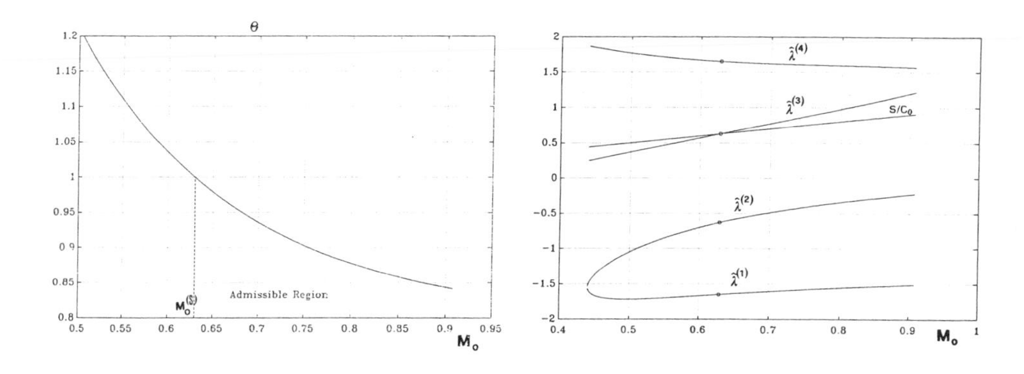

The main interesting result of this statement is the unusual condition that the temperature after the shock is less than the unperturbed one.

The non usual *cold shock* might appear inconsistent with thermodynamics because behind the shock front we have a region of non-equilibrium, i.e. more *disorderly*, and therefore it is reasonable to think that the temperature increases after the passage of the shock. But, as well known, the real measure of disorder is the entropy growth across the shock front and as the density of entropy at non equilibrium depends not only on the temperature but also on the heat flux and the shear stress, it is possible that the entropy increases after the passage of the shock though the temperature decreases. Therefore the cold shock is compatible with the thermodynamics principles and the temperature cannot be considered as a measure of the disorder in the non equilibrium state.

This situation was first discovered in the context of a rigid heat conductor at low temperature [9,10]. In this last circumstance we have a more complicate case because there exists only one shock presenting both the characteristics. In fact for an unperturbed temperature less than a critical one the shock is hot and after this value becomes cold. The transition is strictly related to the second sound physics.

In fig. 6 and 7, we see the temperature ratio and the characteristic velocities as function of the Mach number. In the last figure we observe also in this case an umbilic point in a non admissible region. The upper bound for the Mach number (3.25) is due here to the fact that for this value the maximum compression ($w \to w_{\text{crit}}$) is reached.

4 Conclusions

The Extended Thermodynamic offers a new approach to shock waves when dissipation is present and this procedure is quite similar to the non dissipative case, since the system has the form of a balance laws system. Therefore this analysis gives an alternative methodology with respect to the usual shock structure approach. Moreover the existence of a small set of Mach numbers for which there exists a cold shock may be a challenge for the experimenters to confirm or not the results coming from the modern Extended Thermodynamics.

The complete proofs of the previous results together with the analysis of the characteristic shock (contact shock) not included in the present report, will appear in a forthcoming paper.

References

[1] **I-Shih Liu & I. Müller**, Arch. Rat. Mech. Anal. **83**, 285 (1983).

[2] **I. Müller**, *Thermodynamics*, Pitman, Boston (1985).

[3] **I. Müller & T. Ruggeri**, Eds. *Kinetic Theory and Extended Thermodynamics*, Pitagora, Bologna (1987).

[4] **T. Ruggeri & L. Seccia**, Meccanica **24**, 127 (1989).

[5] **L. Landau & E. Liftshitz**, *Mècanique des Fluides*, pag. 418 Moscow: MIR (1971).

[6] **P. D. Lax**, in: *Contributions to Nonlinear Functional Analysis*, E. H. Zarantonello Ed. Academic Press, New York (1971).

[7] **H. Grad**, Comm. Pure Appl. Math. **5**, 257 (1952).

[8] **A. M. Anile & A. Majorana**, Meccanica **16** (3),149 (1981).

[9] **T. Ruggeri, A. Muracchini & L. Seccia**, Phys. Rev. Lett. *64*, 2640, (1990).

[10] **A. Muracchini, T. Ruggeri & L. Seccia**, *Shock Waves Theory and Changes of Shape on Second Sound Wave in Low Temperature Physics.* To appear. See also the report by A. Muracchini and L. Seccia in the present Lecture Notes.

Department of Mathematics and Research Center of Applied Mathematics
- C.I.R.A.M. - University of Bologna - Via Saragozza 8, 40123 Bologna, Italy.

Part IV

Mathematical Modeling of Experimental Tests on Waves

P BOCCOTTI, G BARBARO AND L MANNINO

An experiment at sea on mechanics of irregular gravity waves: short description

ABSTRACT

The highest waves in storm seas give rise to groups like that of Fig.1. That was predicted as a consequence of "quasi-determinism of rare events in random processes" [1], [2], and now is confirmed by an experiment at sea just in front of the Euromech 270 center.

1. THEORY

Wind generated waves on sea or lake surface may be thought of as the sum of a very large number of small regular cylindrical waves with frequencies and directions such as to form a well characteristic spectrum and phases distributed purely at random. Under those hypotheses both wave elevation and velocity potential, to Stokes's first order, represent stationary Gaussian random processes of time at any point [3], [4].
Provided to know that at a point x_o, y_o within a wind generated wave field (above cited theory) at any time instant t_o there is a wave with a height H very large with respect to the mean, we can predict that that wave, with very high probability, has been produced by the transit of a well defined wave group (the probability approaches 1 if the ratio of given wave height H to the mean wave height tends to infinity) [2]. The expression of the surface elevation and of the velocity potential of the wave group, exact to Stokes's first order, are [2]

$$(1.a)\quad \eta(x_0+X, y_0+Y, t_0+T) = \frac{H}{2} \int_0^\infty \int_{-\pi}^{\pi} S(\omega,\theta)\ [\cos(kX\sin\theta + kY\cos\theta - \omega T) + \\ -\cos(kX\sin\theta + kY\cos\theta - \omega T + \omega T^*)]d\theta d\omega / \int_0^\infty \int_{-\pi}^{\pi} S(\omega,\theta)(1-\cos\omega T^*)d\theta d\omega,$$

$$(1.b)\quad \phi(x_0+X, y_0+Y, z, t_0+T) = g\frac{H}{2} \int_0^\infty \int_{-\pi}^{\pi} \omega^{-1} S(\omega,\theta)[\cosh k(h+z)/\cosh kh][\sin(kX\sin\theta + \\ + kY\cos\theta - \omega T) - \sin(kX\sin\theta + kY\cos\theta - \omega T + \omega T^*)]d\theta d\omega / \int_0^\infty \int_{-\pi}^{\pi} S(\omega,\theta)(1 + \\ -\cos\omega T^*)d\theta d\omega,$$

$k\tanh(kh) = \omega^2/g$, T^* *abscissa of the absolute minimum of* $f(t) = \int_0^\infty \int_{-\pi}^{\pi} S(\omega,\theta)\cos\omega T\ d\theta d\omega$, *whose existence is required,*

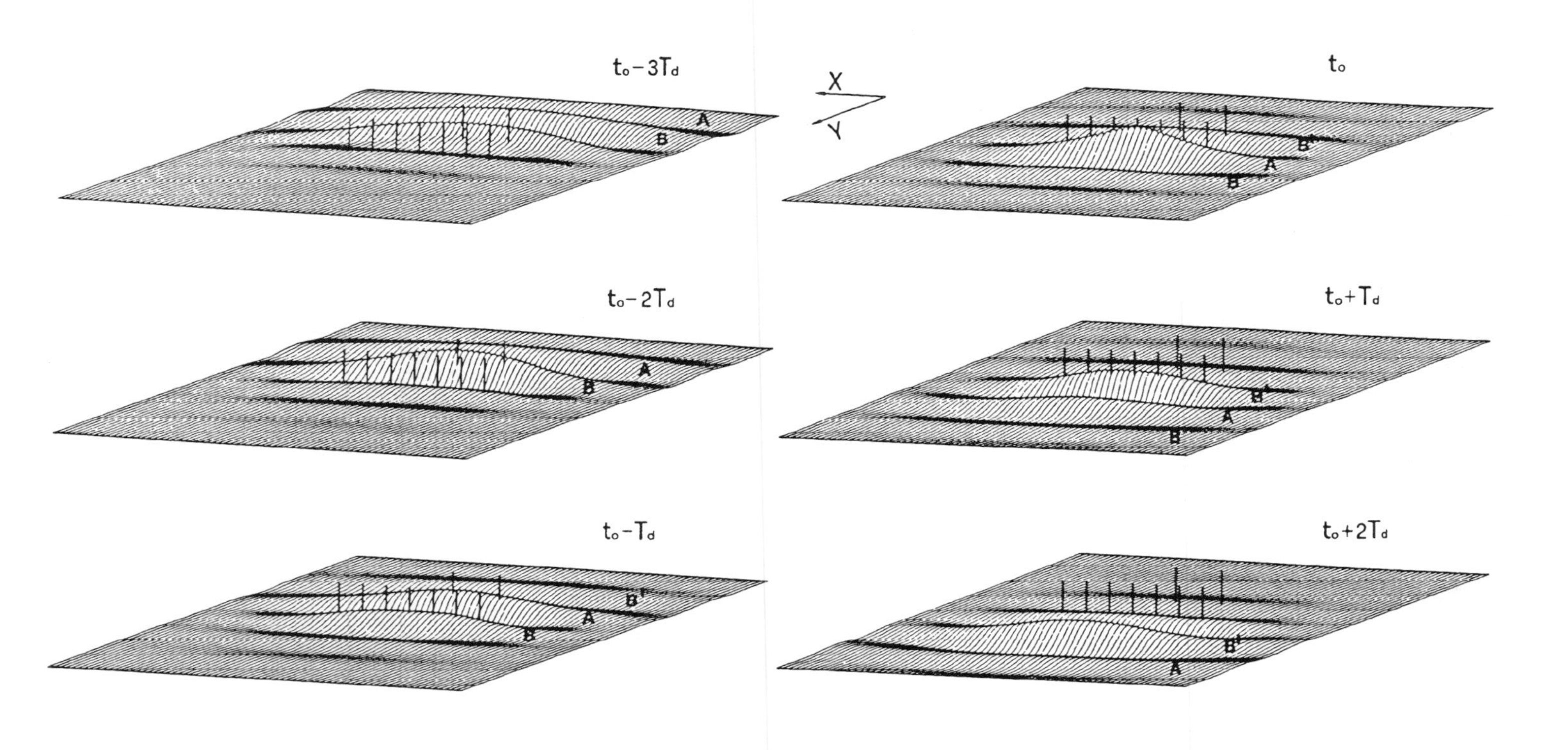

Fig.1 Wave group -eq(1)- on deep water, calculated from classic analytical spectrum [5], [6]. The directional spread of the spectrum is that suggested [6] for the conditions of the experiment: fetch of 10 Km and peak period of the spectrum of 2.0 ÷ 2.5 s.

where t_o is the time instant when the crest of the wave of given height H forms at point x_o, y_o and $S(\omega,\theta)$ is the directional energy spectrum of the random wave field (ω being the angular frequency and Θ the angle of the propagation axis and Y-axis).

Fig.1 shows the wave group on deep water for the classic analytical spectrum [5], [6]. The dominant direction of the spectrum has been assumed to coincide with Y-axis so that, as a consequence, the wave group moves along Y-axis. Point x_o, y_o of given height H is at the center of the framed area and the pictures are taken at time intervals equal to T_d - the peak period of the spectrum -. It may be seen that the wave group has a development stage during which the height of its central wave grows up to a maximum and the front width reduces to a minimum, and a decay stage with the opposite features. Then each wave having a celerity greater than the group, moves along the envelope and undergoes the relevant dimensional variations. Being on deep water, the celerity of a single wave is nearly double than that of the envelope, so that every 2 periods, the envelope center is occupied by a new wave. The great given height H proves to be that at the middle-front of wave A at the center of the group at the apex of its development stage.

2. MEASUREMENTS

Fig.1 shows a set of piles like that we have placed at sea. The X-parallel side of the framed area is 3 wave lengths $L_d (\equiv gT_d^2/2\pi)$ and the distance between pile 1 and pile 7 is 1 wave length L_d (that distance at sea was 7.5 m and also wave length L_d, on average, was of 7.5 m). A quoted plane of the framed area with piles is shown by Fig.2.

Each pile supported an ultrasonic wave probe and a pressure transducer some .5 m below the mean water level. The gauges were connected by submarine cables to an electronic station on an onshore building where data from each gauge were sampled at a rate of 10/s. The double wave measurement (surface elevation and pressure fluctuation at some water depth) permitted to accurately verify the experimental results.

The experiment was executed in front of the Reggio-Calabria beach facing the Messina Straits, on depths ranging from 3 m (piles 1-7) to 4 m (piles 8-9), that is practically on deep water for the waves under examination.

The wind in the area often keeps from NW for several days. After two days of NW wind, usually the southerly swells vanish and the sea states in front of the Reggio-Calabria beach consist of pure wind waves, with significant height typically ranging within .2 m and .4 m and peak period T_d within 2.0 and 2.5 s. The experiment took place during 12 consecutive hours of such a condition. In all 64 sea states each of 9-minutes were recorded.

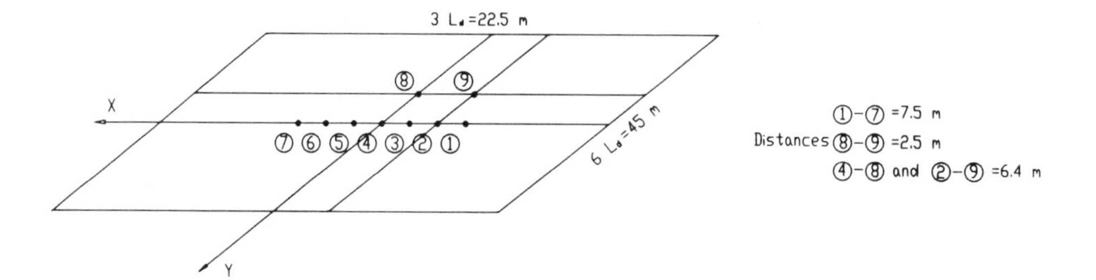

Fig.2 Framed area of Fig.1 and gauges net. The wave length L_d of 7.5 m was the average of the experiment.

3. RESULTS

For each sea state, the maximum wave height recorded by the whole set of gauges was found. In line with the theory, the maximum wave height is denoted by H, the wave is called A and the point (pile) where H has been recorded is called x_o, y_o and is taken as the origin of axis X-Y.

The crest-to-trough height and the length (squared period) of wave A and of preceding wave B on piles 1 - 7 (Y=0) and on piles 8 - 9 (Y=-.8 L_d) were filed for each sea state (in the case that the maximum wave height was recorded by piles 8 or 9 , line 8 - 9 becomed Y=0, line 1 - 7 becomed Y=.8 L_d and wave B' was considered in place of B, indeed, for theory [2], wave B' on Y=.8 L_d is equal to wave B on Y=-.8 L_d). Then, the angle of the wave front and X-axis was extimated from the time shifts of wave A in the records of gauges 1 - 7 . That angle was rather small (15° on average), that is the wave propagation was nearly parallel to Y-axis.

Fig.3 shows the average wave height and length along fronts A and B on lines Y=0 and Y=-.8 L_d (the wave height being related to H and the wave length to L_d). For a comparison, height and length of waves A and B in the theoretical group of Fig.1 are also shown (smoother lines).

Fig.3 apparently displays the alternation of waves A and B at the group center. Wave A grows because goes to occupy the envelope center, and the increase of its height is made possible by a shortening of the length. Both the increase of the wave height and the shortening of the wave length are largest at the front center and are negligible at the front ends, with the consequence that the front gets markedly sharper. On the contrary, the decrease of the height of wave B, which leaves the envelope center, joins with an apparent increase of the wave length.

The phenomenon looks essentially the same in the measured wave groups and in the theoretical wave group, and it exhibits some really striking sizes. It suffices to observe that, on line Y=-.8 L_d, the height of wave A is .7÷.8 times that of wave B whilst, on line Y=0, the height of wave

A reaches 1.8÷2.0 times that of wave B (we refer to the heights at the front center). That means that ratio H_A/H_B of the heights of the two consecutive waves increases nearly by 2.5 times after a course that is shorter than one wave length!

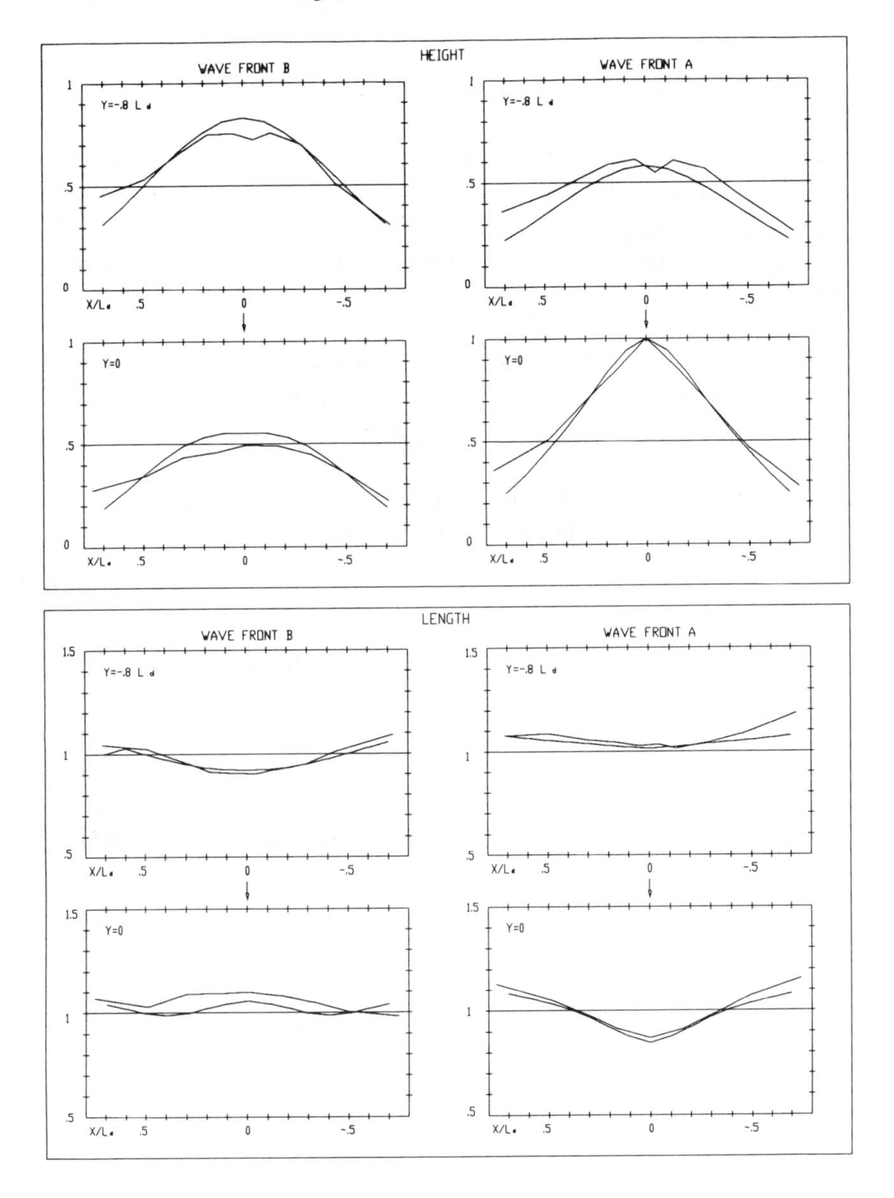

Fig.3 Wave fronts A and B from the experiment and from the theoretical wave group of Fig.1 (the smooter lines clearly refer to the theoretical group).

REFERENCES

1. P. BOCCOTTI, Sea waves and quasi-determinism of rare events in random processes, Atti Acc. Naz. Lincei, Rendiconti, 76, 2 (1984).

2. P. BOCCOTTI, On mechanics of irregular gravity waves, Atti Acc. Naz. Lincei, Memorie, 19 (1989) 111-170.

3. M. S. LONGUET-HIGGINS, The effects of non linearities on statistical distributions in the theory of sea waves, J. Fluid Mech., 17 (1963).

4. O. M. PHILLIPS, The theory of wind generated waves, Adv. in Hydroscience, 4 (1967) 119-149.

5. K. HASSELMANN et al., Measurements of wind-wave growth and swell decay during the Joint North Sea Wave Project (JONSWAP), Deut. Hydrogr. Zeit., A-8 (1973).

6. H. MITSUYASU et al., Observation of directional spectrum of ocean waves using a cloverleaf buoy, J. Phys. Oceanography, 5 (1975).

Paolo Boccotti
Giuseppe Barbaro
Lucio Mannino
Istituto di Ingegneria Civile
Universita' di Reggio-Calabria
Via V. Veneto, 69
89100 Reggio-Calabria ITALY

P LUGER, K HORNUNG AND F HINDELANG

Experiments on dissipative wave propagation in a tube filled with a condensable fluid

ABSTRACT

The propagation of weak shock waves ($1.1 \leq Ma \leq 1.5$) in tubes filled with pure vapour may lead to a significant flow acceleration caused by wall condensation if the tube's walls are heat conducting. Limiting flow conditions are reached much faster than known from conventional shock tube boundary layer theory. The process has been verified by measuring the velocity of shock- and contact fronts by Laser-Schlieren and Mie scattering, post-shock pressure and velocity by piezo probes and Laser Doppler Anemometry respectively.

1. INTRODUCTION

The physics of shock wave propagation in a tube, filled with a saturated vapour is as follows: The shock front suddenly raises the pressure, but the wall temperature essentially remains constant. Therefore strong supersaturation exists near the wall and as a consequence mass flux to the wall occurs, if the wall is heat conducting.

The problem may have various applications, e.g. in cooling circuits, when a nonlinear compression wave is generated in the course of an accident, but it may also be regarded as a phenomenon of principal interest.

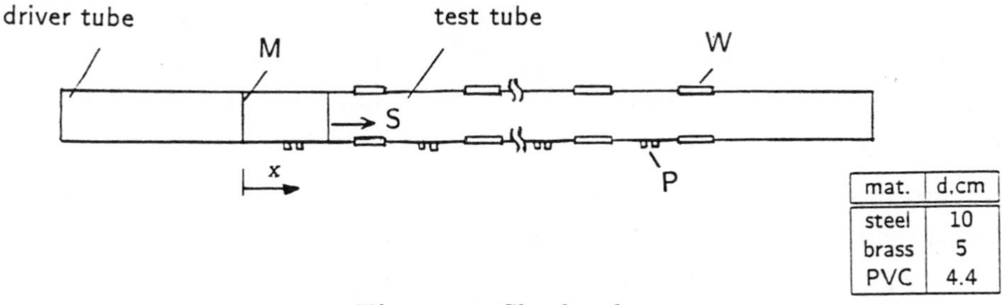

mat.	d.cm
steel	10
brass	5
PVC	4.4

Figure 1: Shock-tube

The apparatus used was a shock tube (fig. 1). For the experiments the driver tube is filled with air at pressures between 100 mbar and a few bar and the test tube with 20 mbar water vapour at room temperature, which means saturation.

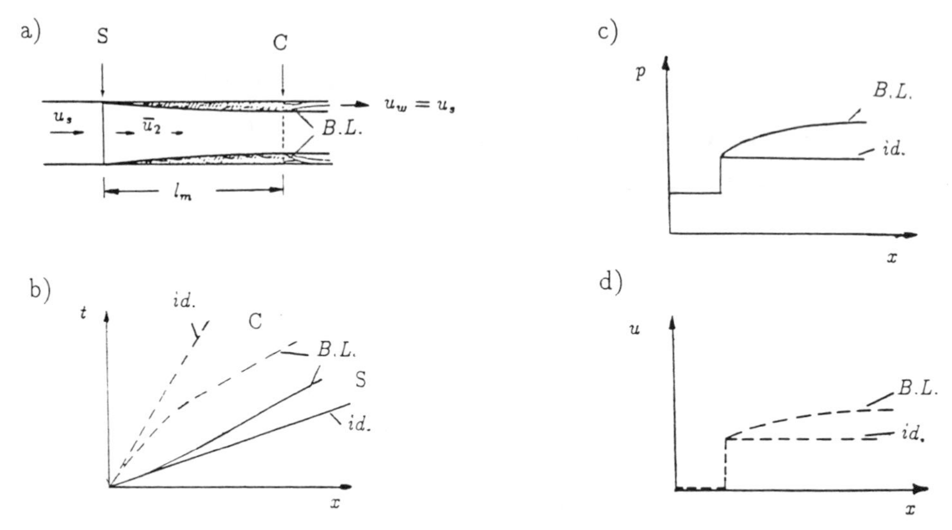

Figure 2: Boundary layer effect on shock-tube flow: principle

After rupture of the membrane M a shock wave S propagates into the water vapour. Observation is done by various pressure probes P and optical windows W. Various shock-tubes have been used in order to investigate the influence of diameter and wall material.

2. INTERPRETATION OF THE WAVE PHENOMENON

There is some similarity with the shock-tube boundary layer problem, treated first in the early sixties ([1],[5],[6]); it's basic features will therefore be summarized briefly in the following.

Figure 2a shows a part of the test tube in the moving frame of reference, the shock front S being at rest. The water vapour comes from the left with supersonic velocity u_s, becoming subsonic ($\bar{u}_{2,s}$) after the front[1]. The wall moves to the right with u_s. A viscous boundary layer ($B.L.$) develops, some of the shocked gas enters into it, attains the wall's velocity and is transported with u_s out of the control volume. The control volume is bounded by the shock front S and the contact surface C, separating driver- and test gas.

After some time a situation is reached, in which the influx of mass through the shock front gets balanced by the boundary layer losses. This is called the "limiting case". The distance between the shock front and the contact surface

[1]A bar above the velocity indicates the moving frame of reference (else laboratory system); the index $2, s$ denotes conditions immediately bedind the shock front.

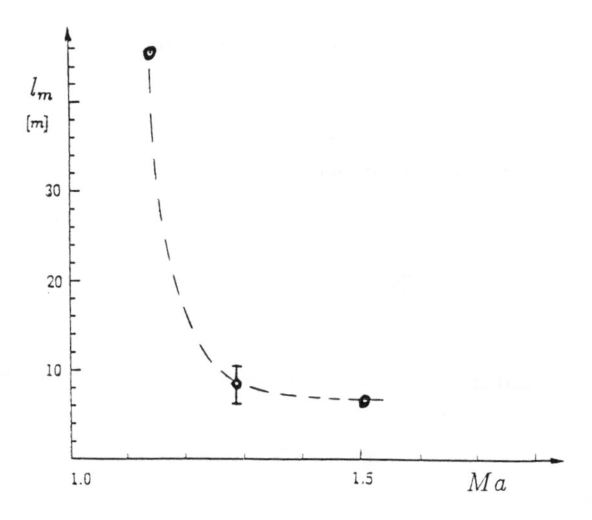

Figure 3: Measured values l_m for a d = 10cm steel tube (water vapour), LDA-data

has then reached a maximum value l_m. This means, that the test gas comes to rest at the contact surface C in the moving frame of reference. In the laboratory frame of reference this means an acceleration of the gas until it even reaches the velocity of the shock front itself.

In the x-t-diagram (fig. 2b) this looks as follows: In the ideal, non-dissipative case the shock S and the contact surface C are diverging on linear paths, whereas in the case with boundary layer they tend to become parallel, which means that they approach the same velocity. There is constant pressure and constant velocity behind the shock in the ideal case and an increase of both in the boundary layer case (see figs. 2c,d). This post-shock increase may even exceed the initial jump, and the whole phenomenon gets the shape of a dispersed wave.

The viscous boundary layer theory, when applied to water vapour, results in very large values for l_m. For $Ma = 1.3$, as an example, l_m has a value of 187m, which is far beyond the scale of fig. 3. Surprisingly flow acceleration occurs much faster here, which means much smaller values of l_m. Therefore a mass loss mechanism must be present which is much more efficient. It turned out, that this mechanism is wall condensation.

It is possible to describe this effect with a simple phenomenological model for the limiting case, which is quasi-stationary in the moving frame, but of course nonstationary in reality .

There is an inflow of mass $\dot{m}_s$ into the control volume, which – in the limiting case – is balanced by the losses $\dot{m}_c$ at the contact surface.

$$\dot{m} = \dot{m}_s - \dot{m}_c = 0 \quad , \qquad \dot{m}_s = \bar{u}_{2,s}\rho_{2,s}A_0 \quad . \tag{1}$$

The difference to the boundary layer case is, that the viscous wall interaction has been replaced by wall condensation

$$\dot{m}_c = \sigma_k(l) d\pi u_s \rho_l \quad . \tag{2}$$

The thickness σ_k of the condensate film increases with the square root of the distance l from the shock

$$\sigma_k = C_k t^{1/2} = C_k \left(\frac{l}{u_s}\right)^{1/2} \quad . \tag{3}$$

This is consistent with recent measurements of Fujikawa [2] in vapours at rest. C_k is a constant, found to only depend on the Mach number for a given tube material.

From C_k the characteristic length l_m is given by

$$l_m^{1/2} = \frac{1}{C_k}\frac{1}{u_s^{1/2}}\frac{d}{4}\frac{\bar{u}_{2,s}\rho_{2,s}}{\rho_l} \quad , \tag{4}$$

and from this follow the profiles of the physical parameters along the centerline of the tube. For this one-dimensional isentropic flow with a fictitious variable cross section A is assumed [5]

$$\frac{A_0}{A} = 1 - \left(\frac{l}{l_m}\right)^{1/2} \quad , \tag{5}$$

which formally models the mass loss within the continuity equation.

This simple model satisfactorily describes our experimental findings (see below). A more thorough theoretical description would require the treatment of the full hyperbolic system for the nonstationary process, including mass loss to the wall and nonstationary conduction of heat of condensation into the interior of the wall.

3. EXPERIMENTAL EVIDENCES

Three types of measurements have been performed: First position-time histories for the shock front and the contact surface by Schlieren optics and Mie scattering respectively. Second pressure by piezo probes and third velocity by Laser Doppler Anemometry.

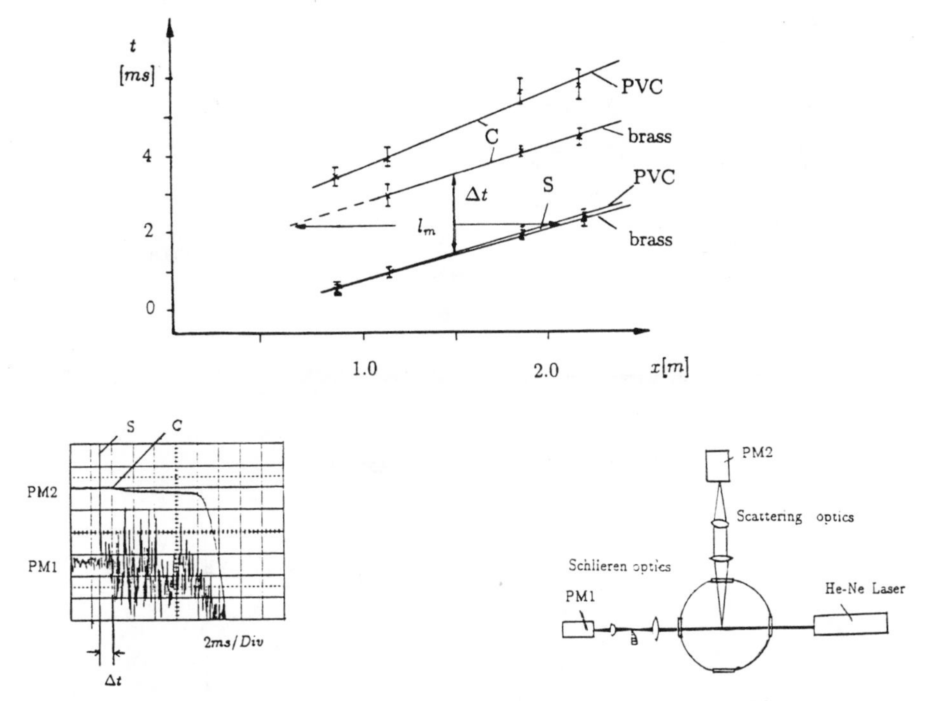

Figure 4: x-t-measurements by Mie scattering and Schlieren

The optics of the position-time (x-t) measurements (fig. 4) is quite simple. A Laser Schlieren system detects the shock front (spike S in signal PM1). A second photomultiplier (PM2) looks under 90 degrees to the center of the tube. It detects the scattered light from those droplets which condense by contact of the water vapour with the cold driver gas. Thus the contact surface is marked by the onset of the photomultiplier signal (point C on signal PM2).

These optical measurements have been performed at different positions along the test tube such, that an x-t-plot for the shock front as well as for the contact surface can be derived.

Figure 4 shows two examples for a 5 cm brass tube and a 4.4 cm PVC tube. The important feature is, that the limiting flow condition is not reached in the plastic tube due to its low heat conductivity (the paths of shock and contact front are not parallel). In the metal tube the limiting case is reached, Δt is the limiting test time and from this the limiting length l_m follows right away.

For the pressure measurements (fig. 5), piezo probes (Kistler 603B) were distributed along 12 positions down the test tube. They had to be isolated thermally by a thin silicone layer, since condensation and subsequent transfer of latent heat to the probe leads to unreal negative pressure signals. The isolated one correctly

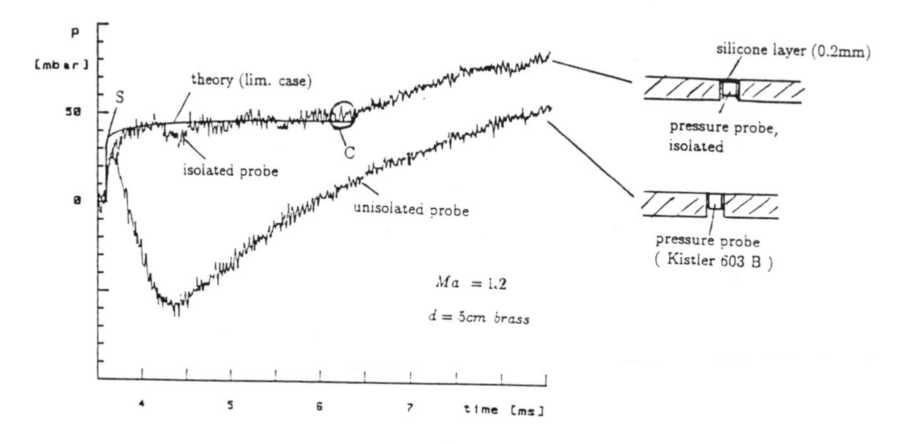

Figure 5: Pressure measurements

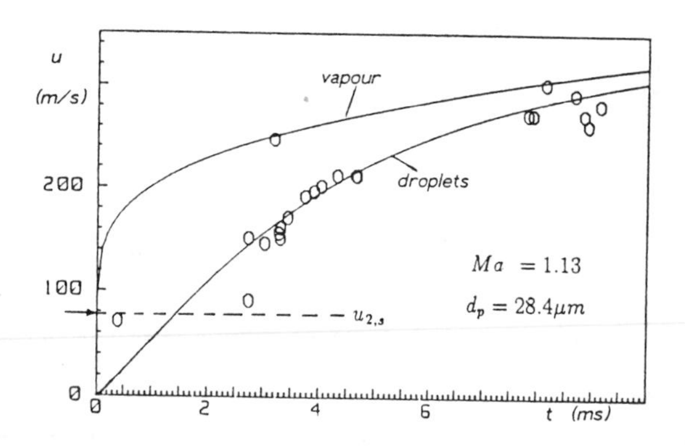

Figure 6: Laser Doppler velocity measurement on tracer droplets

reports the pressure jump across the shock as well as the increase afterwards as predicted by our simple model[2]. The kink at the end (point C) is exactly at that position where the light scattering experiments indicate the arrival of the contact surface. The time difference between S and C again gives a measured value for l_m, using the wave's velocity.

In one of the tubes, the 10 cm steel one, direct velocity measurements were performed in addition. The method used was Laser Doppler Anemometry [3]. For this, small droplets had be added to the water vapour. After some initial velocity relaxation, these droplets attain the vapour velocity. Figure 6 clearly demonstrates the strong post-shock acceleration to values, which exceed the effect of

[2]The superimposed frequency comes from oscillations of the tube's walls and has no relevance to the flow.

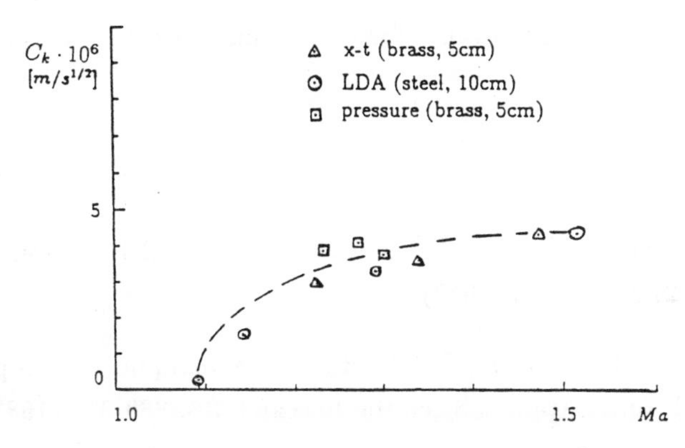

Figure 7: Film growth parameter C_k

the shock front ($u_{2,s}$) by far. The vapour's solid line comes from the simple model described above, and the droplet's solid line is an integration of the equation of motion of one droplet, using an appropriate drag law [4].

Thus, from all three types of measurements, information on the limiting length l_m can be extracted and from this also on the growth of the wall film. Figure 7 shows the dependence of the film growth parameter C_k (equation (3)) on Mach-number. As can be seen, there is no effect below a certain shock strength. Obviously the supersaturation, produced by the shock front is not sufficient to start wall condensation.

On the other hand for Ma above 1.5, C_k seems to approach a certain limit, probably due to the finite ability of the wall to remove the latent heat of condensation out of the liquid film.

This quite good consistency of all three different types of data gives confidence in the – at least approximate – correctness of the phenomenlogical model drawn.

4. POSSIBLE APPLICATIONS

The first thing that excited us when we saw the effect was, that it gives a chance to really observe the limiting case with reasonable tube lengths. This is difficult to achieve with a non-condensable gas, because for this very long and thin tubes and very low pressures are required.

Second: The effect allows an indirect, but rather sensitive measurement of the wall film thickness, which may be of interest for condensation studies.

Third: Whenever weak shock waves travel through vapours near saturation with some walls present, these waves can grow and broaden due to wall condensation. This could be of importance for wet vapour flows in general.

REFERENCES

1. Duff E.D., Shock-Tube Performance at Low Initial Pressure, The Physics of Fluids **2**, 2, 207-216 (1959)
2. Fujikawa S., Akamatsu T., Fujioka H., Studies of liquid-vapour phase change by a shock tube, Applied Scientific Research **38**, 363-372 (1982)
3. Luger P., Hornung K., Hindelang F., Time-Resolved Laser Doppler Anemometry Applied to Shock Waves in Wet Steam, in: Shock Tubes and Waves, Proceedings of the Sixteenth International Symposium on Shock Tubes and Waves, Aachen, W.-Germany (1987)
4. Luger P., Experimentelle Untersuchungen zu Stoßwellen in einkomponentigen Zweiphasengemischen, Diss., UniBw München (1990)
5. Mirels H., *Laminar Boundary Layer behind Shock advancing into stationary Fluid*, NACA TN 3401, Washington (1955); Test Time in Low-Pressure Shock Tubes, The Physics of Fluids, **6**, 9, 1201-1214 (1963); Shock Tube Test Time Limitation Due to Turbulent-Wall Boundary Layer, AIAA Journal **2**, 1, 84-93 (1964); Flow Nonuniformity in Shock Tubes Operating at Maximum Test Times, The Physics of Fluids **9**, 10, 1907-1912 (1966)
6. Roshko A., On Flow Duration in Low-Pressure Shock Tubes, The Physics of Fluids **3**, 6, 835-842 (1960)

P.Luger, K.Hornung, F.Hindelang
Fakultät LRT
Universität der B.W. München
D-8014 Neubiberg, Germany

F MALLAMACE, N MICALI AND C VASI

Experimental study of viscosity in dispersive systems (Role of the Interparticle Interaction Potential)

ABSTRACT

We report the viscosity calculation for disperse solution of charged colloids and microemulsions. The calculation is made in terms of the Maxwell model for viscoelasticity considering in the proper way the interparticle potential. Calculations and experimental data are in good agreement.

1. INTRODUCTION

The viscosity [1], or internal friction, is due to the irreversible transfer of momentum from points where the velocity is high to those where it is small. The transfer momentum due simply to the mechanical transport of fluid particles from one place to another and to pressure forces acting in the system, represents instead a complete reversible process. The stress tensor is represented as:

$$\sigma_{ik} = -p\,\delta_{ik} + \sigma'_{ik}$$

where σ' is the viscosity contribution to stress tensor. It give the momentum flux part not due to the direct momentum transfer with the mass of moving fluid and is:

$$\sigma'_{ik} = \eta\left(\frac{\partial v_i}{\partial x_k} + \frac{\partial v_k}{\partial x_i} - \frac{2}{3}\delta_{ik}\frac{\partial v_l}{\partial x_l}\right) + \zeta\,\delta_{ik}\frac{\partial v_l}{\partial x_l}$$

where v is the velocity vector, the constants η and ζ are know as viscosity coefficients and x the position vector. In a incompressibile fluid the viscosity is determined only by the coefficient η as:

$$\sigma'_{ik} = \eta\left(\frac{\partial v_i}{\partial x_k} + \frac{\partial v_k}{\partial x_i}\right) = \eta\, 2\,\dot{u}_{ik} \quad ; \qquad \dot{u}_{ik} = \frac{1}{2}\left(\frac{\partial v_i}{\partial x_k} + \frac{\partial v_k}{\partial x_i}\right)$$

2. MICROSCOPIC CALCULATION

As it is well known that the molecules motion is reversible whereas the transport processes, observed in a

macroscopic scale, are clearly irreversible [2]. In a gas the single molecule spends most of its time in the empty space between one collision and another; hence the trasport of momentum is accompained by each molecule that carries its own contribution. Instead, in a liquid each molecule is enclosed in a cage formed by its nearest neighbours, and only occasionally migrates to a neighbouring cage. It spends most of its time in this environement and is therefore under the influence of the intermolecular interactions with its next-neighbours. Consequently the momentum is transferred through the space from the faster to the slower molecules.
The laminar flow in a simple Newtonian liquid is :

$$P_{xy} = \sigma'_{xy} = \eta \frac{\partial v_x}{\partial x_y}$$

where P_{xy} is the shear stress in the x-direction (perpendicular to the y-axis) and $\partial v_x/\partial y$ is the velocity gradien and represent how the x-component of velocity varies in the y-direction.
Considering a unitary volume V, the rate of the x-momentum transfer in the y-direction is numerically equal to the shear stress (magnitude of momentum transferred · velocity of transfer) : $F\partial t \cdot \frac{a}{\partial t} = P_{xy}\, V$. Therefore we easily obtain the rate of the momentum transfer for the generic molecule, that is:

$$F_x \cdot y = F(r) \cos\alpha \cdot y = F(r) \frac{x\, y}{r}$$

where α is the angle between the x-axis and the line joinin the considered molecules and the generic molecule of the cage, and F(r) is the force between two molecules at the distance r . For the unit volume we have :

$$P_{xy} = \Sigma\, F(r) \frac{x \cdot y}{r}$$

and in this terms the summation must be performed over the $N^2/2V^2$ pairs of the unit volume . In order to perform the sum we need the radial distribution function under non-equilibrium conditions. The equilibrium distribution function is proportional to the radial distribution function $\rho(r)$ that represents the probability that one particle is in the r coodinate when another is in the origin. This function is experimentally determinable by means of spectroscopic techniques. For non-equilibrium systems such a function has to be moltiplicated for a correction term $u(r)$. On the base of the above considerations the viscosity in a liquid system is [2,3,4]:

$$\eta = \frac{2\pi}{15} \cdot \frac{N}{V} \cdot \int \frac{d\phi(r)}{dr} \rho(r)\, u(r)\, r^3\, dr$$

where the sum is subtituted by the integration over the distribution function ρ , $\phi(r)$ is the intermolecular potential function and $u(r)$ represents the radial distorsion caused by non-equilibrium condictions produced by the velocity gradient.

The classes of liquids that possess elastic and viscous properties are called viscoelastic. If they are sheared are able to store elastic energy and their the viscosity stress tensor can be written as [5] :

$$\sigma'_{ik} = G_\infty \cdot 2\, u_{ik} \quad ; \qquad u_{ik} = \frac{1}{2}\left(\frac{\partial x_i}{\partial x_k} + \frac{\partial x_k}{\partial x_i} \right)$$

where u_{ik} represent the strain tensor.
The Maxwell's model of viscoelasticity [6] allows to calculate the viscosity of a fluid of interacting particles as the product of the high frequency shear modulus G_∞ and the relaxation time τ. The relaxation time τ corresponds to the time taken by the system to relax when a shear strain is applied [5]. From a lattice model of viscoelastic liquids the shear modulus (rigidity) is obtained from the interaction potential $\phi(r)$ and from the mean separation distance between the particles R [7] :

$$G_\infty = \frac{\alpha}{r} \left(\frac{\partial^2 \phi(r)}{\partial r^2} \right) \qquad \text{calculated for } r=R$$

3. EXPERIMENTAL EVIDENCE

Let us consider two systems: a) latex particle dispersion

and b) microemulsion system.

a) Polystyrene particle dispersion (latex): the viscosity in a two fluid model can be written as:

$$\eta = \eta_O + \eta_1 + O(\phi)$$

where η_O is the solvent viscosity (water in our case) and η_1 the additional contribution due to the interaction among the colloidal particles while the third term represent the hydrodynamic contributions. In such a model the total viscosity is due to a contribution of the interacting particles imagined to be a separate fluid plus the solvent contribution and the hydrodynamic effects. The η_1 contribution is viscoelastic and is obtained from the Maxwell's model using a proper interaction potential $\phi(r)$. For our analysis we use the DLVO (Derjaguin, Landau, Verwey and Overbeek) theory [8,9]:

$$\phi(r) = 2\pi\epsilon\psi_\circ^2 \ln(1+\exp(-2\kappa R_0(r_O-1))) \quad \text{for } (R_O\kappa \gg 1),$$

$$\phi(r) = \frac{4\pi\epsilon\psi_\circ^2 R_O^2}{r} \exp(-\kappa r_O) \quad \text{for } \kappa R_O \ll 1$$

where $r_O = r/2R_0$, r is the distance among the centers of the particles, ϵ is the dielectric constant of the solvent medium, $1/\kappa$ is the Debye screening length):

$$\kappa = \left(\frac{2e^2 N 10^3}{\epsilon k_B T} I \right)^{\frac{1}{2}}$$

where e is the electron charge, N the Avogadro number, I the ionic strength calculated including the total concentration n of the free-state (diffusible) cation and anion in the solution, $\psi_\circ$ is the particle surface potential related to the charge Z, due to the ionizable sites, on the colloidal particle that for small values of κ ($\kappa R_O \leq 6$) can be written as:

$$\psi_\circ = \frac{Z}{4\pi\epsilon R_O(1 + \kappa R_O)}$$

The corresponding modulus is:

$$G_\infty = 4\pi\alpha\epsilon\psi_\circ^2 R_O^2 \left(\frac{\kappa^2 r^2 + 2\kappa r + 2}{r^4}\right) \exp(-\kappa r_O) \quad \text{for } \kappa R_O \leq 3$$

used experimental condictions is a system macroscopically homogeneous and in a thermodynamical stable phase [10, 11]. Again in term of the two fluid model its viscosity is

$$\eta = \eta_0 + \eta_1 + \eta_2 \cdot$$

the interaction potential $\phi(r)$ is repulsive hard core [12]

$$\phi(r) \sim \left(\frac{\sigma}{r-\sigma}\right)^n$$

the corresponding contribution η_1 to the viscosity is

$$\eta_1 \sim \eta_0 \left(\frac{\phi}{\phi_m}\right)^{(n+1)/3} \left(1 - \left(\frac{\phi}{\phi_m}\right)^{1/3}\right)^{-n}$$

the relative viscosity is

$$\eta_S = 1+\frac{1}{\eta_0}(\eta_1+\eta_2) = (1+2.5\ \phi) + A \left(\frac{\phi}{\phi_m}\right)^{(n+1)/3} \left(1 - \left(\frac{\phi}{\phi_m}\right)^{1/3}\right)^{-n}$$

The comparison with the experimental viscosity data at 10 °C furnishes n=2 (fig. 2)

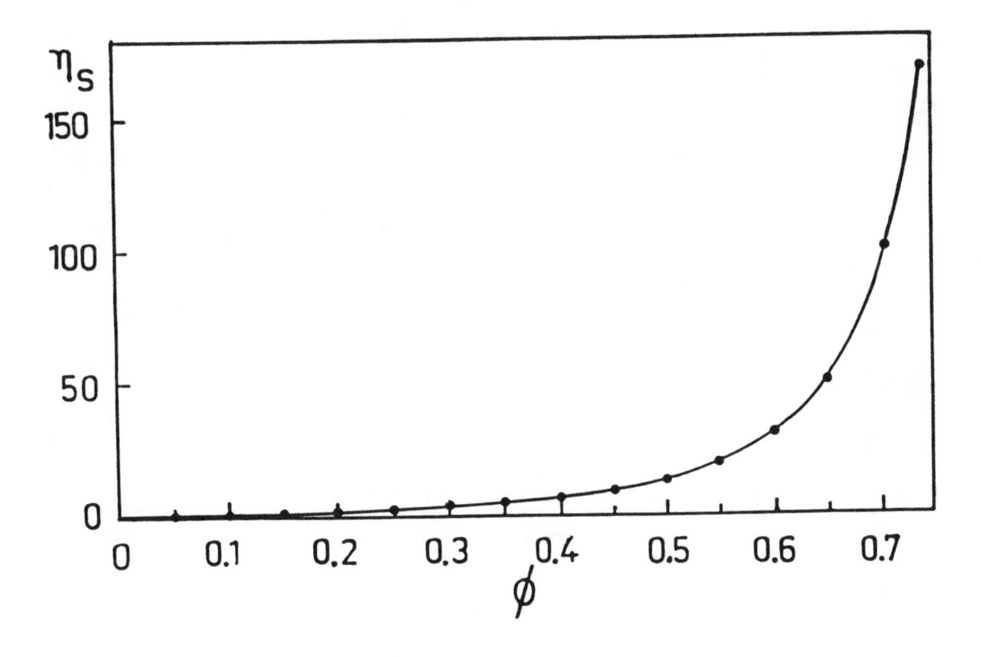

FIG 2 Dot experimental viscosity, full line calculated viscosity with n=2

used experimental condictions is a system macroscopically homogeneous and in a thermodynamical stable phase [10, 11]. Again in term of the two fluid model its viscosity is

$$\eta = \eta_0 + \eta_1 + \eta_2 .$$

the interaction potential $\phi(r)$ is repulsive hard core [12]

$$\phi(r) \sim \left(\frac{\sigma}{r-\sigma}\right)^n$$

the corresponding contribution η_1 to the viscosity is

$$\eta_1 \sim \eta_0 \left(\frac{\phi}{\phi_m}\right)^{(n+1)/3} \left(1 - \left(\frac{\phi}{\phi_m}\right)^{1/3}\right)^{-n}$$

the relative viscosity is

$$\eta_S = 1+\frac{1}{\eta_0}(\eta_1+\eta_2) = (1+2.5\ \phi) + A \left(\frac{\phi}{\phi_m}\right)^{(n+1)/3} \left(1 - \left(\frac{\phi}{\phi_m}\right)^{1/3}\right)^{-n}$$

The comparison with the experimental viscosity data at 10 °C furnishes n=2 (fig. 2)

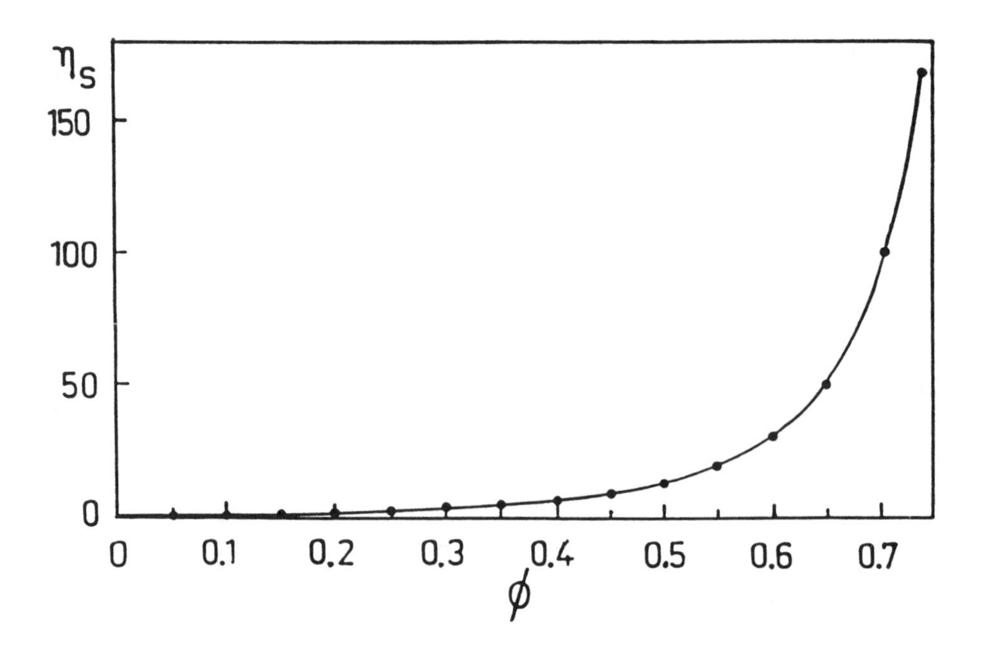

FIG 2 Dot experimental viscosity, full line calculated viscosity with n=2

The relaxation time in such a case results proportional to the inverse of the diffusion coefficent of the droplets In both the systems in terms of a two fluid model and by using the Maxwell law of viscoelasticity we are able to calculate the measured η from the interacting potential among the particles.

REFERENCES

1 L. D. LANDAU , E. M. LIFSHITZ , Fluid Mechanics (Pergamon Press LTD 1959)

2 H. N. V. TEMPERLEY , D. H. TREVENA , Liquids and Their Properties (Ellis Horwood Limited , Chichester 1978)

3 P. A. EGELSTAFF, An Introduction to the Liquid State (Academic Press London 1967)

4 C. A. CROXOTON , Liquid State Physics (Cambridge University Press 1974)

5 L. D. LĀNDAU , E. M. LIFSHITZ , Theory of Elasticity (Pergamon Press LTD London 1959)

6 MASON , Physical Acoustics volume II B (Academic Press Inc New York 1965)

7 R. BUSCALL , J. W. GOODWIN , M. W. HAWKINS , R. H. OTTEWILL , Viscoelastic Properties of Concentrated Latices , J. Chem. Soc. Faraday Trans. I, 78 (1982) 2873

8 E. J. VERWEY , J. TH. G. OVERBEEK , Theory of the Stability of Lyophobic Colloids (Elsevier, Amsterdam, 1948)

9 D. MAJOLINO , F. MALLAMACE , S. VENUTO , N. MICALI, Viscoelastic Properties of Charged Colloids Phys. Rev. A. , 42 , (1990)

10 J. H. SHULMAN , D. P. RILEY , X-Ray Investigation of Structure of Transparent Oil-Water Disperse System , J. Colloid Sci. 3 (1948) ,384

11 K. MITTAL , B. LINDMAN , Surfactant in solution (Plenum, New York, 1984)

12 F. MALLAMACE , N. MICALI , C. VASI , Viscosity Measurements in Dense Microemulsions , Phys. Rev A , 42, (1990)

F. Mallamace[o] , N. Micali* , C. Vasi*

o Dipartimento di Fisica dell'Universita' di Messina

* Istituto di Tecniche Spettroscopiche del C.N.R. Messina

Salita Sperone 31, 98166 S.Agata, Messina, Italy

M A WHEEL AND P S LEEVERS

Modelling of the high speed double torsion test

1. INTRODUCTION

The High Speed Double Torsion (HSDT) Test has been developed to study the phenomenon of rapid crack propagation (RCP) in various polyethylene (PE) pipeline materials and to compare their dynamic fracture resistances, R, (the work required to produce a unit increment of fracture surface area). The test sample consists of a 6mm thick plate measuring 100x200mm, which is supported at four points in the test rig, as shown in figure 1. A 1mm deep 'V' groove is machined on the underside of the sample along the major axis and then a prenotch of length a_0=40mm is machined part way along this axis. The sample is impacted normally close to the notched end, which causes the material lying either side of the notch to be deformed as two rectangular sectioned torsion beams rotating in opposite senses. The resulting torsional disturbance, that travels along each half of the specimen, loads the notch tip causing a crack to initiate at some time t_f after the impact event and then run along the axis at speeds, å, of up to 240m/s. The load history is measured by piezo-electric load cells under the load plane supports while the crack length is monitored in 10mm increments during the test, using a pattern of conductive and resistive strips painted onto the lower surface of the sample.

2. A LINEAR ELASTIC MODEL OF THE HSDT TEST.

Each half of the sample can be modelled as a torsion beam that is unrestrained behind the crack tip (x<a) but restrained by a torsional elastic foundation ahead of the crack tip (x>a). The restoring torque applied by this foundation at any particular section is then proportional to the rotation, $\phi(x,t)$, of that section at that time. Kanninen [1] developed a similar model of the double cantilever beam sample. The model for the HSDT test is:-

$$\mu K \phi_{xx} - \rho I \phi_{tt} - H(x-a)\sigma\phi = 0 \qquad (1)$$

where H is the Heaviside step function:-

$$H(x-a) = 0, \quad x < a \qquad\qquad H(x-a) = 1, \quad x > a \qquad (2)$$

The torsional rigidity of the beam section is μK, the polar moment of inertia of the section per unit length is ρI, σ is the torsional stiffness per unit length of the foundation and ρ is the density. The shear modulus, μ, is found using ultrasonic methods which give an estimate of the modulus under high strain rate, low strain loading conditions. The torsional wave speed, C_T $\{=(\mu K/\rho I)^{1/2}\}$, is typically 250m/s. The initial conditions are:-

$$\phi(x,0) = 0 \qquad\qquad \phi_t(x,0) = 0 \qquad (3)$$

which state that the sample is undeformed and at rest prior to loading. At the load plane the striker velocity is measured just prior to impact and this is assumed to remain constant throughout the test. The design of the striker and support then ensures that the load plane rotation rate, $\phi_t(0,t)$, remains almost constant throughout the test so the load plane boundary condition can be expressed as :-

$$\phi(0,t) = \phi_t(0,t)\, t \tag{4}$$

Crack growth is simulated by allowing the foundation boundary to translate at the crack tip speed after fracture initiation, that is, by setting:-

$$a = a_0 + H(t - t_f)\, å\, (t - t_f) \tag{5}$$

in equation (1), which can then be solved numerically by integrating along the characteristics, allowing the sample deformation to be determined as a function of time. Figure 2 shows how discontinuities will propagate in the x-t domain. The line OPQ represents the position of the initial loading wave that emanates from the load plane while the bold line gives the crack tip position. When the initial loading wave reaches the notch tip at point P, it is partially reflected along the characteristic PR and a further reflection takes place on returning to the load plane at point R. Upon crack initiation at time t_f an unloading wave is emitted from the crack tip which propagates along the characteristic ST and is reflected from the load plane at point T. Clearly Figure 2 shows the beauty of the HSDT test :- none of the discontinuities will interfere with the crack tip motion during the crack propagation phase. The test simulation is normally terminated when the initial loading wave reaches the end of the sample.

In order to check the validity of the model represented by equations (1) to (5) HSDT samples were marked and instrumented and the tests filmed using a high speed camera. Figure 3 compares the rotation profiles predicted by the model at three particular times, to those actually measured from the photographs. Excellent agreement is obtained.

R can then be determined at each timestep in the integration procedure by calculating the imbalance between the rate at which external work is being supplied to sample, dU_W/dt, and the rates of increase of kinetic energy, dU_K/dt, and strain energy, dU_S/dt, within the specimen :-

$$R = \frac{1}{B_C å} \left\{ \frac{dU_W}{dt} - \frac{dU_S}{dt} - \frac{dU_K}{dt} \right\} \tag{6}$$

where B_C is the width of the fracture surface.

Figure 4 shows the computed variation in R with crack length for a typical test. The main feature of this figure is that R remains approximately constant as the crack length increases and hence a mean value of R can be determined for each test. However, as Figure 5 shows, when mean R values are computed using experimental data obtained from a series of tests performed on one PE grade and are plotted as a function of crack velocity, considerable scatter in the resistance results is seen. This scatter is systematic, at a given crack velocity, a higher R value is yielded by a test performed at a higher impact speed. Another feature of the model is that the predicted applied loads are much greater than those actually measured during the tests, which suggests that inappropriate modulus values are being used in the simulations.

3. DETERMINATION OF SHEAR MODULUS VALUES FOR THE HSDT TEST

In order to determine shear modulus values appropriate to the HSDT test a technique that makes use of the HSDT geometry has been developed. Two independent rectangular sectioned torsion beams are clamped together at one end using a paper fastener, to produce a sample with the same dimensions as the fracture test piece. These samples are then loaded by impact using a range of striker speeds in the HSDT test rig, with the striker twisting each beam at its unrestrained end. Figure 6 shows the assumed deformation of the sample during the test. The front of the disturbance is assumed to propagate at a torsional wave speed that depends on the ultrasonically measured low strain shear modulus, μ_0, because the region ahead of the disturbance is unstrained. The torque applied to each beam, T, can be determined from the mean measured load while the twist induced in the disturbed region, ϕ_x, is given by :-

$$\phi_x = \frac{\phi_t(0,t)}{C_T}, \qquad C_T = \sqrt{\frac{\mu_0 K}{\rho I}} \tag{7}$$

The twist can be varied simply by changing the impact speed. The effective shear stress, τ_{EFF}, and maximum shear strain, γ_{MAX}, in any rectangular section are defined by :-

$$\tau_{EFF} = T\frac{B}{K}, \qquad \gamma_{MAX} = B\phi_x \tag{8}$$

where B is the sample thickness. An 'effective' secant shear modulus, μ_{SEC}, can then be defined as :-

$$\mu_{SEC} = \frac{\tau_{EFF}}{\gamma_{MAX}} = \frac{T}{K\phi_x} \tag{9}$$

This allows μ_{SEC} to be found and plotted as a function of the twist. Figure 7 shows how μ_{SEC} varies with twist for the PE grade on which the fracture tests were previously performed. The ultrasonically measured shear modulus value is also plotted at zero twist.

A linear approximation:-

$$\mu_{SEC} = \mu_0 - m\phi_x \tag{10}$$

is then used to represent the decay in secant shear modulus. A simple analysis gives the 'effective' tangent shear modulus, μ_{TAN}, defined by $d\tau_{EFF}/d\gamma_{MAX}$, as:-

$$\mu_{TAN} = \mu_0 - 2m\phi_x \tag{11}$$

4. A NON LINEAR ELASTIC MODEL OF THE HSDT TEST

When the variation in modulus with twist given by equation (11) is included within the model of the HSDT test the non-linear hyperbolic equation :-

$$(\mu_0 - 2m\phi_x)K\phi_{xx} - \rho I\phi_{tt} - H(x\text{-}a)\sigma\phi = 0 \tag{12}$$

results. It has been assumed that the foundation stiffness remains linear elastic, that is, σ remains constant. The initial and boundary conditions given by equations (3) and (4) remain valid. Equation (12) can then be solved numerically by integrating along the characteristics, using Hartree's method [2], modified for second order equations.

Figure 8 compares the sample deformations predicted by the linear and nonlinear models in a typical test simulation. Like the linear model, the nonlinear model predicts that the rotation varies linearly in the free beam region ($x<a$) and decays non linearly ahead of the crack tip in accordance with the photographic results. Figure 9 shows how the mean R values vary with crack velocity when the non linear model is used to analyse the experimental data considered previously. This figure shows considerably less scatter in the range of R values at any particular crack speed than does Figure 5. However, Figure 9 does suggest that R falls with increasing crack velocity, a conclusion which is not supported by fracture surface studies, which have shown that the surface features remain unchanged as the crack velocity rises.

One further improvement that can be introduced is to consider the suppression of free warping of the rectangular beam sections, coupled with the appearance of axial stresses, in the region where the twist varies [3]. The model becomes:-

$$2(1+\nu)\mu_{TAN}\lambda^2 I\phi_{xxxx} - \mu_{TAN}K\phi_{xx} + \rho I\phi_{tt} + H(x\text{-}a)\sigma\phi = 0 \tag{13}$$

where ν is Poisson's ratio and $\lambda/B \approx 0.3$ for the test piece. This inclusion produces an effective stiffening of the beams ahead of the crack tip causing the loading wave to disperse in this region. One additional boundary condition stating that the axial stresses are zero must be applied at the load plane so:-

$$\phi_{xx}(0,t) = 0 \tag{14}$$

while at the far end of the sample two conditions which state that both the axial and shear stresses are zero :-

$$\phi_{xx}(L,t) = 0, \qquad \mu_{TAN} K\phi_x - 2(1+\nu)\mu_{TAN} \lambda^2 I\phi_{xxx} = 0 \qquad (15)$$

must be applied. Equation (13) has been solved by employing an explicit finite difference scheme in which the values of the nonlinear coefficients are computed at the start of each timestep. These are then used to compute the rotations at the next timestep.

Figure 8 also depicts the rotation profile of the beams predicted by the enhanced nonlinear model. Once again a linear variation in rotation in the free beam is seen while the profile in front of the crack tip decays sinusiodally as the wavefront disperses. Figure 10 shows how the mean R values vary with å when the tests discussed previously are once again resimulated. Not only do these results show less scatter than those given in figure 5 they also indicate R is independent of å , which agrees with the fracture surface studies, and implies that R is a property of the material.

REFERENCES

[1] M.F. KANNINEN, Int. J. of Fracture, Vol.10 No.3, 1974

[2] W.F. AMES, Nonlinear Partial Differential Equations in Engineering, Academic Press, New York, 1965

[3] J.M. GERE, J. App. Mech., Vol. 21, pp 381-387, Dec. 1954

M.A. Wheel and P.S. Leevers

Mech. Eng. Dept., Imperial College, London, SW7 2BX, U.K.

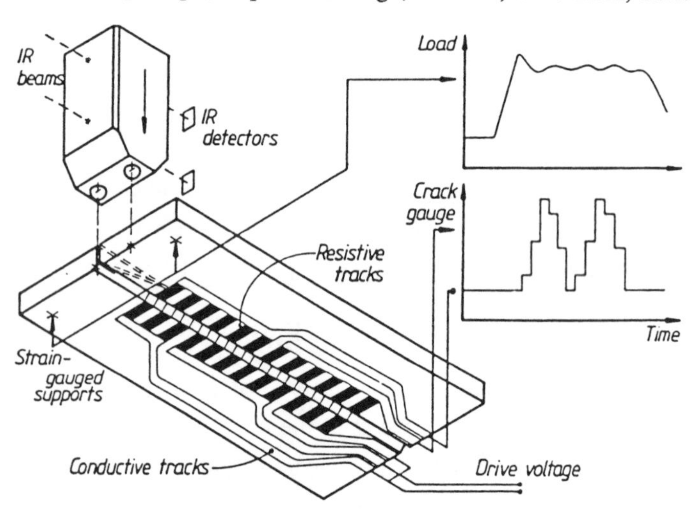

Fig. 1 The High Speed Double Torsion Test

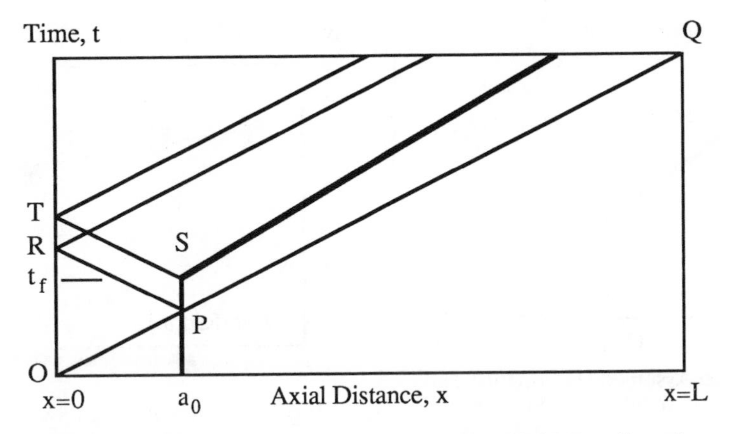

Fig.2 Propagation of Discontinuities in the HSDT Test Sample

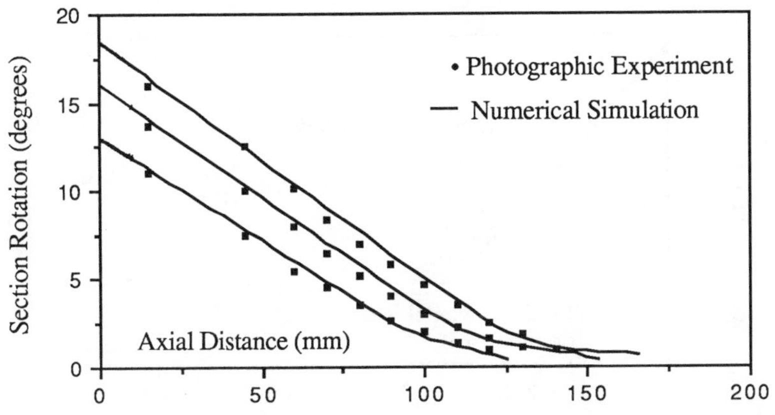

Fig.3 Comparison Between Computed and Experimentally Determined Profiles

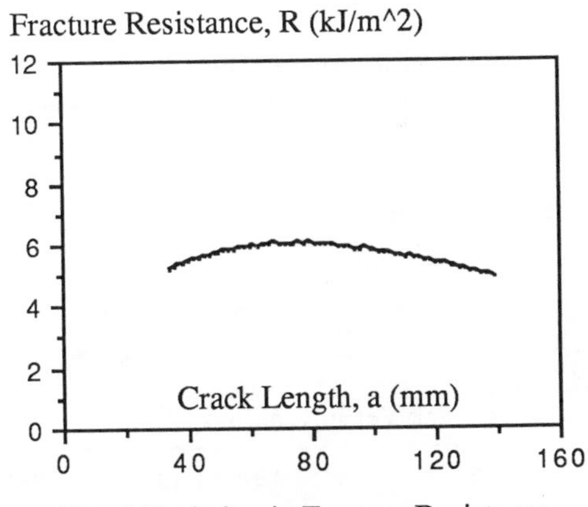

Fig. 4 Variation in Fracture Resistance with Crack Length

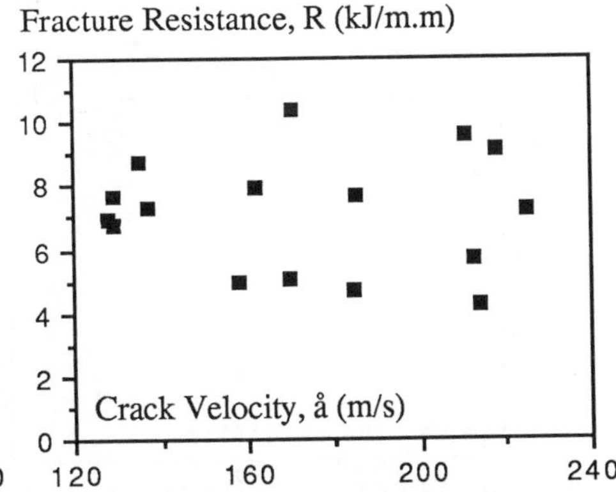

Fig. 5 Variation in Fracture Resistance with Crack Velocity (Linear Model Eqn. 1)

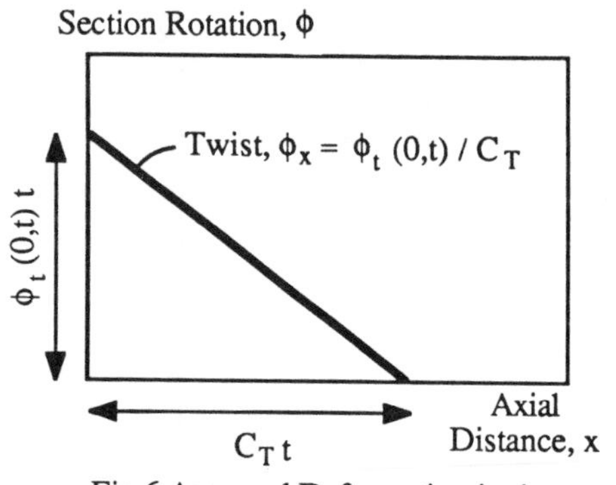

Fig.6 Assumed Deformation in the Modulus Test Sample

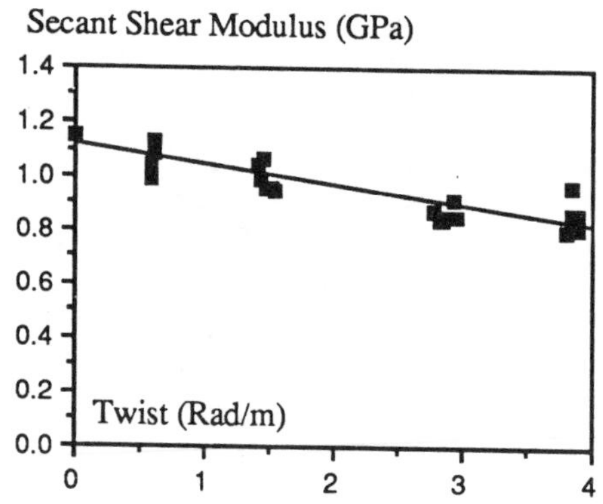

Fig.7 Variation in Secant Shear Modulus with Twist

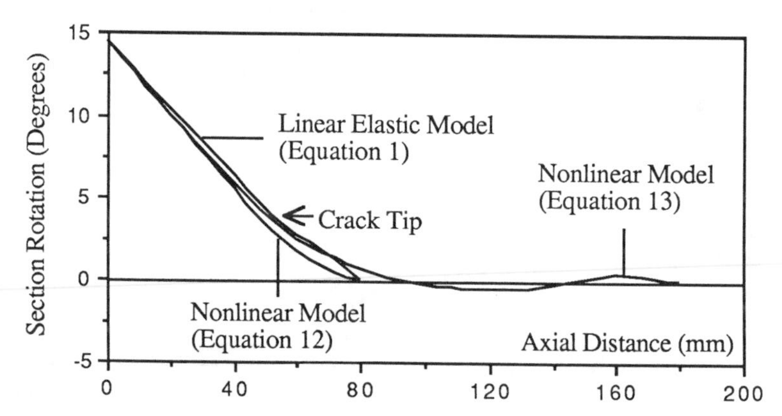

Fig. 8 Comparison of Rotation Profiles Predicted by Linear Elastic and Nonlinear Models

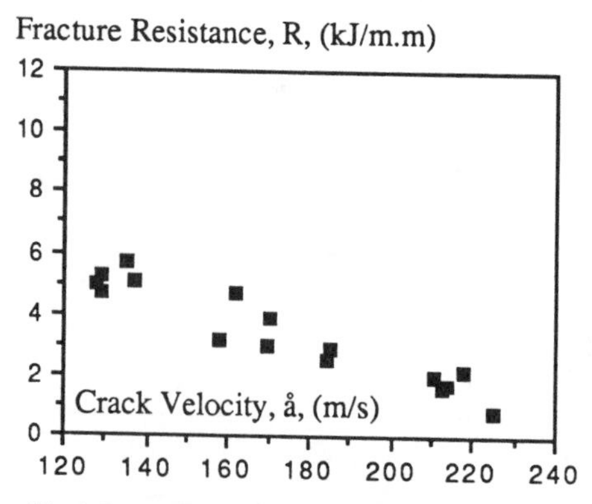

Fig.9 Variation in Fracture Resistance with Crack Velocity (Nonlinear Model Eqn.12)

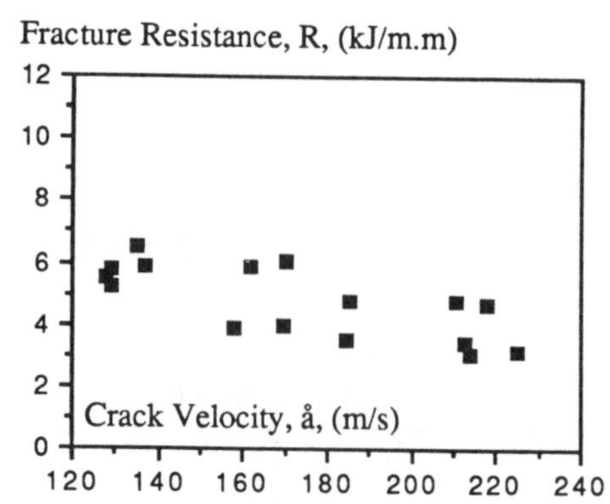

Fig.10 Variation in Fracture Resistance with Crack Velocity (Nonlinear Model Eqn.13)